Port

and Development

Portable Electronics Product Design and Development

For Cellular Phones, PDAs,
Digital Cameras,
Personal Electronics,
and More

Bert Haskell

McGraw-Hill

New York   Chicago   San Francisco   Lisbon
London   Madrid   Mexico City   Milan   New Delhi
San Juan   Seoul   Singapore   Sydney   Toronto

Cataloging-in-Publication Data is on file with the Library of Congress.

ISBN 978-0-07-163402-1

The sponsoring editor for this book was Steve Chapman and the production supervisor was Pamela Pelton. It was set in Century Schoolbook by International Typesetting and Composition. The art director for the cover was Margaret Webster-Shapiro.

This book was printed on recycled, acid-free paper containing a minimum of 50% recycled, de-inked fiber.

McGraw-Hill books are available at special quantity discounts to use as premiums and sales promotions, or for use in corporate training programs. For more information, please write to the Director of Special Sales, McGraw-Hill Professional, Two Penn Plaza, New York, NY 10121-2298. Or contact your local bookstore.

Contents

Preface

Since the introduction of portable transistor radios in the 1950s, portable electronics have captured the imagination of the general public. Today, concepts such as the Dick Tracey wristwatch-TV-phone no longer seem like science fiction, as compact video phone products are appearing in the marketplace. Portable devices such as these are capturing the imagination of people all over the globe as these new technologies change the way people work and live.

Invisible to most users of these products, however, is the business and technology infrastructure that is required to make these products successful. Breakthrough products, such as the Palm Pilot or the Motorola Flip-Phone, tend to garner more than their fair share of acclaim in an industry that is really about evolutionary, rather than revolutionary, change. Nor can these products be considered the result of single technology breakthroughs. The invention of integrated circuit technology and liquid crystal display technology in North America was critical to the emergence of portable electronics, but so too were the mass production techniques and electronics packaging technologies developed in Japan.

Modern portable electronics has come into being through a process of evolution, with product designers retaining the successful technologies and features from past products, while injecting new innovations on an ongoing basis. Quite often, these new innovations fall by the wayside as the Darwinian forces of the marketplace render their judgment. The judgment is made by consumers who are quick to marvel at the latest electronic wizardry but slow to part with their cash when the product actually hits the market. The design of a portable electronic product is really about delivering the benefits of technology to the individual. Fail to do so and the product in question will not be successful.

So it was when I entered into this exciting segment of the industry in 1985. I had no idea at the time that my education in mechanical engineering and robotics would serve me so well in the field of portable electronics. But it turns out that portable electronics design is truly a multidisciplinary field. I have found that the greatest contributors to product design in this field are those who have a systems engineering perspective and a good understanding of human factors.

This book is not about the details of electronics design. It is about the creation of the product concept and the technology infrastructure needed to deliver technology to the individual. Any individual with basic engineering training and a strong interest can aspire to design portable electronic products. This book is intended to encourage and enable multidisciplined thinkers to become portable electronics designers.

Structure

The structure of this text is divided into 13 chapters. The first chapter reviews the process of portable electronic design. This chapter includes a discussion of the overall product development process, an introduction to the critical factors that affect portable electronic design, and a walkthrough of the system design process.

Chapter 2 is a brief overview of digital and analog processing, which enables the basic functionality of portable electronic products. In no way intended as a design guide for integrated circuits, this chapter is designed to familiarize the multidisciplined thinker with a basic understanding of the role of signal processing and memory in a portable electronic system.

Chapter 3 covers electronic packaging in some detail. The portable electronic designer is likely to be directly involved in the technical design of the electronic packaging since this is the means by which integration of the system components takes place. Electronic packaging design, as has been demonstrated by Japanese manufacturers, is a decisive tool in creating advanced portable electronic products.

Chapter 4 discusses the critical role of display technology to portable electronic products and introduces the reader to some key terms used in the characterization of display technology. The discussion is mostly in the context of LCD's, but OLED and other emerging technologies are mentioned briefly.

Chapter 5 is titled Power Sources, but from a practical standpoint it is about batteries. Numerous examples of battery implementations are depicted to familiarize the reader with different methods of battery implementation. A brief discussion of battery life and power management is also included.

Chapter 6 addresses mechanical design. Product housings, electrical shielding, thermal management, mechanical integration, and manufacturability are all discussed. Considerable attention is paid to thermal design in notebook computers since this situation represents one of the more intricate mechanical design issues in portable electronics today.

Chapter 7 covers both software and communications. The product designer will generally drive the software and communications requirements through the functional specification and through the generation of use-case scenarios. Nevertheless, it is important for the portable electronic designer to understand the commonly used abstractions for describing the software and communications elements of the system. Related sections on systems I/O and wireless standards are also included in this chapter.

Chapters 8 through 11 cover four major areas of portable electronics, namely, cellular phones, portable PCs, PDAs, and digital imaging products. Each of these product areas is discussed in the context of a product case study taken from the database of Portelligent, Inc. These examples are intended to rapidly familiarize the user with the typical design practices used in contemporary portable electronic products.

Chapter 12 discusses the economics of portable electronic products. Learning curve theory is discussed in some detail, as it is a fundamental and powerful tool that can have a major impact on product design decisions. Likewise, product platform management is discussed to help the product designer increase the profitability of a product line.

Chapter 13 speaks to the past, present, and future of portable electronics. A brief history of the portable electronics industry is presented to give the reader a sense of the factors that have shaped the current state of affairs. Some advanced product concepts are also discussed and an opinion on the future direction of the industry is presented.

Effective portable electronic product design requires knowledge of design processes, key technologies, infrastructure, economics, and a vision of the future. Taken together, these chapters are intended to provide a unique and comprehensive perspective on the design of portable electronic products.

Acknowledgments

I would like to thank Dave Carey and the entire team at Portelligent, without whom this book would have been impossible. Almost all of the examples and photographs and much of the analysis in this book are derived directly or indirectly from Portelligent intellectual property. Any errors in interpretation of their excellent work are my responsibility alone.

I met Dave Carey when I went to work for the Microelectronics and Computer Technology Corporation (MCC) in 1989. Initially I was assigned there as an employee of Eastman Kodak to work on multicompany research in advanced electronic packaging. By 1994, I had become a direct employee of MCC and Dave and I were working together on the *Portable*

Electronics Project. Over the next six years, we built an evolving team with many outstanding contributors, including Tom Hunter, Bill Weigler, Masako Robertson, Phil Gilchrist, Ed White, Chris Windsor, Claude Hilbert, Howard Curtis, Rick Nolan, Scott Anderson, and John Stockton. Much of this book comes from what I learned as a result of being a part of this team. Tom, Bill, Masako, Chris, and Howard left MCC in 2001 with Dave, to start Portelligent, where they improved their methods further and continue to provide the best information on portable electronic design available to the industry. I have continued to work closely with the Portelligent team as a member of their Board of Directors.

I would also like to thank a number of people who supported me over the years, as either loyal customers or as managers, in my efforts to learn more about portable electronics. These include Larry Smith of MCC, who granted me the budget for my first product teardown; Terry Clas, David R. Smith, and Cathy Olenick of Kodak; John Thome, Sanjar Gahem, Iwona Turlick, and Bob Hurley of Motorola; Larry Marcanti of Nortel; Seppo Pienemma of Nokia; Denny Hammill of 3M; Lance Mills of HP; and John McRary, who gave us the necessary trust and independence to effectively run the Portables program at MCC.

Finally, I would like to thank my wife, Tina, for her editing assistance and moral support in the preparation of the manuscript.

1

The Portable Electronic Design Process

To be successful, the portable electronic designer must view the product design process as a part of a broader process of product development. While the insights and creativity of the individual designer are critical in creating a successful product, the designer must be prepared to engage through every phase of product development if the product is to maintain its intended characteristics. The successful product designer will be skillful at shepherding the product through the entire product development process.

1.1 The Product Development Process

The development of new technology-based products is a complex process and involves multiple disciplines. Product design, marketing, engineering, manufacturing, and procurement organizations all have a role in the development of a new product. As such, the design of a portable electronic device is only a part of the overall product development process. The portable electronics designer should understand the role of product design in product development. The phases of a typical product development process are shown in Fig. 1.1.

This view of product development was formulated to allow outside consultants to communicate with managers of large corporations. The phases shown here roughly correspond to the functional divisions that exist within a large corporation. Smaller companies may find this formalism too constraining. Regardless, it is important that the product designer understand all the conceptual phases of product development, since the latter phases are impacted by the product design.

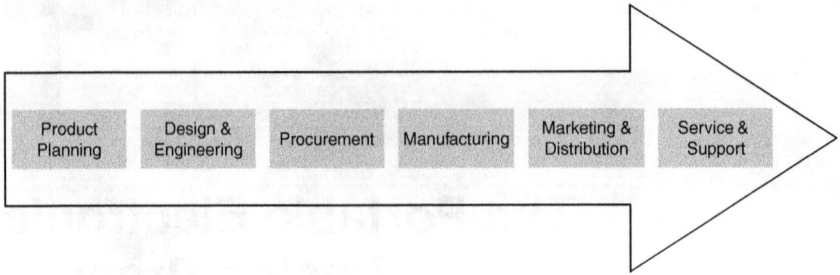

Figure 1.1 A typical product development process.

1.1.1 Product planning

Product planning is the process of evaluating the feasibility, profitability, and initial business plan for bringing a new product to market. Completion of the product planning phase results in management approval (or venture capital financing) of the design and engineering resources necessary for the next phase of development. Product planning must contemplate all other phases of the product development to arrive at a credible business plan.

The product design begins during this phase. The product concept is articulated, defining the target market and enabling an assessment of the total available market for the product. The product concept is reduced to a set of product requirements. These product requirements are not a detailed technical specification, but rather a high-level set of functional requirements that enable the product designer to establish technical feasibility and approximate cost to produce.

1.1.2 Design and engineering

The design and engineering phase is the most highly leveraged portion of the design and development process. According to the Aberdeen Group, 70 percent of the cost of a product is constrained during the design and engineering portion of the product development cycle. Only 30 percent of the cost can be impacted once the product is released to manufacturing. This is due to the fact that the cost of a design change increases exponentially as the device advances through the product development process. Once manufacturing tooling is committed and purchasing contracts are in place, certain types of design changes will become cost prohibitive with respect to the business plan. Thus, the design and engineering phase will define the manufacturing infrastructure and the supply chain required to support the product.

The first task in the design and engineering phase is to create a *product specification* derived from the product requirements. The product designer must work carefully with engineering and marketing to ensure that the product requirements are met without injecting unnecessary cost adders into the design.

Next, *system architecture* must be developed which enables an efficient product design, while enabling all system specifications to be met. This architecture must also accommodate other aspects of the product development process by enabling the product to be upgraded, modified, optimized for manufacturing, etc. The product designer must then select the major system components that will comprise the system. Trade-off analysis is performed to determine the optimal combination of components and system partitioning for meeting the product specifications. In practice, the results of a trade-off analysis may cause the designer to modify the system architecture as well. The ability to quickly trade off the benefit of various architectures and components is essential to effective portable electronic product design. This iterative process of architectural definition and trade-off analysis of available components ultimately results in an optimal system architecture that can be used to drive the *circuit design*.

Once the system architecture definition and circuit design is complete, the *physical design* may take place. Physical design involves a major expenditure of engineering and design resources and will result in a complete definition of the product needed for manufacturing.

1.1.3 Procurement

With a physical design established, the procurement organization must develop contracts to provide the necessary components to build the product. In some cases, certain components may be supplied internally, but most components (and major subassemblies) will likely be sourced globally.

While actual procurement will generally take place after the product design is finalized, the procurement organization is often called upon to provide certain information during the component selection phase. Accurate estimates of component pricing and availability are critical in component selection. Commodity components such as chip resistors are available from multiple suppliers and basic supply information can be quickly provided with a high degree of certainty. Highly specialized components or custom components require more effort to gather the desired price and availability information and, once gathered, this information is subject to greater uncertainty. The product designer must work closely with procurement to manage the risk associated with this uncertainty. This generally involves getting quotes from the component vendors and conducting due diligence to assess the credibility of the information provided in the quotes.

1.1.4 Manufacturing

Once a product is ready for manufacturing, a significant amount of resources has been invested in creating the product design. Even so, the decision to commit manufacturing tooling to a specific revision of the product design is one of the most critical financial decisions in the product development process.

Manufacturing tooling includes items such as plastic injection molds, special fabrication tooling for various metal parts, mechanical and electrical test fixtures, creation of control programs for automated manufacturing equipment, and custom semiconductor masks for ASIC devices. Prototyping techniques, while expensive on a per unit basis, are typically employed to verify the product design prior to a commitment to high-volume tooling.

The present trend is for manufacturing to be performed by contract manufacturing houses, which are very efficient at executing manufacturing operations.

The product designer must be very familiar with the choices and capabilities of various manufacturing processes. An entire discipline, known as *design for manufacturing* (DFM), has been developed to ensure that product designs can be manufactured cost effectively. Most design tools enable the designer to constrain a design to comply with the manufacturing process limitations of a specific manufacturer. These constraints are commonly known as *design rules*. Careful adherence to design rules during the physical design phase will limit the number of design changes required once the product enters manufacturing. On the other hand, accepting some technical risk and pushing the limits of process capabilities often result in a product that is very competitive.

1.1.5 Marketing and distribution

Effective marketing and distribution of a portable electronic device should begin very early in the product development process. Portable electronic devices generally require high volumes of shipments in order to be a commercial success. Therefore, large numbers of potential customers need to be made aware of the product benefits. Effective advertising and promotional campaigns must be implemented to create strong demand at the time of product introduction. Failure to do so can result in large inventories in the distribution channel. These inventories can threaten the commercial viability of an otherwise successful product. Product launch could be (and often is) initiated with a relatively low volume of supply in the distribution channel. This reduces financial risk, but leaves the door open for nimble competitors to respond with like products and capture market share.

Marketing activities are not limited to the generation of end user demand. Retailers and wholesalers must be persuaded to stock and promote the product. Infrastructure partners must often be convinced to provide support technologies and services. Indeed, suppliers may even need to be convinced that their resources are best spent supporting the product rather than some other customer project. The world of portable electronic devices revolves around infrastructure, and the product designer must be prepared to support marketing efforts aimed at developing that infrastructure.

1.1.6 Service and support

The product designer can minimize the service and support costs by creating a reliable and easy-to-use product. Low-cost products may simply be replaced if they develop a defect. More expensive products should be designed so that key components can be quickly and easily replaced by a service technician. For products that are connected to a network, the ability to remotely troubleshoot and manipulate the system software is a powerful way to provide service and support. The very best products, however, are designed not to break!

The reader should notice that most of the actual design of a portable electronic product takes place in the product planning and in the design and engineering portions of product development. The key phases of product design that we are concerned with are product concept, product requirements, system architecture, trade-off analysis, circuit design, and physical design. These are shown in Fig. 1.2.

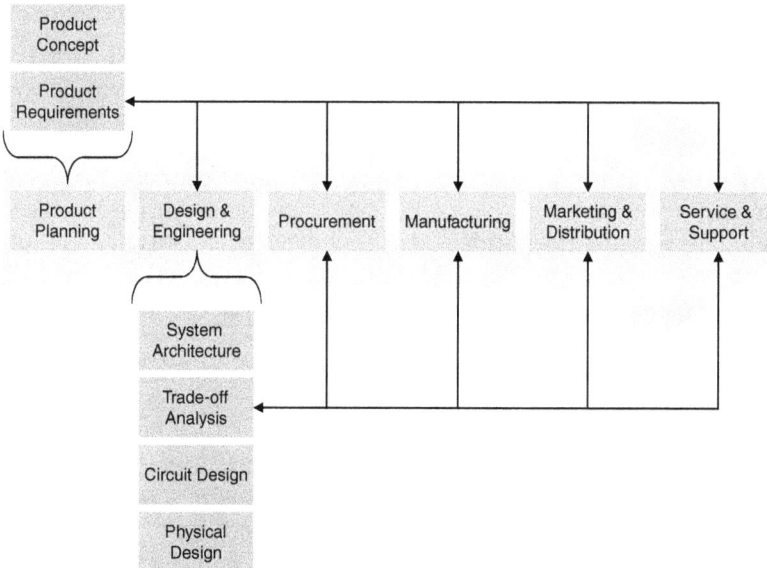

Figure 1.2   The key phases of product design within the product development process.

1.2   Portable Electronic Design Factors

The principal factors that shape the design of a portable electronic device are as follows:

- Functionality
- Performance
- User interface
- Form-factor
- Battery life
- Cost
- Time-to-market
- Reliability

The priority of these factors depends on the product concept and the market segment that is being targeted. While trade-offs between these different factors are a natural part of the design process, maintaining a clear distinction between these factors is essential in developing the product concept. While there are other ways to organize the elements of a portable electronic device, this particular set of factors is particularly important because it covers all dimensions of the product, and each factor corresponds to a quality that is directly experienced by the end user.

1.2.1   Functionality

In almost all cases, functionality is the most important factor in the design of a portable electronic product. To design a successful product, the functionality must be clearly defined, focused, and not confused with the other design factors. Unfortunately, many product designers succumb to the temptation to add additional functional features because the technology enables additional features at no extra cost. The result has often been an unfocused product concept that allows the individual to perform too many functions. All too often, such products do not perform any single function very well.

While many portable electronic devices do have multiple functions, these functions are often secondary to a primary product functional concept on which the product is based. These secondary functions may be well received if they complement the primary function, but the designer must take great care that these additional features do not obscure or dilute the primary purpose of the device.

In his book, *The Evolution of Useful Things*, Vintage Books, New York, 1994, Henry Petroski examines the characteristics of extremely successful tools and devices. Petroski demonstrates the extent to which certain designs have evolved over the years to a point of near perfection. Commonplace tools, such as paper clips, forks, and screw drivers, have been fastidiously modified over many decades, to perform a single function very well.

Tools such as the screwdriver and the chisel perform distinctly different functions and yet are very similar in form and appearance. It might be tempting for a designer to make a chisel that could also function as a screwdriver, but the result would be a device that performed neither function very well. Adding functionality that causes the product's primary function to become nonoptimal is, in general, a poor design choice.

The modern PC is, of course, a direct contradiction of this concept. The PC has been the preferred platform for delivering new types of functionality for the past 20 years. One must keep in mind, however, that from a historical perspective, we are still in the early stages of a new exploiting electronics technology. Furthermore, the PC is in essence a type of physical user interface, consisting mostly of a keyboard and a display. The products are actually the individual software packages that perform a specific set of functions. In his book, *The Inmates Are Running the Asylum*, Sams Publishing, Indianapolis, 1999, Alan Cooper takes the technology industry to task for producing software that confuses the user by offering too much unnecessary functionality.

As the PC industry matures from its embryonic stages, we are starting to see segmentation of the market into a broader range of specialized product types. Current types of PCs include servers, workstations, desktops, desktop replacement notebooks, full size notebooks, thin and light notebooks, and ultraportables. Workstations tend to be used for high-end engineering, science, or financial applications and simulations in a stationary office environment. Ultraportables tend to be used mostly for mobile connection to e-mail and applications that support business communication and analysis. The PC is slowly but surely being refined into a range of specialized products that will have very focused functionality. Just as large rocks were the first tools for the early humans, the PC is the blunt instrument of the emerging information age.

Examples of well-defined functionality may be observed in a number of successful portable electronic product concepts. Garage door openers (Fig. 1.3) are an almost unnoticed element of modern suburban life. Most people never give their garage remote a second thought; that is, until it stops working! The utility of the garage door opener is unquestionable. Pushing a single button performs an essential function. The user does not even need to select which button to pick since the same button opens and closes the door. The device relieves the user of a laborious chore and yet requires almost no mental or physical effort on the

Figure 1.3 Picture of garage door opener.

part of the individual. To the extent that the user doesn't even have to think about pushing the button, the effect is almost telepathic.

The television remote control is another very successful product that has a very well-defined set of functions. A typical remote controller, like those pictured in Fig. 1.4, will turn the TV on and off, will allow the user to scan up and down through the channels, and will allow for volume control and direct selection of a particular channel number. Users have come to expect these basic functions and will typically make little use of any other functions on the controller. Universal remotes have been developed recently to control an expanding list of home entertainment components. These devices are much more difficult to use and as a result most people opt to use a collection of individual remotes that are each optimized to control a different home entertainment component.

The wristwatch is perhaps the world's most familiar portable information appliance. These devices have evolved over the years to provide an essential function with a minimum of user effort. Indeed, the wristwatch has become such an essential tool that it becomes an extension of the individual's personal identity.

The Sony Walkman is a prime example of a portable electronic product that embodies well-defined and valuable functionality. The Walkman was designed to enable an individual to listen to high-quality music whenever and wherever desired. This well-defined notion of functionality has

Figure 1.4 Picture of television remotes and universal remotes.

made the Walkman one of the most successful portable electronic devices ever created.

To design a successful portable electronic device, it is important not only to define the product functionality clearly, but also to avoid confusion between functionality and other factors such as performance and user interface. Lack of clarity around these factors can undermine the designer's ability to meet his or her objectives. Table 1.1 lists a number of portable electronic devices and the functionality provided by each.

In general, the product functionality should be clearly articulated as the primary purpose of the product. Technical metrics or jargon should not be necessary in describing the functionality of the device.

1.2.2 Performance

Performance is a quantitative description of how well a product performs its function. Performance describes the level of functional satisfaction experienced by the user. There are three regimes of performance. First, there is a minimum performance required for the product to effectively deliver the desired functionality. Second, higher levels of performance may allow for product differentiation in a competitive market. Third, there are performance levels above which the user is no longer able to perceive an improvement in functionality. Figure 1.5 shows the different regimes of product performance.

TABLE 1.1 Primary Functionality of Some Common Portable Electronic Devices

Device	Primary functionality	Comments
Cellular phone	Enables remote voice conversation	Smart phones are gradually adding secondary functionality
Walkie talkie	Enables remote voice instant messaging	Multicast is useful secondary function
Pager	Delayed text messaging	Being made obsolete by cellular phone messaging and voice mail
Watch	Displays time (& date)	Secondary functions are mostly for niche markets
Calculator	Mathematical calculations	Well-focused functionality, *not* made obsolete by the PC or the calculator watch
Garage door opener	Opens and closes garage door	Highly focused functionality
Television remote	Controls TV	Turn TV *off* and *on*; scan channels; direct access channel selection
Universal remote	Controls multiple AV devices	TV, VCR, audio system, set-top box, etc.
Handheld game platform (like GameBoy)	Play video games	Sega tried to add TV tuner—didn't sell
Walkman	Private enjoyment of high quality music	Volume, bass, FWD, reverse, pause—what more could you want?
Laser pointer	Presentation pointing	
Transistor radio	Shared and private broadcast music and news	Made obsolete by Walkman and boom-box
Electronic organizer	Personal contact and calendar information	Being made obsolete by palm top PCs
Palm top PC	Platform for portable *Outlook*	Outlook function made popular on the desktop PC, transferred to a smaller form-factor
Notebook PC	Platform for windows applications	Multipurpose platform for a wide range of software product functionality
Handheld GPS	A smart map	Will probably be more popular as a secondary function than as a dedicated device

There may be a number of metrics that describe performance for a particular product. Optimizing different performance metrics may be important in positioning the product into different market segments. The relationship of these performance metrics to product functionality is by no means linear and is often quite unpredictable. The interaction between different performance metrics can also be quite surprising.

Performance Regimes

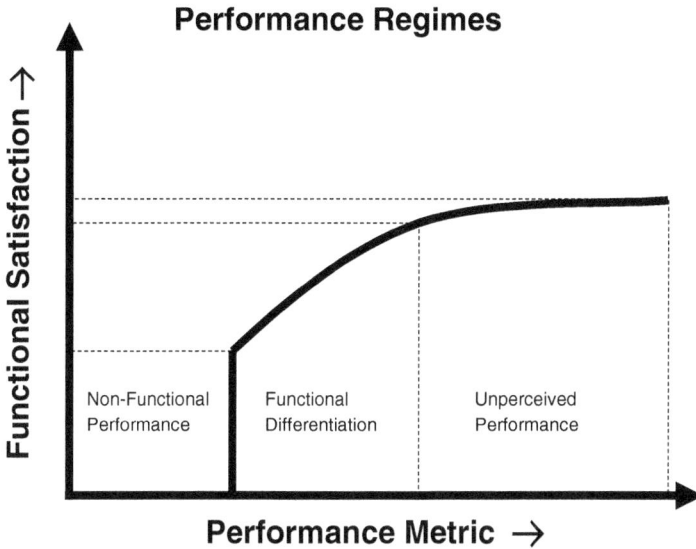

Figure 1.5   Product performance regimes.

Furthermore, the value of performance is often defined in the eye of the beholder. Some users will accept a complex or slow user interface, or high maintenance requirements, in order to gain performance. Often these compromises are a necessary condition of high performance.

Improving certain dimensions of performance can make a product more competitive up to a point. Once a certain level of performance is achieved, however, further improvements may become unperceivable to the end user. This is certainly the case today with microprocessor speeds in the personal computer industry. As processor clock speeds have moved above 1 GHz, most users have not noticed a meaningful improvement in the product functionality. In the case of personal computers, the users' perception of functionality is now limited by other aspects of performance such as the delay associated with accessing peripheral devices, network related delays, and of course the unexplained "hour glass" waiting periods imposed by the operating system. The response time of an application to a user's interaction is what the user actually notices. Further improvements in PC microprocessor speeds, which do not noticeably improve application response times, are about as useful as developing a consumer automobile that can go 250 mi/h.

Audio fidelity is an interesting performance metric. Obviously, great performance gains have been made in the audio performance of electronic devices from the invention of the gramophone until the development of Dolby audio systems. Again, we can examine the example of the Sony Walkman. Prior to the creation of the Walkman, it was certainly possible to plug a low-quality monaural earphone into a transistor radio or a handheld tape recorder, but the experience was not compelling for most individuals. Sony's products broke through the minimal audio performance levels required to deliver the desired product function.

In light of this example, it is interesting to ponder the potential effects of audio fidelity relative to cellular phone products. Even under ideal conditions, the audio performance of contemporary cellular phones leaves much to be desired. To what extent would truly high fidelity sound impact the world's largest portable electronics market segment?

The line between user interface and product performance may often be unclear. In the case of a portable DVD player, for example, the video resolution of a display is an important performance metric that describes the level of product functionality. In the case of a cellular phone, however, the display resolution is not a critical performance metric, since the display is used as a user interface and not to deliver the primary product functionality. That is not to say that higher display resolution cannot be a differentiator in a cellular phone product. Indeed, the user interface may be greatly improved by clearer font renderings and the use of display graphics. But it is important for the product designer to delineate the performance metrics that directly impact the primary product functionality.

The bottom line is that *performance* should reflect how well the user perceives the product perform its primary function. Table 1.2 gives a number of portable electronic product functions and their associated performance metrics.

TABLE 1.2 Performance Metrics for Various Portable Electronic Devices

Device	Primary performance metrics
Cellular phone	Range, audio fidelity, latency
Walkie-talkie	Range, audio fidelity
Pager	Range
Watch	Accuracy
Calculator	Precision, number of functions
Garage door opener	Range, false signal rejection
Television remote	Range
Handheld game platform (like GameBoy)	Processing speed, graphics/video fidelity
Walkman	Audio fidelity
Transistor radio	Range, audio fidelity
Electronic organizer	Response time, content storage capacity, graphics capability
Palm top PC	Application response time, application initiation/transition speed, content storage capacity, graphics/video fidelity, network communications bandwidth
Notebook PC	Application response time, application initiation/transition speed, content storage capacity, graphics/video fidelity, network communications bandwidth
Handheld GPS	Accuracy, response time

1.2.3 User interface

The user interface for a portable electronic device is the means by which a user extracts the functionality of the device. The user interface encompasses both the user inputs to the device and the outputs from the device that are sensed by the user.

Output elements of the user interface include visual, audio, and tactile actuators that the user can interpret. These correspond to the senses of sight, sound, and touch. Thus far, no one has developed a way to communicate taste or smell from a portable device. Input elements of the user interface include visual, audio, and physical sensors that enable the user to control the device. Table 1.3 gives examples of user interface mechanisms for portable electronic devices.

There is another element of the user interface which goes beyond the physical elements of the device. Portable electronic devices have architecture of logical sequences, which enable the user to interact with the device. This architecture defines the sequence of buttons or other physical device manipulations that must occur to achieve a particular result. As the user manipulates the physical controls, the device changes its logical states. Even in relatively simple portable electronic devices, a user may become confused as to the actual state of the system. Alan Cooper coins the term *cognitive friction* to describe the confusion that a user can experience when using computer technology.

Cognitive friction results from a failure to focus on the needs of the user. Good products should deliver their primary functionality to the user with ease. Interactions with the product should be intuitive to the user. Problems often arise when the product designer adds "features" to a product concept that do not add value to the primary functionality of the device. This situation is known as *feature creep* and it tends to result in a more confusing user interaction. The decision to add features means that the product can be put into a greater number of states. The user is then forced to deal with the complexity of these

TABLE 1.3 User Interface Mechanisms for Portable Electronic Devices

Human sensory mode	User interface mechanisms	
	Input	Output
Visual	Image sensor, scanner, photo-cell	Display, indicator light, gauge
Audio	Microphone	Speaker
Tactile	Button, switch, keyboard, dial, joystick, mouse, trackball, touch-pad, touch-screen, inertial sensor	Vibration feedback
Taste	—	—
Smell	—	—

additional states even if he or she is not interested in the features that they provide.

Thus, the user interface comprises both the physical elements that facilitate input and output to the device as well as the user interaction as defined by the system's logical architecture. The user interface of a portable electronic device should be optimized around the device's primary functionality. Extraneous user interface features which address secondary functions or noncritical performance metrics can detract from the user's satisfaction. In short, well-designed portable electronic devices should not require an instruction manual.

1.2.4   Form-factor

Form-factor encompasses the size, weight, and shape of a portable electronic device. Form-factor is critical to any portable electronic design. Indeed a small form-factor determines whether or not a product is considered portable. As with performance, breakthroughs in form-factor can enable entirely new product concepts and functionality. Achieving the desired form-factor is often the most critical engineering challenge in developing or differentiating a portable electronic device.

While miniaturizing the system components is a common challenge in portable electronic design, the minimum size is limited by user-interface considerations in many cases. Figure 1.6 shows how the display and keypad define the frontal area of a portable electronic device. Until a foldable display technology becomes available, at least one plane

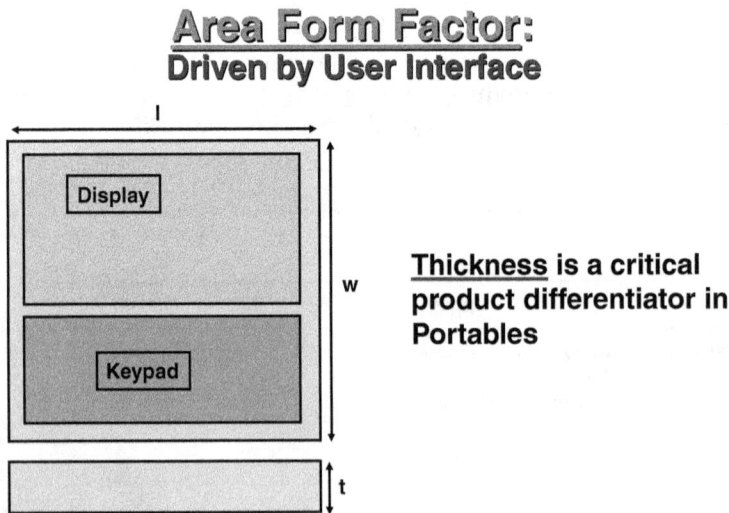

Figure 1.6   User interface drives form-factor.

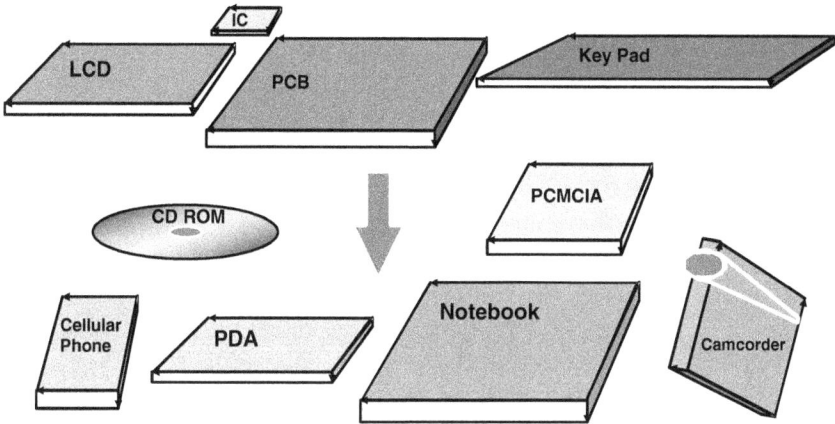

Figure 1.7 Two-dimensional interfaces drive flat products and components.

of the product must have an area equal to or greater than the area of the display module. Indeed, a larger display is seen as an advantage when comparing two otherwise identical products.

So, the desire for an ample display and the desire for small form-factor are at odds. The result is that thickness is often an important differentiator among portable products and the constituent component technologies (Fig. 1.7). Minimal thickness enables minimal volume in the face of a fixed user-interface area. Flat devices also tend to be more easily stowed in a pocket or purse than thicker devices that occupy the same total volume.

There are a number of categories of form-factors into which the world of portable electronic devices may be logically divided. These form-factors are named after common personal objects that have well-understood portability parameters. These form-factors are by no means standardized and in many cases they overlap. Understanding the way a product will be used or carried, not just its size, is important in determining the proper form-factor category.

Tool kit. As the name suggests, the tool kit form-factor includes the largest products that could be considered *portable*. Products with this form-factor can be quickly packed up and moved although they may be rather heavy and bulky. Portable military communications gear, early "luggable" PC products, boom boxes, and a wide range of specialized medical and industrial instruments fall into this category. See Fig. 1.8 for examples of products with this form-factor.

Notebook. The notebook form-factor is meant to imply a device that is roughly the size and shape of a notebook. This form-factor has actually

Figure 1.8 Products with a tool kit form-factor.

been extended to cover any product that can be carried around like a book, even if that book might be a rather hefty encyclopedia! Notebook products are expected to be carried from place to place by hand or in a briefcase but are typically set onto a table or flat surface when in use. In some cases, such as a notebook tablet, the device may be used while being cradled in one arm.

The notebook computer market is currently divided into a number of subcategories of form-factors. These include desk-top replacement, full size, thin and light, and ultraportable and tablet. Figure 1.9 shows the approximate size and weight ranges for these different form-factors. One of the shortcomings of notebook computer designs is that they do not achieve the weight and ruggedness of similar-sized books. As a result, while notebooks may be carried in hand or in a briefcase similar to a book, the portability of these products is not as comfortable as it should be. Also, their fragility means they must be handled with care. They cannot be carelessly dropped on the floor or tossed onto a table top. Figure 1.10 shows some specific product examples.

Notebook PC Form-Factor	Max. Length (cm)	Min. Length (cm)	Max. Width (cm)	Min. Width (cm)	Max. Thickness (cm)	Min. Thickness (cm)	Max. Weight (kg)	Min. Weight (kg)	Max. Volume (L)	Min. Volume (L)	Range of Specific Gravity	
Desk Top Replacement (DTR)	38.1	30.48	30.48	25.4	6.35	3.81	3.6	2.7	7.37	2.95	0.49	0.92
Full Size	33.02	27.94	27.94	22.86	4.572	3.175	2.7	1.8	4.22	2.03	0.65	0.89
Thin & Light	33.02	22.86	27.94	19.05	3.175	1.778	1.8	0.9	2.93	0.77	0.62	1.17
Ultra Portable	22.86	17.78	21.59	16.51	3.175	1.27	0.9	0.5	1.57	0.37	0.58	1.22

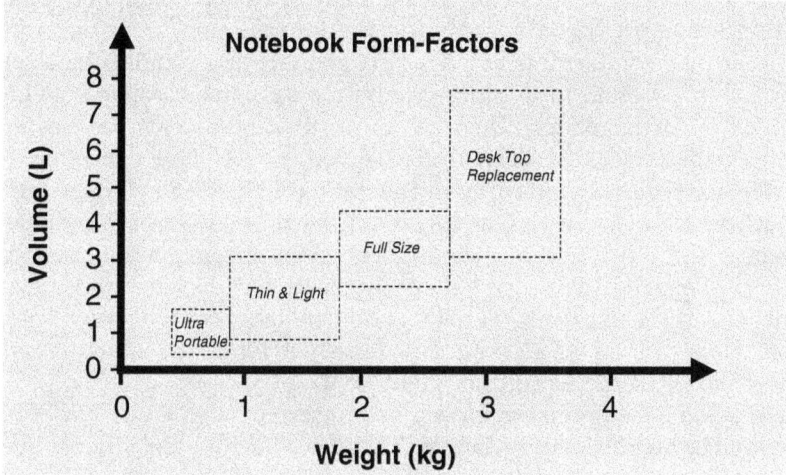

Figure 1.9   Notebook computer form factor parameters.

- DTR
- System weight: 7.6 lbs.
- Dimensions (inches): $13 \times 10.7 \times 1.8$

- Ultra-Portable
- System weight: 1.8 lbs.
- Dimensions (inches): $1.2 \times 7.3 \times 5.5$

- Thin & Light
- System weight: 3.5 lbs.
- Dimensions (inches): $0.9 \times 10.5 \times 9.5$

Figure 1.10   Notebook computer form-factor examples.

Palm top. The palm-top form-factor is one of the most common and important form-factors for portable electronic devices. It is the form-factor of choice for calculators, personal organizers, palm-top PCs, and some TV remote controllers. Typically, palm-top devices are carried in a pocket or briefcase when not in use. They are usually held in the hand during use or occasionally set onto a flat surface next to the user.

Creating a comfortable balance between size and weight (density) is a frequent problem in designing palm-top systems and portable electronic devices in general. Devices that contain many electronic components or batteries often end up being very dense. This can make the device uncomfortable to carry in the hand and also awkward to put in a pocket, since it will weigh the pocket down and deform the shape of the user's clothing. Some in the industry have suggested that a specific gravity of around 1.0 is the ideal density for a portable electronic device. Since this is the density of water, it closely matches the density of human flesh; therefore an object with this density is likely to feel like a natural extension of the hand that wields it.

Palm-top devices tend to have a user interface with a number of buttons and/or an information display. The idea of holding the product in the palm, while in use, implies that the functionality of the device involves significant user interaction. Calculators and palm-top PCs certainly illustrate this idea (see Fig. 1.11).

Figure 1.11 Examples of palm top form-factor devices.

Figure 1.12   Examples of cigarette pack form-factor devices.

Cigarette pack (shirt pocket). The cigarette-pack form-factor significantly overlaps the lower end of the palm-top form-factor in terms of size. The cigarette form-factor is driven more by the desire for compactness than by the way the product is used. This form-factor is intended to enable the device to be carried in a suit pocket, a front trouser pocket, a small purse, or in some cases a shirt pocket. Many cellular phone products, cameras, and Walkman-type products strive to achieve this general form-factor. (Current social trends may someday require the renaming of this form-factor.) Figure 1.12 shows several cigarette-pack form-factor devices.

Pen. Pen form-factors are driven by a desire to carry the device in an extremely unencumbering fashion and/or by a desire to take advantage of the handwriting posture for the user interface. Also, a pen form-factor is ideal for clipping into a shirt pocket. This is an extremely useful form-factor but it is limited in functionality due to current technological limits. Product designers have tried to push the envelope of this form-factor with mixed results (see Fig. 1.13).

Key chain. Key chain form-factors are driven by a desire for extreme portability. Since most people have learned to keep track of their keys, connecting a small device to a key chain is a good way to ensure that it

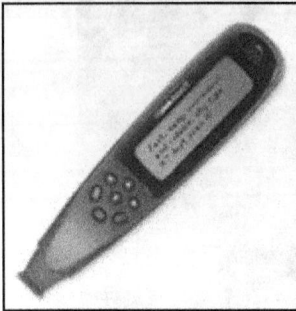
Scanner/Translator

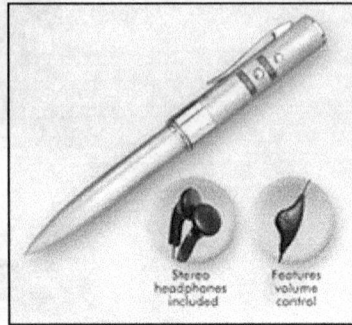
Radio

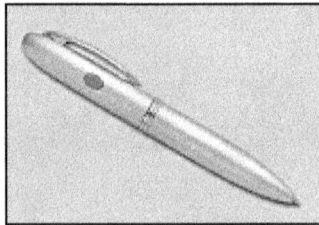
Laser Pointer

Figure 1.13 Examples of pen form-factor devices.

is not lost or left behind unintentionally. Electronic automobile access and other security devices are an obvious functionality for this form-factor. Mass storage USB drives have become available in this form-factor and are proving to be quite popular, as shown in Fig. 1.14.

Credit card. The credit card form-factor is also driven by the desire to achieve transparent portability. Most credit cards can in fact be considered portable electronic devices, since they utilize a magnetic strip to store information. Smart cards are an evolved form of credit card that uses an embedded chip to store larger amounts of information. Other more functional devices have also sought to achieve the credit card form-factor. These include calculators, secure ID cards, electronic scales, and even personal organizer devices such as the Rex. Figure 1.15 shows several examples of this form-factor.

The credit card form-factor can be carried in a wallet if sufficiently rugged. More fragile versions can be carried in a shirt pocket, a briefcase, or a purse. The surface area permits a reasonable size display or a number of buttons to be incorporated to the user interface. The credit

Figure 1.14   Examples of key chain form-factor devices.

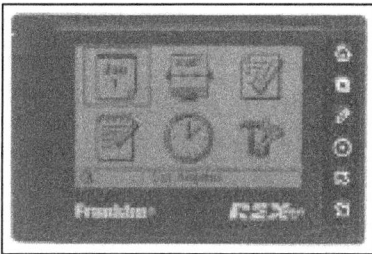

Personal Information Device

Calculators

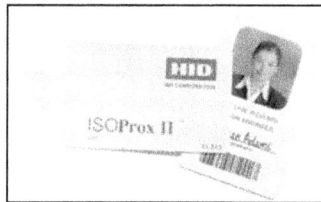

Security Badge

Figure 1.15   Examples of credit card form-factor devices.

card form-factor is one of the most important form-factors for future portable electronic devices.

Wearable form-factors. Wearable electronics was a very popular research theme in the mid 1990s, attracting much attention because of the "WOW" factor associated with "wearing" a computer. Most of the devices that were showcased during this era were too awkward in appearance for all but the most socially challenged. Small bricklike battery packs worn on the belt, or binocular-sized head-mounted displays may be fine for the soldier, but they are unlikely to be popular with the average consumer and business person. Fortunately, there are some successful wearable form-factors that have been developed over the course of human history.

The wristwatch represents the first widely adopted portable information device. Its widespread use was driven by military officers in the nineteenth century who required instant access to the time in order to conduct effective military maneuvers. By keeping the timepiece strapped to the top of the wrist, chronological information is readily accessible even when the hands may be occupied with other tasks. Due to the popularity of mechanical wristwatches, digital watches were quickly adopted as one of the first successful portable electronic devices. Specialty watches that provide additional information beyond the primary function of telling time, such as stopwatch, pulse monitor, and altimeter, have been moderately successful. Other products have attempted to overextend the functionality of the form-factor and have been less successful.

Some recent examples of wristwatch form-factor devices are shown in Fig. 1.16.

Calculator

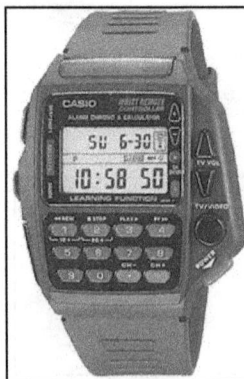
TV Remote

Digital Camera

Figure 1.16 Examples of wristwatch form-factor devices.

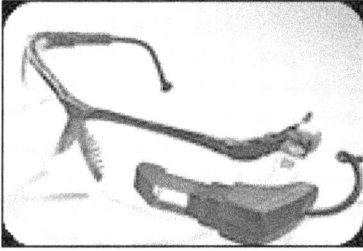
Eyeglass mounted display

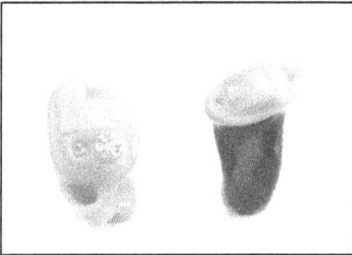
Belt Mounted–Voice Activated
Computer

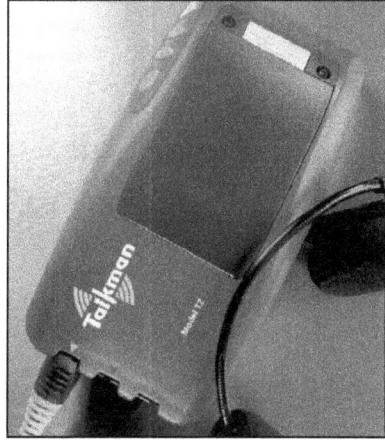
Hearing Aid

Figure 1.17   Examples of *other* wearable form-factor devices.

Other items whose form-factors may be emulated for portable electronic devices include eye-glasses, hearing aids, rings, necklaces, belts, buckles, and clothing (Fig. 1.17).

Other comments on form-factors.   The form-factors discussed in this chapter are for illustrative purposes. Certainly there are no formal constraints on how form-factors should be classified. Portable consumer camcorders designed in the early 1990s exhibited a wide range of custom form-factors as designers searched for an optimal design. It is of critical importance that portable electronic designers carefully select a form-factor that is consistent with the way a product is intended to be used and carried.

It is useful to observe that the form-factor of most portable electronic devices is determined by the frontal area required to accommodate the user interface, or by the size of some required storage media such as a cassette or a DVD. The product thickness then tends to be the dimension by which many designers will differentiate the product. This is certainly exemplified in the case of notebook computers where a considerable premium is afforded products which are less than 1 in thick. A summary of portable form-factors is presented in Fig. 1.18.

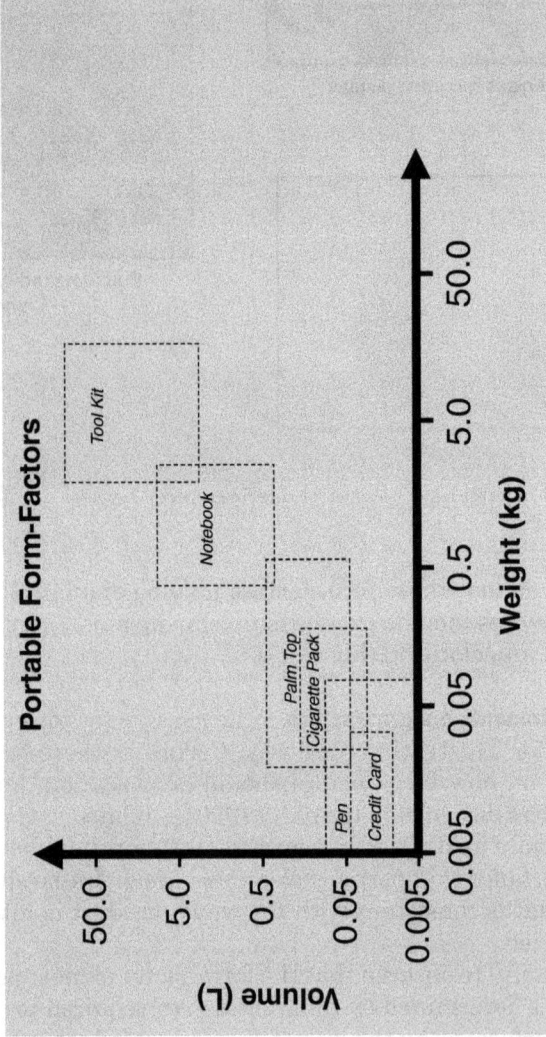

Portabe Form-Factor	Max. Length (cm)	Min. Length (cm)	Max. Width (cm)	Min. Width (cm)	Max. Thickness (cm)	Min. Thickness (cm)	Max. Weight (kg)	Min. Weight (kg)	Max. Volume (L)	Min. Volume (L)	Range of Specific Gravity	
Toolkit	61	20	30	15	30	10	18.144	2.268	56.634	3.146	0.32	0.72
Notebook (Computers)	38	18	30	17	6	1	3.629	0.454	7.374	0.373	0.49	1.22
Palm Top	11	8	10	7	4	1	0.567	0.014	0.442	0.027	1.28	0.50
Cigarette Pack	10	8	6	4	3	1	0.227	0.023	0.164	0.043	1.38	0.53
Pen	18	9	2	0.64	2	0.64	0.091	0.005	0.065	0.004	1.41	1.27
Credit Card	10	8	8	5	0.32	0.25	0.045	0.005	0.025	0.010	1.85	0.46

Figure 1.18 Portable form-factor parameters.

1.2.5 Battery life

Battery life is an important factor in the design of a portable electronic device because existing power sources (usually batteries) are quite limited. Battery life describes how long a product may be used in an untethered and truly portable mode. A short battery life is in direct conflict with the functionality of some products but has relatively little impact on others. In most cases, users have learned to adapt the way they use a product to match its battery life. In almost all cases, a longer battery life would be a valuable product differentiator. Some day, high-energy batteries may make it possible for the user to be completely oblivious to power source issues.

The success of early cellular phone products was hampered by short battery life. Improved battery technology and reduced power consumption were quickly developed, and now most cellular phones can operate for an entire day without being recharged. Heavy users still run into battery life problems, and while most users have become accustomed to plugging in their phone every night, the ability to completely ignore recharging issues would be welcomed by all.

Some products, such as electric toothbrushes and electric shavers, are used only for a short period of time each day and can be stowed in a charging cradle so that the user never perceives a battery life issue. Even with these products, however, increased battery life could reduce the counter or shelf space required to store the product and eliminate the need to transport charging hardware when traveling.

Notebook computers are highly sensitive to battery life issues. The demand for untethered usage of small form-factor notebook computer products is on the rise. Battery technology, power consumption, and power management are not yet able to provide the battery life that users would really like. Corporate users must still keep an eye open for a conference room outlet just in case their notebook runs out of juice during an important meeting. Extra batteries and power adaptors must be carried by road warriors if they are to ensure their ability to be productive while traveling. Consumers want to be able to sit on the couch with a notebook or tablet PC and not have to contemplate stringing a power cord across the living room floor.

On the other hand, there are portable electronic devices for which battery life has ceased to be an issue. Devices that have many months of battery life include calculators, watches, and television remote controls. Solar calculators have an unlimited battery life as long as there is ambient light available to provide power.

Engineers and designers talk about battery life in terms of continuous usage and standby mode. A TV remote is always in standby mode and may only be transmitting control signals for a few seconds per day. A cell phone may be capable of being turned on and ready to receive calls for several days without a recharge, but a single 2-h phone call can drain the battery. The user of a notebook PC is not impressed if the product has a 30-day battery life in standby mode, particularly if it shuts down in the middle of a working session after a couple hours of heavy usage.

While extending the standby time of a product can be a worthy goal, it is certainly no substitute for extending the operational time of the product's primary functionality. Running out of power when the product is most needed can frustrate a user. It is unavoidable but ironic that most cell phones, when down to their last couple of hours of standby time, do not have enough power to receive or initiate a phone call. Would it not be better to shut the phone down completely, while there is still enough power left for a 2-min call?

There are three main elements that the designer must comprehend in order to develop a product with the appropriate battery life. They are session duration, session frequency, and standby conditions.

The session duration is the length of time that a single usage of a portable electronic device is expected to last. A product designer would certainly want to have a thorough understanding of the distribution of session times that might occur (minimum, average, maximum).

The session frequency is the number of times a product might be used in a given interval of time. Again, this is not a fixed number, but a distribution.

Standby conditions can vary greatly depending on the type of device. Some products are simply turned off by the user when not in use. Others turn on and off automatically when the user pushes any button. Electronic wristwatches are on all the time. Notebook PC designers have implemented an elaborate set of standby states in a brave attempt to extend system battery life as much as possible.

Providing a docking cradle that recharges the product is one useful way to smooth a user's battery life concerns. Domestic cordless phones are an example where a recharging cradle masks a relatively short battery life. Individuals are happy to return the phone to the cradle frequently so that they can find it when the phone rings. A recharge cradle will not, however, mitigate the battery life problem for all product concepts.

Ultimately, the designer's job is to thoroughly understand the desired patterns of usage and come up with a battery life strategy to meet those demands. Battery life is an important factor in the trade-off analysis aspect of system design discussed in Chap. 3.

1.2.6 Cost

As a general rule, the designer of a portable electronic product should seek to meet all the product requirements at the lowest possible cost. Product designers should have a cost model at their disposal that allows them to estimate system cost based on the current state of the design. Interaction with marketing, procurement, and manufacturing teams is critical in maintaining awareness of tactical and strategic issues which will impact system cost.

Understanding the true cost of a product is much more difficult than it may first appear. Looking at the cost of the product bill of materials (BOM) and the assembly cost does not paint the entire picture. Development costs can comprise a very large percentage of total system cost if the production volumes are relatively low. If the product involves a significant amount of market risk, it may be imprudent to invest heavily to reduce the manufacturing cost until demand for the product is proven.

Since it takes the sale of hundreds of thousands of units to justify the cost of developing a custom integrated circuit, many designers will choose a more expensive (on a per unit cost basis) programmable IC device to create the first generation of a product. System-on-chip (SOC) solutions, however, are making it easier for designers to use highly integrated, low-cost parts in an initial system implementation, but only when the system production volumes justify the SOC cost.

Large companies that manufacture portable electronics have a significant advantage in developing low-cost products. Besides being able to leverage purchasing power across multiple products, these companies can also minimize the development costs of products by reusing the designs of key components and subsystems. Nokia recently introduced a low-cost cellular phone product, for example, that uses the same electronics as a high-end phone designed over two years earlier and no longer in production. Such reuse, known as *waterfalling*, allows companies to amortize development costs to a very low percentage of product cost.

While there is no doubt that competitive pressures drive such companies to seek the lowest possible cost solutions for many products, there are also instances of strategic cost injection, which enable the manufacturer to realize a disproportionate advantage in the marketplace. Strategic cost injection can be used to improve a product platform's flexibility for future reuse, or it can be used to increase the product appeal and attract higher margins. It may also increase market share, driving learning curve related cost reductions to yield what is ultimately a less expensive product.

1.2.7 Time to market (TTM)

The old adage "Timing is everything" is an understatement when it comes to the introduction of a new portable electronic device. Being first to market with a new product concept or a major improvement to any of the critical product factors can have an immense impact on the product success. Missing the start of the holiday sales cycle, for example, could cost a company millions of dollars in lost revenue. Missing a market deadline by 12 weeks could relegate a company to a nonparticipant status in a hot new market. As with cost, there are many complex issues around TTM. The product designer must work closely with other business functions to make sure that TTM goals are met.

Large established companies that are leaders in the portable electronics market, such as Sony, do not need to be first to market to enjoy product success. For these types of companies, they will often adopt a *fast-follower* strategy, meaning that they will let others take the market risk and will respond with overwhelming force if a new opportunity is exposed. While Sony has done well in the innovation of new products such as the Walkman, it has also had great success in areas where it elected to be a fast follower, such as electronic still cameras and camcorders.

1.2.8 Reliability

Reliability describes the ability of a device to provide the intended functionality whenever needed, for as long as needed, under reasonable environmental conditions. A portable electronic device does not need to be broken to be considered unreliable by the user. It need only fail to perform its intended function at the moment it is required.

Different market segments have different reliability standards. Certain military electronics, for example, are required to function under extreme environmental conditions, during exposure to severe shock and vibration and after long periods of storage in adverse conditions. (This is one reason why military hardware tends to be very expensive.) Reliability specifications, such as Military Standard 883C, call out in detail the specific reliability testing that must be conducted in order to produce acceptable military products.

Much of what is known about physical reliability was developed by the military electronics and automotive electronics industries. Fortunately, much of what has been learned has been leveraged by commercial component suppliers to ensure a basic level of reliability with off-the-shelf components and subassemblies.

For the designer creating a product for a high-reliability market, the appropriate reliability specifications for that industry must be consulted and dealt with. For many consumer and commercial applications, the reliability requirements are not well defined. The product designer must review the individual component specifications to ensure that no unexpected reliability problems are created.

System stability is also an important element of reliability in portable electronic devices that have complex functionality. Extensive testing must be performed to ensure that an unexpected sequence of user inputs do not cause the system to stop functioning correctly. This is a major problem in the notebook computer industry and one that causes users a great deal of frustration.

Consumers tend to have rather high expectations of device reliability. Familiar portable consumer electronic devices such as the Walkman, the cordless phone, the cellular phone, and the calculator are fairly rugged and reliable. These devices turn on instantly, the operating systems do not crash, they can be used in a wide range of temperatures, they can withstand occasional exposure to moisture, and they can be dropped without breaking.

The product designer must develop system architecture and physical design that are fundamentally sound in order to ensure system reliability. The extent to which the user will be annoyed by an unreliable product should not be underestimated.

Since the reliability of individual mass-produced components is guaranteed by the manufacturer, it is the job of the portable electronic product designer to ensure that the system which integrates all these components is reliable with respect to the intended usage paradigm. Portable electronics are exposed to a very high level of abuse when compared with nonportable consumer electronics devices. Portable electronics, ideally, should be able to operate in any environment that is survivable by a human being. In addition, they should be able to survive being dropped, tossed, and sat upon.

Temperature extremes can cause havoc with portable electronics. Outdoor temperatures above 90°F can cause some LCD panels to lose contrast and stop functioning. The layers of optical material inherent to flat panel displays are subject to the green-house effect when exposed to solar energy. The result is that on a warm sunny day, an LCD left in the sun can heat up quite a bit higher than the ambient air temperature, causing temporary failure of the display.

Cold temperatures can also interfere with LCD function. Temperatures below 40°F can slow or stop response of the liquid crystal transition, rendering the display nonfunctional. Moving from a cool, air-conditioned car or building into a hot, humid environment causes condensation which can interfere with LCD function. Moisture condensation on system electronics can also cause the product to stop functioning temporarily.

Perspiration damage is a very real problem for portable electronic products. Even a single exposure to the dissolved electrolytes in perspiration can create a migration path which may result in a permanent short circuit. Repeated exposure to perspiration can cause corrosion of system components.

Increased product functionality tends to decrease product reliability. This is intuitively obvious, since greater complexity creates greater opportunity for failure. Notebook computer users are conditioned to treat portable computers with the utmost care since the mechanical fragility of a large flat panel display is self apparent. TV remotes, on the other hand, are flung across the room and rarely stop working. Difficulty can arise, however, when the user's expectation of ruggedness exceeds the actual product ruggedness. If a designer creates a portable product that looks and feels like a credit card, he should also ensure that the product can withstand the physical abuse of a credit card (which can be extreme) or else take significant steps to educate the user as to the actual level of ruggedness.

1.3   System Design

The system design is a subset of the product development process but it represents most of the functions performed by the product designer. Figure 1.19 shows system design process steps from product concept through physical design, which result in the creation of manufacturing and procurement documentation.

1.3.1   Product concept

The concept for a new portable electronic device may come from many different sources. The concept may be the brainchild of a single individual or it may be the combined output of a small team or large organization. The concept may be developed by someone with deep technical

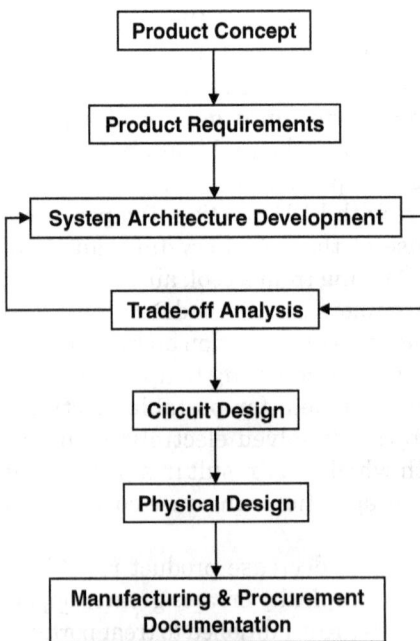

Figure 1.19 System design process.

knowledge in the field of electronics, or it may be the idea of someone with no technical expertise whatsoever. A successful portable electronic product concept must be built on the fulfillment of a specific need in the context of a specific usage paradigm.

The development of a new product concept is a creative process and as such is more of an art than a science. Technical knowledge can play an important role in concept development but insight into human factors is even more valuable. In this respect, the study of anthropology may be as relevant to developing a product concept as the study of electronic engineering.

Producing a product concept is a highly creative activity and is often performed without the benefit of a well-defined process. There is no proven formula for systematically creating an innovative and successful product. That being said, this book will set forth an approach that can be very beneficial in improving the chances of conceiving and developing concepts for an innovative portable electronic device. This approach is composed of three phases: *innovation, creation*, and *communication*.

1.3.2   Innovation

A new product concept is generally driven by a spark of an idea that is then developed into a well-defined concept. This idea may be a small improvement in an existing concept, or it may be a radical departure from existing products. This idea to make a difference in the product, from what has existed before, may be referred to as the *innovation*. If this difference is large, the product may ultimately be judged to be *innovative*.

The vast majority of new products are actually slight modifications of existing product concepts. An example of this would be a new notebook computer design based on the design of a previous generation. Perhaps the modification would be nothing more than upgrading the CD drive to a CD RW drive. In these types of situations, the product concept is not really new since it is based on an existing *platform*.

The product concept for a new product platform will involve different levels of innovation. The concept may be a relatively conservative variation of existing product concepts or it may involve a radically new functionality and usage paradigm.

For example, a company such as Nokia might decide to develop a cellular phone product that is 20 percent less costly to manufacture, in order to penetrate a new market segment. While such a product might require significant technology innovation in order to meet the cost objective, the product concept does not necessarily require a high degree of innovation. Such a product addresses a well-understood functionality and usage paradigm. An innovative product concept may even be avoided to ensure that the technology investment required for cost reduction is not exposed to the risk of product acceptance in the marketplace.

Alternately, Nokia might decide to develop a product that is half the size and weight of the nearest competitor and is operated entirely through voice activation. This product concept achieves the functionality of existing cellular phones but alters the usage paradigm significantly.

Furthermore, Nokia might decide to begin developing a credit card–sized device for wirelessly exchanging real time video and audio between individuals. This concept would entail new functionality (wireless personal video transmission) and a new usage paradigm (ultra-portable video acquisition).

So the level of innovation achieved by a product concept is a rather subjective measurement and is determined by comparing the functionality and usage paradigm to those of existing products.

Product concept innovation should be distinguished from technology innovation, and innovative technology does not guarantee innovative products. A very innovative flat panel display technology, for example, could be used to produce a new PDA product. If the display does not alter the functionality or usage paradigm of the PDA, then the technology innovation has not resulted in a product innovation. This is not to say that technology innovations cannot be useful in such a situation. If the new display technology results in a product with a slightly different visual quality or price point, for example, the technology may serve to *differentiate* the product in the market place.

Of course, innovative technology can be an enabler of innovative products if correctly applied. In the early 1990s, Casio used low-cost, high-density electronic packaging technology, and emerging sensor and LCD modules, to create a compact, low-price digital camera known as the QV-10. Although digital cameras had been used by photographers prior to this time, these types of devices had been out of the reach of consumers. Once consumers got a taste of instant photography that allowed immediate sharing of images on the LCD, and a computer ready email-able image format, the market for digital still cameras erupted. Thus, technology innovation enabled a new price point and a corresponding usage paradigm that was extremely innovative in the consumer market.

There are many innovative products, on the other hand, which have been developed without the use of innovative technologies. An excellent example is the Sony Walkman. Prior to the development of the Walkman, Sony had developed compact stereo tape recording products. Separately, they had also developed a set of lightweight headphone products. It was not until the late 1970s when Sony made the decision to combine these two existing technology elements that a new usage paradigm was created: a personal, portable, high fidelity audio system. The Sony Walkman was born. Today we have portable CD and MP3 players, which use innovative storage technologies, but are based on the same usage paradigm as the original Walkman products.

An innovative product concept can also be problematic when it comes to educating the customer about the product functionality and usage paradigm. Consumer reactions can be difficult to assess. In general, consumers' reactions to new product description can be very misleading. Initial surveys regarding the Sony Walkman indicated that consumers were not interested. Consumers could not visualize the usage paradigm and subsequent benefit of portable high fidelity audio. Fortunately for Sony, consumers caught on quickly.

TiVo is an excellent example of an innovative electronic product that has taken a long time to reach its potential. TiVo found it very difficult to describe the benefits of the product in a way that consumers could readily comprehend. It is the experience, not the description of the experience, which is compelling. A high price point also made it difficult for a large number of consumers to experience the TiVo personal video recorder (PVR) functionality first hand. As a result, TiVo started to realize its true potential only when the PVR functionality was embedded into cable and satellite set-top boxes, which had a much larger customer base.

Again, most products do not exhibit high levels of product innovation. This is understandable since most new product designs are actually new iterations of proven product concepts. For well-established product concepts, the job of the system designer is greatly simplified. A well-established product platform may be in place, so that the designer need only be concerned about adding one or two new features, leaving the rest of the design unchanged.

The first order of business then, is to come up with an idea for creating a new product concept. Often these ideas appear to be the result of a random thought or a brilliant insight. In reality, innovative ideas can also be generated by a fairly systematic process and used to seed the creation of a product concept.

As we have discussed, an innovation of a portable electronic device represents a difference in the functionality and/or the usage paradigm of the product relative to existing devices. So, if we can systematically generate differences in product functionality or usage paradigm, we can systematically generate the seeds of an innovative concept.

Making a list of available (or newly conceived) functionalities and usage paradigms is one way for a product designer to begin this process. Putting various lists into a matrix format allows the designer to contemplate various combinations of functionality and modes of use. This approach can be a good starting point for triggering an innovative idea.

The product functional enabler matrix is given in Table 1.4. This is an example of a matrix that can be used to generate innovative ideas for new products. The primary functional enablers that exist in various types of products are marked under the column for each product. Existing product types are shown in this matrix to validate the mapping process.

TABLE 1.4 Product Functional Enabler Matrix

Functional enablers	Notebook computer	Digital still camera	Camcorder	Cellular phone	Walkie-talkie	One-way pager	PDA	Voice memo capsule	Concept A	Concept B	Concept C
					Product						
Text display	X	X	X	X	X	X	X		X	X	X
Image display	X	X	X				X		X	X	X
Video display	X		X							X	
Audio speakers	X		X	X	X			X			
Audio capture (microphone)	X		X	X	X			X			
Still image capture		X									
Video capture			X								
Mass data storage	X	X	X				X	X			
Wired network connection	X										
WLAN connection	X								X		
WWAN connection				X	X	X			X		
Global positioning									X	X	
Pen text input							X				
Keyboard text input	X										
Voice processing											X

TABLE 1.5 Functional Description and Usage Paradigm for Cellular Phone and Walkie-Talkie

	Cellular phone	Walkie-talkie
Description of functionality	Enables remote voice conversation	Enables remote voice instant messaging
Usage paradigm	Peer to peer connection; private conversation	One to many broadcast; mission critical communication

The last three columns represent new product concepts. The set of functional enablers indicated for each of these concept products is different from the set of enablers for any of the existing products. This difference implies that there may be a potential for an innovative product concept. These functional enablers for concept products can be randomly selected or they could be driven by some creative idea on the part of the innovator. In this case, by selecting an unusual combination of functional enablers, we hope that we might be led into thinking of an innovative product concept.

These functional enablers are only a part of the equation. The completed product concept will also involve a higher level description of the product functionality and usage paradigm. Furthermore, a single set of functional enablers could be associated with multiple product concepts where the functional description and usage paradigm are different. In Table 1.4 the example is the walkie-talkie and the cellular phone. Notice that a walkie-talkie product could be built with the same functional enablers as a cell phone but the functional descriptions and usage paradigm are different, as shown in Table 1.5.

Different form-factors can cause products with the same functionality to have different usage paradigms. Consider the example of a boom box and a Walkman. Both provide very similar functionality, but the small form-factor of the Walkman results in a significantly different usage paradigm for that product.

So, by formulating new combinations of functional enablers and form-factors, a product innovator can create ideas for new product functions and usage paradigms.

1.3.3 Creation

An innovation idea alone, however, does not constitute a product concept. A description of product functionality and usage paradigm must be synthesized based on the innovation. This synthesis represents the actual *creation* of the product concept.

Let us consider the product Concept A in Table 1.4. This product has a display which can render text and images (but not video). It has a wireless wide area network (WWAN) connection, global positioning capability, and pen text input. Having fixed this combination of functional enablers, we can now think creatively about what valuable function such a device might perform.

Obviously, the presence of a global positioning capability causes us to think of applications where determining exact location is important. We might start with the general concept of a person in an urban setting who needs to work with continuously changing locations. Since the person's location can be determined by the GPS function, the display could then be used to display a map. The network connection could be used to provide the map data as well as some specific information related to the task at hand. Many job functions could be envisioned that would exploit this general type of device: a police officer, a delivery person, etc. A good portable product design might examine many of these applications in looking for an opportunity to appeal to a broad market. The best products will look at a very specific application, however, and will be designed to fulfill that particular purpose very well.

We will select a U.S. government census worker as a very specific target customer. The census worker starts her shift by turning on the device. Already we envision a tablet form-factor. The screen displays a color street map covering a two-block radius of the worker's current location. The worker's current location is marked on the display with a red dot.

Upon powering up, the tablet makes a wireless network connection to the Census Bureau server which indicates the worker is now beginning her shift and is ready for an assignment. The nearest unsurveyed residential address is identified by the server and sent to the tablet that displays the address on the worker's map. The worker is notified, and she may have to zoom out if the location is not close enough. To zoom out, she taps a zoom button displayed on the tablet. A green dot on the map denotes the location of the target address.

The worker accepts the assignment by tapping on the green dot. The best walking route is immediately displayed on the map and the worker can zoom in and out as needed to observe the necessary detail. A small compass, mounted on the housing and flush with the display, can be used to help orient the tablet so that the map directions correspond to physical directions. The worker proceeds on foot to the target address. If the tablet is light and rugged, the worker does not need a briefcase to carry or protect the device.

Once the worker arrives at the address, the tablet calls up the census form and the worker can interview the resident. All data are entered on the tablet by pen input. Once the data have been completed and reviewed by the worker, the data are downloaded to the Census Bureau server and the address is marked as "surveyed." The next closest unsurveyed address is then displayed on the screen.

This example of a Census Pad demonstrates the first pass at the creation of a product concept. We have described a way in which a particular set of functional enablers might be used. We have described the product functionality, to a first order. We could certainly describe several other functional models for how the census bureau worker might use the device, or we might choose to explore other types of applications. This type of narrative is sometimes called a *use-case scenario*. Use-case scenarios can be quite succinct or they can be very elaborate documents, examining every relevant situation with which the user might be confronted.

Even if we decide that the census application is the most promising, we still have work to do on the product concept. It is incumbent on the product designer to validate the concept well enough to determine its feasibility. This validation process may look very much like an abbreviated business case that includes a first-order estimation of engineering feasibility.

1.3.4   Validation

The process of validation not only allows the product designer to test the feasibility of the concept, but also enables the concept to be refined. There are a number of aspects of the product concept that must be validated, including use-case scenario, competitive analysis, market opportunity, technical feasibility, cost feasibility, support infrastructure, and marketing/distribution feasibility.

The use-case scenario can be validated by researching the characteristics of the target user. Informal interviews with target users can be a good way of providing useful information. As mentioned earlier in the text, a truly innovative product concept may be difficult for the target user to grasp if a prototype is not available. In this case, the designer will have to have some special insight into why the product might still be successful, in order to validate the concept.

It is well worth the product designer's time at this point to engage in some initial competitive analysis. The designer should learn if there are any similar products already available. If so, the designer should become familiar with the functionality of these products so that he or she is able to judge whether the concept has any advantages with respect to the use-case scenario. Discovery of competing products is not necessarily a negative event for a product concept. Existing products serve to validate and clarify the market conditions that will govern the success of the product, reducing the risk of product acceptance. The product concept must show some distinct advantages or improvements over the competition, however.

Validating the market opportunity can be fairly straightforward at this phase of the product development process. Market research data and demographic data can help create an estimate of the total available market (TAM). Also, pricing data for similar devices or pricing based on product value needs to be estimated. In our census example, the total available market might be something like 100,000 people for a one year period at the end of each decade. If we determined that $500 per unit is a reasonable sales price, then our market opportunity is roughly $50 million in revenue. This should be enough information to determine whether the size of the opportunity is sufficient for the organization in question.

Technical feasibility should also be addressed to a first order. The use-case scenario will shed light on basic requirements such as battery life, form-factor, functionality, and performance requirements. First-order approximations are all that are necessary at the concept phase. Oftentimes, technical feasibility may be self-evident if the envisioned properties are similar to existing products. If the product concept relies heavily on an aggressive form-factor, an innovative type of functionality, or an unusual level of performance, then the concept may involve considerable technical risk. If this is the case then the feasibility may not be fully understood until the product development activity enters the trade-off analysis phase.

Cost feasibility is closely related to technical feasibility. The product cost at this stage is estimated by comparison to the cost of similar devices or by adding up the cost of known key components. Of course the entire point here is determining that the cost is likely to be something less than product price to the end-user. The manufacturing cost of a portable electronic device is typically one-third to half the retail price.

Infrastructure feasibility is critical to almost every portable electronic product concept. The desired support infrastructure may exist, but does it perform adequately for the use-case envisioned? The most predominant infrastructure issue today is that of wireless communications. Will the network provide the communications bandwidth at the desired locations and at a reasonable cost? In the census example, one might ask: Does satellite-based GPS work reliably in a dense urban setting?

Marketing and distribution deserve at least a cursory look at this phase of product development. The designer should at least be able to articulate how the product *might* be marketed and distributed in a way that is consistent with the use-case and target audience. For example, it may be sufficient at this juncture to simply state that the product can be sold through retail consumer electronics outlets. In the census example, it might be necessary to identify a specific government contract, which would have to be awarded in order for the business to move forward. It is also useful to ask: What existing interests, if any, are threatened by the development of this product? Legal injunctions from the music or film industry, for example, could put a damper on the product success.

Once the concept has been validated in the mind of the designer, it must be communicated to others in order for the project to move forward.

1.3.5   Communication

The product concept must be communicated to others in order to become a reality. Effective communication of the product concept is critical if the project is to be successfully financed and executed. A presentation must be developed which effectively conveys the product concept and the results of the validation.

One important tool in communicating the product concept is a narrative of the use-case scenario. This narrative can be made more effective in the form of a storyboard that graphically illustrates how the target user would interact with the product. PowerPoint presentation software is a very useful tool for developing these types of storyboard presentation.

PowerPoint can also be used to quickly and cheaply create an on-screen product simulation. To do this the designer creates a set of successive views of the product as it changes states in response to various user inputs (like the pushing of a button). Since the designer has scripted the precise inputs and responses, it will appear that the device on the screen is functioning in accordance with the product concept.

More sophisticated tools are also used for developing a product concept. Altia, Inc. is one example of a company that makes such tools. Altia's products enable a designer to create an on-screen functional model of a portable electronic device. The device is not limited to a scripted set of user inputs, but can provide appropriate functional responses to a variety of user inputs. The Altia product even generates embedded code automatically that can later be used to drive the actual product. This type of product simulation takes more time to produce than the PowerPoint method and thus may be overkill for the initial product concept communication. Regardless of how it is accomplished, some type of product simulation is invaluable in communicating the product concept to others.

Some form of business case is also necessary to secure the resources needed for the next phase of product development. This business case can be a very high-level description of the market opportunity, the technical/infrastructure feasibility, and a statement of how the product might be marketed and distributed.

Assuming the concept is successfully communicated and judged to be worthy of investment, the concept may be funded and the rest of the design process can move forward.

1.3.6   Product requirements

Product requirements are the first step in taking the product from concept to a product design. Too often, the product requirements stage is glossed over and engineers immediately begin creating the product specification. This can be a costly mistake. A good product requirements document can keep engineering efforts properly directed and greatly reduce the cost of designing the product.

Product requirements should address each of the product design factors listed in Chap. 2, namely, functionality, performance, form-factor, battery life, user interface, cost, time-to-market, and reliability. Product requirements should maintain a pure reflection of the use-case scenario around which the product concept is based. Product requirements should focus on describing the required product functionality without making engineering judgments about how to achieve those requirements. Product requirements that include engineering judgments will unnecessarily constrain trade-offs that need to be made in later phases of the product development process.

The use-case scenario paints a picture of the way the product concept works, but it does not necessarily provide all of the product requirements explicitly. Additional insight and judgment must be applied on the part of the product designer to create a comprehensive set of product requirements.

Let us again consider our example of the Census Pad to illustrate the product requirements.

Functionality. The device must have some form of graphic display to provide the user with an image of a map. The device must have a pen input mechanism so that the user can interact directly with portions of the image being displayed. The device must have some method of determining its current location. The device must have instant access to several types of data and communications.

Analysis of these types of data is important. First, the Census Pad must have access to the required map image data. Some research is required to find out the geographic range of map data that might be accessed on a particular device. We will assume that a Census Pad needs to access the map data for a single county during a given work shift, but that county could be anywhere in the United States on a given day.

The device must also have instant and mobile (wireless) access to real-time address assignment data from the Census Bureau server. The device must also be able to communicate the survey information back to the Census Bureau server as soon as it is entered and verified.

The observant reader will notice that these conclusions about functionality closely correspond to the functional enablers initially used to seed the idea for this Census Pad concept (see Concept A, Table 1.4). If this approach of a functional enabling matrix is used to initiate the product concept, then this result should not be surprising. Usually, however, product concepts are not derived this way, and so the functional requirements might be less obvious.

Thus, the functionality requirements might be explicitly stated as follows:

1. Display a zoomable street map covering an entire county.
2. Provide a built-in compass as a means of orienting the map.
3. Acquire and plot the device's current location on the street map.
4. Acquire and plot the target address on the display.
5. Automatically determine and apply an appropriate zoom to simultaneously plot current and target locations.
6. Calculate and plot the most efficient route to the target address depending on the mode of transportation selected.
7. Track and plot the positional progress en route to the target address.
8. Display alphanumeric address and estimated time of arrival simultaneously with map image.
9. Capture the survey data through a pen input mechanism.
10. Send the survey data to the server in short order.
11. Capture and send data regarding rejected assignments or failed attempts to conduct the survey.

The functionality requirement should represent not only the minimum functional requirements of the device but also the maximum functional requirements. With technology-based products such as portable electronic devices, designers are often tempted to add "free" functionality simply because the hardware platform may easily support these additional functions. Additional functions should *not* be added during product development without reexamining the entire product concept, use-case scenario, and overall usage paradigm. Let us create a list of functions that could potentially be supported by the Census Pad hardware platform:

1. Email
2. Two-way paging
3. Web browsing
4. Time card submittal
5. Census Bureau phone directory
6. Access to employee benefit information
7. Ability to run Microsoft office applications
8. Cellular telephony
9. MP3 player
10. Etc.

The list could go on. Some of these functions might seem like reasonable additions based on the purpose of the device. In reality, these additional functions will dilute the real value proposition of the product and will have negative impact on battery life, form-factor, user interface, and overall productivity as envisioned in the use-case scenario.

It is also important to take note of the fact that defining functionality does not include defining the user interface. We have simply analyzed what the product does; we have not yet discussed requirements regarding how the user should interact with the system, other than the functional requirement of accepting pen input, as defined by the use-case scenario. We will emphasize again that maintaining a clear distinction between user interface and functionality is essential to keeping the product concept focused.

Form-factor.   The form-factor for the Census Pad is likewise extracted from the use-case scenario. A *tablet* or *clipboard* usage paradigm is envisioned. An 8.5 in × 11 in sheet of paper has proven over the years to be a familiar and useful size for reading and writing information onto a clipboard. This roughly corresponds to the notebook form-factor that we defined in Chap. 2. Given our knowledge of current notebook computer products, however, we might reasonably conclude that the Census Pad should be considerably lighter and more rugged than a notebook computer if it is to be a comfortable and productive tool for the user. Human factors experts may be consulted and/or some experience carrying around weighted clipboards or wooden mock-ups might lead the product designer to conclude that a 12 in × 9.5 in × 1 in pad weighing 2 kg or less is a comfortable form-factor for the user, which could also accommodate the 8.5 in × 11 in display area. A minimum size requirement is probably not needed for this product, given the obvious display size requirement.

Setting product requirements for form-factor is always fraught with peril. Establishing a requirement that is unnecessarily aggressive may result in the engineering conclusion that the product concept is not feasible. Allowing the form-factor to expand beyond certain maximum parameters, however, can result in an unsuccessful product. The margin that separates failure from success can be surprisingly thin when it comes to form-factor. An extra quarter inch here or an extra 200 g there can cause a product concept to fall apart.

Battery life.   Additional research into the Census Pad user profile might indicate that a typical work shift lasts 12 h. It is not unreasonable to expect that the worker will plug the device in overnight for recharging. Requiring that the device be recharged during a lunch hour may be out of the question. Managing spare batteries may introduce a complexity and resulting loss of productivity that is unacceptable to the paying customer (U.S. government).

In a dense population area, the user may be interacting with the device on a nearly continuous basis. The requirements for battery life should include a quantitative description of the functional workload required from the device (also known as the duty cycle). A statement of battery life without reference to the duty cycle is meaningless.

Our battery life requirements for Census Pad will be structured around a worst case scenario. We will require that the device be capable of providing a 12-h battery life under the heaviest usage conditions. (This requirement may be driven by the insight that the competing product is a traditional notebook computer with a battery life of only 8 h under light usage conditions.) Let us assume that time study data from the Census Bureau indicates that a top performing worker in a dense urban area would survey up to four residences per hour. Each hour this would require four map references of 30 s each, four target address downloads from the server, four survey data uploads to the server, and an average of 40 min per hour entering survey data into the device (10 min per residence).

User interface.   Fundamental to the Census Pad product concept is the requirement that the user interact with the device through pen input. Based on the use-case scenario and some additional creative design thinking, we will create a list of user interface requirements as follows:

1. When the device is turned *on*, a map is displayed with the worker's current location marked in red.
2. The map automatically zooms to a magnification level that enables both the current location and the nearest eligible target address to be displayed simultaneously.
3. The target address is marked by a green dot.
4. The user touches an icon of either a car or a walking person to indicate the intended mode of travel to the destination.
5. The most efficient travel route to the target address is plotted.
6. The alphanumeric street address is displayed on the periphery of the map along with the estimated transit time.
7. The user accepts the target address assignment by touching the corresponding green dot on the map.
8. If the user wishes to reject the assignment she may do so by touching a rejection icon.
9. A rejection icon will cause the display to present a form requesting the reason for rejecting the assignment.
10. Once the rejection reason is input, the user touches the send icon and the rejection data are sent to the server.

11. The device returns to the state described in step 2.

12. Upon acceptance of an assignment, the target address flashes green.

13. Red dot moves along the route in correspondence with the user's transit progress.

14. To conserve battery life, the display may shut down automatically whenever there is no user interaction for more than one minute.

15. The display can be turned back on instantly with a touch of the pen.

16. When the user arrives at the target address, she knocks at the door.

17. If the resident agrees to an interview, she touches a survey icon with the pen, the survey form is displayed, and the user conducts the survey.

18. When the survey is completed, the send icon is touched and the survey data are sent to the server.

19. If the resident is not home, or rejects the interview, the user touches the rejection icon and inputs the reason for rejection, touches the send icon and the data are sent to the server.

20. The device returns to the state described in step 2.

To clarify the user interface requirements, a flow chart is helpful. A flow chart for the Census Pad user interface requirements is shown in Fig. 1.20.

Based on this description of functional interaction, the product designer should be able to create a user interface that is true to the original product concept. Notice that this requirements document says very little about exactly how the user interface should look and feel. Even the colors specified for tagging locations are provided only for the sake of example. The actual colors could be changed depending on human factors, display characteristics, etc. The user interface requirements should reflect a usage paradigm that is consistent with the product concept. The specifics of the interface will be designed later.

As with functionality, the user interface requirements should not only be considered a minimum requirement but a maximum requirement as well. This notion is important in avoiding feature creep, which can undermine a product's effectiveness at achieving its primary functionality. Without changing the basic functionality, the designer of the user interface might be tempted to add features such as the following:

1. An alphanumeric display of the time and date

2. The ability for the user to adjust the idle tie that triggers display shutdown

3. The ability to *personalize* some attributes of how the information is displayed

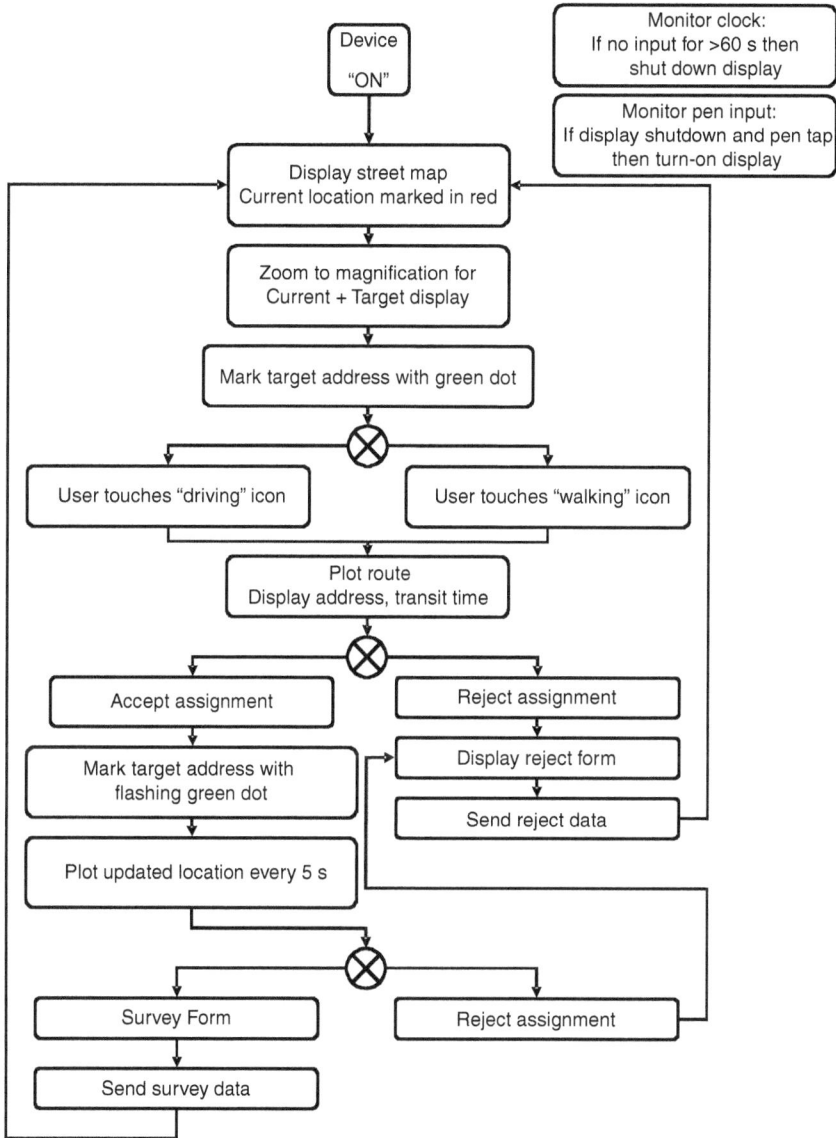

Figure 1.20 Flow chart for Census Pad user interface requirements.

4. The ability to set a default transportation mode of "always walk" or "always drive"

5. The ability to analyze the user's personal performance statistics

6. The ability to display the entire route throughout the shift

7. Etc.

A creative software engineer could extend this list indefinitely. Such features can be easily added in software but should not be added unless there is specific alteration to the product concept or customer feedback indicating that a specific new feature would add significant value. The user interface requirements therefore indicate what should, and by omission what should not, be encompassed in the user interface.

Cost (price).   The product cost requirements are driven by the expected sales price of the portable electronic device. The product designer should work with marketing to develop a range of expected sales prices and corresponding sales volumes. Based on this matrix, a target manufacturing cost can be developed.

The cost requirements should specify the maximum cost for the product bill of materials (BOM) as well as the cost to assemble and test the product. Other expenses, such as packaging and marketing, will be assumed to be subtracted from the remaining gross profit margin.

In the Census Pad example, we can rationalize a $500 per unit price point for the following reasons:

1. Productivity gains resulting from real-time scheduling of census workers would provide the basis for a rapid return on investment.

2. The price point is significantly below the cost of a notebook computer, which would be the most likely competitive product.

Time-to-market.   Time-to-market (TTM) requirements for a portable electronic device should be specified in terms of specific market cycle objectives. Therefore, it is appropriate to say "the product should be on retail shelves for the Christmas 2003 shopping season." The TTM requirement should not be *nine months*. An unambiguous statement of the target market cycle will drive all aspects of the product development cycle and cause each part of the organization to work backward from the target date to determine its own schedule.

It is also useful to quantify the lost revenue opportunity if the product launch is delayed by one or more market cycles. This information can be very useful in helping various product development teams justify additional resources if the project starts to drift off track.

For the Census Pad we could state the following TTM requirements:

1. Deliver 50 prototype units to U.S. Census Bureau by May, 2003.

2. Deliver 100,000 units to U.S. government in January, 2004.

3. Produce 10,000 units per month for U.S. and international customers from January, 2004, until end of life.

Reliability. Reliability requirements for portable electronic devices should be specified in terms of environmental constraints as well as physical stresses and system stability. Compliance to industry reliability standards may also be required for certain market segments.

We will establish the following reliability requirements for the Census Pad:

1. The device must be able to operate indoors and outdoors at all times in all regions of the United States.

2. The device must withstand being dropped repeatedly from a chest-high position onto concrete.

3. The device must not enter an unstable state (crash) more than once per 100 h of use, at which time it must be able to reset within 20 s and must be able to access any survey data that were stored but not yet downloaded to a server.

4. Eighty percent of the devices must remain in service for at least 18 months of usage (60 h/week).

Performance. Review of use-case scenario will reveal the context in which a performance requirement must be developed. Going from the use-case scenario to a performance specification is also a creative process that incorporates insight into the business issues and human factors which govern the product use. A minimum level of performance may be necessary to achieve the task described in the use-case scenario. Higher levels of performance may change the end user's perception of the product, however.

Cost thresholds for various levels of performance are not obvious at this phase of the product development. It is most important that the product designer not inadvertently create expensive performance thresholds. The product performance requirements should be stated in terms of minimum levels of performance required to successfully realize the product concept. Additional performance can be added later if the cost trade-offs justify the benefits.

Performance requirements should be stated in human and operational terms, not in quantitative technical terms. Quantifying the performance into technical parameters occurs as the product develops through the system architecture and trade-off analysis culminating in a product specification.

Performance requirements must be generated for each aspect of the product functionality. For the Census Pad, the following type of performance specification could be developed in conversation with the potential customer.

It is helpful to start with the list of functional requirements and determine the performance of each on an individual basis. Thus, we have the following:

1. Display a zoomable street map covering an entire county
    a. Easily readable indoors and outdoors
    b. Easily readable in bright light and low light levels
    c. Multicolor (not necessarily full color)
    d. Comfortable viewing angle for user, assuming pad will be used like a clipboard
2. Provide a built-in compass as a means of orienting the map
    a. Accurate (±10°) when pad is semihorizontal
3. Acquire and plot the device's current location on the street map
    a. Accurate to within one-fifth of a city block
    b. Requires less than 10 s to acquire and plot
4. Acquire and plot the target address on the display
    a. Accurate to within one-fifth of a city block
    b. Requires less than 10 s to acquire and plot
5. Automatically determine and apply the appropriate zoom to simultaneously plot current and target locations
    a. One second or less
6. Calculate and plot the most efficient route to the target address depending on the mode of transportation selected
    a. One second or less
7. Track and plot the positional progress en route to the target address
    a. Update every 10 s
    b. Accurate to within one-fifth of a city block
    c. Requires less than 10 s to acquire and plot
8. Display alphanumeric address and estimated time of arrival simultaneously with map image
    a. Easily readable indoors and outdoors
    b. Easily readable in bright light and low light levels
    c. Multicolor (not necessarily full color)
    d. Comfortable viewing angle for user, assuming pad will be used like a clipboard
9. Capture the survey data through a pen input mechanism
    a. Instant visual response on screen from pen
    b. Must be able to read neatly printed character correctly 99.99 percent
    c. Must be able to digitally store free format input in some fields
10. Send the survey data to the server in short order
    a. Must upload survey data within 15 min of completion

    b. Must store one shift of survey data and allow for wired upload if wireless network is unavailable

11. Capture and send data regarding rejected assignments or failed attempts to conduct the survey

    a. Instant visual response on screen from pen

    b. Must be able to read neatly printed character correctly 99.99 percent

    c. Must be able to digitally store free format input in some fields

    d. Must upload survey data within 15 min of completion

    e. Must store one shift of survey data and allow for wired upload if wireless network is unavailable

The above list represents a set of market-driven performance requirements for the product. Notice that the numbered items are descriptions of functionality and the lettered sub-bullets describe the performance level in operational terms. There are no preconceived notions at this point regarding the CPU clock speed, the amount of DRAM required, the hard drive capacity, etc.

These operational performance requirements must next be systematically converted into technical metrics, which are used to drive the development of the system architecture.

1.3.7 System architecture development

System architecture development is the most important phase of product engineering. The system architecture will create the constraints within which all other aspects of system development must occur. System architecture development can be a highly creative process for complex system designs. In these cases, there is a trade-off analysis that must take place as alternative architectures are compared. So moving from the system architecture phase to trade-off analysis phase is not strictly a sequential process but can involve a number of iterations. It is through experience and technical insight that product designers and architects are able to limit the amount of iterations and converge on successful product architecture.

As a creative process, there are many ways that product architecture can come into being. The process described here is presented as a systematic option for developing product architecture but is by no means the only approach.

The first step in architectural development is to look at the functional requirements of the system. Each functional requirement can be translated into a technical requirement for an input signal, some type of signal processing, and an output signal. This notion is supported by the concept that for anything useful to happen, an input must be provided and a response given. By creating and categorizing a complete list of input signals, processing requirements, and output signals, the system architect can begin to assess the total processing requirements for the system. These processing requirements will determine what types of processors are needed and where processing resources can be shared between functions.

Functional requirements such as a requirement to display graphic data carry with them the implication that the system has a display. While the display is technically a part of the user interface, the display function may be so central to the product concept that it must be treated as a functional requirement.

High-level block diagrams are initially created, which describe the product functionality in terms of inputs, processing sequences, and outputs. These individual block diagrams are then mapped onto various system architectures to test how they might share the system resources.

So, the sequence of steps to develop system architecture is as follows:

1. Translation of system functionality requirements into inputs, processing requirements, and outputs (IPO)

2. Analysis of system resource options that could meet each IPO requirement

3. Formulation of an efficient architecture that can meet the aggregated IPO requirements

4. Analysis of the performance limitations, if any, resulting from the envisioned resource sharing scheme of each IPO

5. Physical design strategy (partitioning)

6. Evaluation of architecture with respect to critical factors

7. Modify architecture and repeat evaluation

Integrated circuits are the building blocks of system architecture. Portable electronic products rely on microprocessors, memory, logic devices, analog devices, microcontrollers, DSPs, and other specialized silicon to meet the full set of IPO requirements. It will often be the case that a microprocessor or microcontroller will be identified as the most efficient approach to achieving some or all of the product IPO requirements. The set of key silicon components delineated by a particular system architecture is known as the *hardware platform.*

The product designer may start by generating combinations of key components and associated block diagrams in an effort to identify the most efficient combination of components needed to deliver the product. Assessing computational bandwidth for various functional blocks and data transfer bandwidth requirements between blocks will lead to selection of a workable set of components. Product development tools are also available, which can select an appropriate hardware platform based on a functional product description encoded by the product designer.

For portable electronic products it is important that a physical design strategy also be developed during the architectural stage. This includes defining the system partitioning and electronic packaging architecture. It is a common error to ignore this part of the system development until the physical design phase. Since form-factor is so critical to portable electronics design, neglecting the physical design strategy until later in the process may result in nonoptimal system architecture.

Examining the I/O specifications for available components is the first step in considering a physical partitioning strategy. A first-order analysis of the number of I/O that connects the major functional blocks is used to assess whether additional silicon might be an option. This analysis of I/O also drives the printed circuit assembly (PCA) partitioning strategy.

1.3.8 Trade-off analysis

Trade-off analysis for a portable electronic product is different from that of other electronic products because it tends to be centered on the critical factors of form-factor, battery life, cost, and user interface. Functionality should not be a trade-off factor given our stated design approach of limiting the functionality to what is necessary to fulfill the product concept; no more, no less. Similarly, the performance level of that functionality should meet the minimum level set in the product requirements. Providing higher levels of performance than those required should only be done if it has no impact on the other critical factors. Reliability and time-to-market are not normally given great weight during the trade-off analysis stage unless new technologies or components are required by the product design, or unless the product is targeted at a high reliability application. Use of off-the-shelf components carries with it the implication that those components have an established reliability and availability.

The trade-off analysis is integral to the system architecture development. Since there are many variables involved in optimizing the system design, it is practical to consider a fixed hardware platform while varying other system elements. Thus there are really two levels of trade-off analysis—one to optimize the hardware platform and one to optimize the other elements of the system.

So, for a given hardware platform (set of silicon), the first pass of a trade-off analysis requires the selecting candidates to fulfill the undetermined aspects of the system including batteries, displays, electronic packaging, thermal management components, external connectors, RF hardware, buttons, keyboards, mass storage devices, sensors, optics, and any other major components.

Form-factor is a good place to start the trade-off analysis for a portable electronic product. The physical dimensions of each of the known system elements can be obtained from a spec sheet for a typical component. The dimensions of a simple three-dimensional primitive can be specified to represent the volume required to accommodate the presence of each component within the product. The primitives for all of these major components can be manipulated on paper to estimate the feasibility of the required form-factor. If simple prismatic primitives are not complex enough to provide a reasonable model, or if form-factor feasibility is marginal and requires a high-resolution estimate, a 3D-mechanical CAD system may be used to perform a more precise analysis.

With many product concepts the form-factor area will be determined by user interface requirements such as the display size and the area required for a keypad. The thickness then becomes the critical dimension in maintaining a small form-factor. Conducting a component placement analysis with particular attention paid to the stacking conditions of various components is a useful exercise.

In the case of a notebook computer system, the component placement strategy is centered on the identification of thickness-limiting components. In these products, the thickness-limiting components include the display and backlight assembly, the battery pack, the minimum plastic thickness, the keyboard, the hard drive, the motherboard, and the system thermal solution. The goal is to limit the thickness of the product to the thickness of the thickest component, plus those components which must be stacked along with the thickest component.

Figure 1.21 shows how the thickness limiting stack-up in the lower enclosure of a notebook computer might be arranged. The thickest component is the hard disk. Since the hard disk must be contained within the lower enclosure, the ideal minimum thickness for the lower enclosure would be the thickness of the hard drive plus the thickness of the plastic that forms the enclosure.

In this example the hard drive is not stacked on the motherboard because it would increase the product thickness unnecessarily. The space

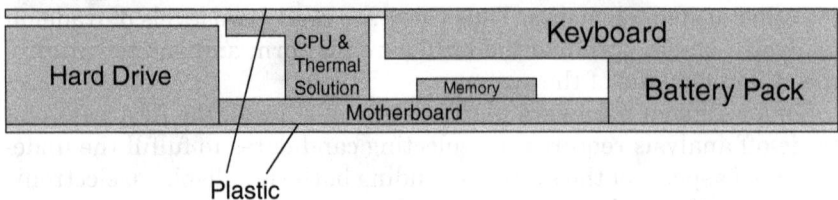

Figure 1.21 Stack-ups for lower notebook enclosure.

limitations do require that the keyboard be placed over 45 percent of the product depth. The motherboard is positioned so that the CPU thermal solution is not underneath the keyboard. This allows the thermal solution to be designed as thick as possible, minimizing thermal resistance, without imposing a limit to the minimum product thickness.

The memory module might have been mounted on the bottom of the motherboard and accessed through a panel in the bottom plastic. This would have created thickness-limiting stack or at the very least would have impacted the maximum thickness of the CPU thermal solution. Instead, the memory module is positioned under the keyboard so that RAM can be easily upgraded by removing the keyboard module to access the SODIMM connector.

The designer is left with no other choice but to stack the keyboard and battery pack if the desired product length and width are to be maintained. Stacking these two components plus the plastic enclosure becomes the limiting factor for the minimum product thickness.

Estimating the form-factor for the product's PCAs is a primary activity for the portable electronic product designer. While most system components will have form-factors specified by the manufacturer, the PCAs are probably the largest contributor to product form-factor over which the product designer has direct control. Thus, selecting an optimal electronic packaging strategy is of primary importance in designing a portable electronic product.

One method of developing an optimal electronic packaging design would be to perform the complete physical design process with all the available packaging technologies. Each design could then be sent out for quotes to determine the cost of implementation. Given the many permutations of IC packages and substrate technologies that are available, this approach is not feasible since it would require excessive design resources to explore every design possibility. Even if the design resources were not an issue, the time required to complete these designs is unacceptable during this early phase of product development.

A more feasible method is to make reliable estimates of the cost and form-factor that would result from the use of different packaging strategies. Some computer-aided design companies have created "design advisor" software to help the portable product designer identify optimal packaging solutions, by estimating the outcome of a complete design process. These solutions have the disadvantage, however, of requiring detailed information about the system components that is not readily available to the designer when the trade-off analysis portion of the design process is being conducted.

Another approach is one based on simple algorithms and comparative metrics based on the attributes of the system silicon. This is a preferred method provided the designer is confident that the metrics and algorithms yield a reasonably accurate result. The ability to rapidly assess electronic packaging form-factors is essential to the portable electronic designer.

Based on an initial selection of system components, the designer can estimate the battery life of the product. This exercise consists of adding up the power consumption of each system component and comparing this figure to the energy capacity of the battery that has been selected.

In practice, the actual duty cycle of each component needs to be understood, as low duty cycles can improve battery life significantly. A simple and conservative power management scenario should be developed by the designer during the trade-off analysis phase, to support system optimization.

The contribution of each system component to the overall system cost must also be accounted for. Here again a simple cost model based on high-level system attributes is preferable to a model that requires a lot of detailed inputs. While cost quotes for key system components will be available from the component manufacturer, assessing the lowest available costs without price negotiations may require experienced input from an electronics buyer, an understanding of learning curve theory, a comparative cost database, or perhaps all three. In addition, the cost for the system PCA is highly dependent on the electronic packaging strategy. While PCA costs comprise a low percentage of cost in some classes of electronic products, they tend to be more critical in highly miniaturized products that require leading-edge electronic-packaging technology.

Figure 1.22 summarizes how this trade-off analysis process works for a portable electronic product. Notice that after the initial analysis, the designer must asses whether the product requirements have been met.

If the product requirements are not met, the designer must begin to make trade-offs. For example, if the form-factor is too large but there is still room for a cost increase, the designer might select a more expensive substrate technology for the electronic packaging. This might result in a sufficient decrease in form-factor to meet the product requirements. Alternatively, the designer might look for a thinner display module or battery cell. This effort might be successful, or it might lead to the conclusion that the form-factor goals simply cannot be met without a new type of battery technology. The designer might then begin looking at some emerging high-density battery solutions—a new component technology. If none of these options pan out, the designer may conclude that different platform architecture (system silicon) needs to be considered. Many such iterations may take place during the trade-off analysis process. For a competitive solution, the designer should not seek to merely meet the product requirements but rather come up with an optimal solution in which cost, form-factor, or battery life objectives are exceeded, if this can be done without a significant increase in risk.

```
┌─────────────────────────────────────────┐
│ Select candidates for key system components │◄──────┐
└─────────────────────────────────────────┘       │
                    │                              │
                    ▼                              │
      ┌───────────────────────────┐               │
      │ Estimate PCA form factor  │               │
      └───────────────────────────┘               │
                    │                              │
                    ▼                              │
  ┌───────────────────────────────┐   ┌──────────────────────────────┐
  │ Aggregate form-factor primitives │  │ Change:                      │
  └───────────────────────────────┘   │                              │
                    │                  │ Electronic Packaging         │
                    ▼                  │ Component Selection          │
      ┌───────────────────────────┐   │ Component Technology         │
      │ Estimate Battery Life     │   │ Platform Architecture        │
      └───────────────────────────┘   └──────────────────────────────┘
                    │                              ▲
                    ▼                              │
        ┌───────────────────────┐                 │
        │ Estimate Cost         │                 │
        └───────────────────────┘                 │
                    │                              │
                    ▼                              │
  ┌───────────────────────────────┐               │
  │ Requirements met? Optimized?  │───────────────┘
  │ {Cost, Form-Factor, Batterylife} │
  └───────────────────────────────┘
                    │
                    ▼
      ┌───────────────────────────┐
      │ Trade-off analysis complete │
      └───────────────────────────┘
```

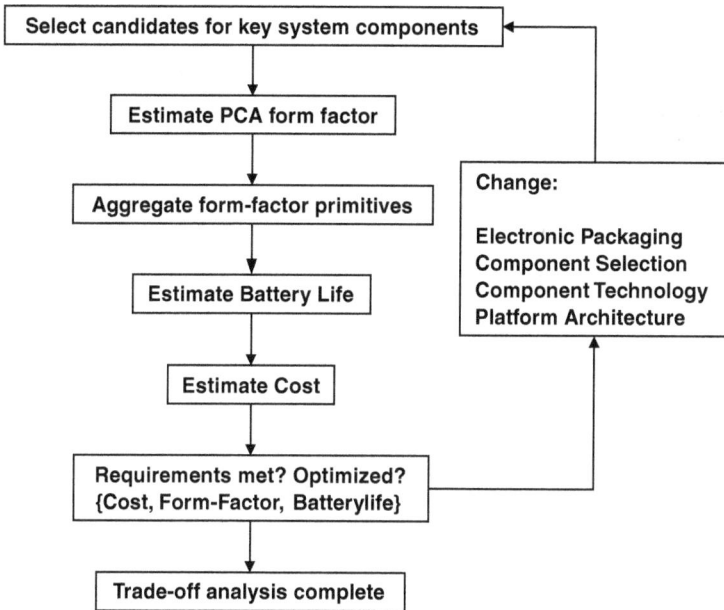

Figure 1.22   Trade-off analysis process.

Comparative metrics and cost models, such as those employed by Portelligent, Inc., provide actual product data which can be used to infer an optimal packaging solution for a product that has yet to be designed. This approach is very effective because it is based on data from actual implementations and, as such, also provides insight into the competitive posture of the product with respect to an aggressive form-factor. Metrics that can be used to estimate product PCA form-factors are discussed in Chap. 9.

1.3.9   Cost modeling discussion

Cost modeling is different from the cost estimating process and is used by the portable electronic designer for different purposes. Cost modeling is used in the preparation of the initial business case and during the trade-off analysis portion of the product design. It is also used in conducting competitive analysis once a product has been produced.

A cost estimate, in contrast, is an attempt to determine the actual manufacturing cost of a completed physical design. The cost estimate is the basis for production budgets and performance management. It is based on the product bill of materials (BOM) and reflects firm quotes provided by component suppliers and manufacturing operations. Compiling an accurate cost estimate is a time-consuming process because it requires confirmation of much detailed information.

Cost modeling can be performed more rapidly than cost estimating and without the benefit of a complete physical design. Cost modeling relies on an awareness of cost sensitivity to different variables in the system design. It also relies on heuristic shortcuts, which enable certain cost boundaries to be established based on system level metrics rather than detailed design parameters. A compendium of such cost modeling techniques was developed at the Microelectronics and Computer Technology Corporation (MCC) throughout the 1990s. These cost-modeling methods were developed specifically for use with portable electronic devices and are now used by Portelligent, Inc. (Austin, TX). Portelligent conducts cost modeling of portable electronic devices as a part of the information services that they provide to the electronics industry. Below is Portelligent's explanation of how they effectively conduct their cost-modeling activities. It is included here because it represents a definitive example of a disciplined approach to cost modeling.

Portelligent Cost Estimation Process—Overview and Discussion

Cost modeling is tricky business. Multiple variables affect the actual production costs a manufacturer will experience, including development expenses, unit volumes, supply-and-demand in component markets, die yield-curve maturity, OEM purchasing power, and even variations in accounting practices.

Different cost-modeling methods employ different assumptions about how to handle these and other variables, but we can identify two basic approaches: one which seeks to track short-term variations in the inputs to the production process, and the other which strives to maintain comparability of the output of the model across product families and over time.

Portelligent's philosophy in cost modeling is to emphasize consistency across products and comparability over time, rather than to track short-term fluctuations. During the past eight years, we have developed an estimation process that, while necessarily lacking an insider's knowledge of the cost factors that impact any one manufacturer is reasonably accurate in its prediction of unit costs in high-volume production environments.

We do not claim that the model will produce the right answer for your firm's environment. However, Portelligent does give customers a key analytical tool with a complete set of data in our bill of materials (BOM). The BOM allows readers to (1) scrutinize the assumptions behind our cost model and (2) modify the results based on substitution of their own component cost estimates where they have better information based on inside knowledge.

Our estimation process decomposes overall system cost into three major categories: electronics, mechanical, and final assembly. We begin by creating a complete electronics BOM. Each component from the largest ASIC to the smallest discrete resistor is entered into a BOM table with identifying attributes such as size, pitch, I/O count, package type, manufacturer, part number, estimated placement cost, and die size (if the component is an IC). Integrated circuit costs are calculated from measured die area.

Using assumptions for wafer size, process type, number of die per wafer, defect density, and profit margin in combination with die area, an estimate of semiconductor cost is derived.

Costs for discrete components and interconnect are derived from assumption tables, which relate BOM line items to specific cost estimates by component type and estimates for part placement costs are included.

For LCD display costs, we employ a model, which tabulates expected cost from measurements of glass area, LCD type, and total pixel resolution. When market costs are available from alternative sources, LCD panel costs are taken from and referenced to these sources.

Costs of mechanical enclosures and fasteners along with the cost estimates of final system assembly and integration are modeled using Boothroyd Dewhurst design-for-assembly (DFx) tools. Other system items, such as optics, antennae, batteries, and so on, are costed from a set of assumption tables derived from a combination of industry data, average high volume costs, and external sources. In effect, we rebuild the torn-down product, tabulating final assembly costs as the process of reconstruction proceeds.

The three major categories for system cost contributors can be broken down into the subcategories of ICs, other electronics parts, displays (as appropriate), printed circuit boards, electronics assembly, mechanical/housing elements, and final assembly.

By adding the cost estimates for each of these subcategories, an overall estimated cost is derived for the system under evaluation. Product packaging and accessories (CDs, cables, etc.) are also documented and estimated for their contribution to total cost as appropriate. We believe our cost estimates generally fall within ±15 percent of the right answer, which itself can vary depending on the market and OEM-specific factors mentioned earlier.

While the Portelligent cost model is imperfect, it yields important insights into technology and business dynamics along with good first-order contributions to system cost by component type. Additionally, the consistency of approach and gradual modification to assumptions (smoothing out frequently shifting pricing factors) hopefully yields a credible, but user-modifiable, view of OEM high volume cost to produce.

All the cost analyses in this book are based on the Portelligent cost model. The primary purpose in using this type of cost model is to compare the relative cost of different design options. By emphasizing comparative modeling, it is possible to optimize the cost of a new design concept and to conduct competitive analysis regarding how a product's cost structure might be positioned.

Actual manufacturing costs are affected by the actual production volume. Structuring the cost model to account for different production volumes can make its usage significantly more complicated. Using a cost model in conjunction with the learning curve model is a better way to deal with high volume production issues. The cost model can be targeted at developing an initial production cost for, say, the first 10 K units produced. The learning curve model discussed in Chap. 12 can then be used to project what the costs might be in higher volumes. Studying the learning curve dynamics of key cost adders is a very useful exercise in developing an overall product strategy.

Both cost modeling and learning curves are useful during product concept and trade-off analysis. Of course, once the product moves into production, the actual cost to manufacture will be determined by a host of factors, including the actual negotiated component prices, overhead costs, production throughputs, yields, etc.

1.3.10   Circuit design

Once the trade-off analysis is complete and optimal system architecture has been developed, the detailed circuit design phase may begin. The completed circuit design will take the form of schematics, which detail all the electronic circuits in the system.

The system architecture is one or more block diagrams that show the major functional blocks and signal paths within the system. The *blocks* in these block diagrams may represent individual integrated circuits, or they may represent entire subsystems consisting of many components.

The circuit designer begins by analyzing the signals that must pass between functional blocks. In doing so, the circuit designer defines the *interface* to each functional block. Additional infrastructure, such as power and clock signals, must also be included in this interface description.

Once functional interfaces are defined, the circuit designer looks inside individual functional blocks. Further subdivisions may take place and the corresponding interfaces can be analyzed. At some point the functional block in question corresponds to a manageable design problem and a detailed schematic may be developed for that block. It is by this general process of decomposition that the system design is divided and conquered.

The simplest possible block diagram would be a single block corresponding to a single integrated circuit component. The circuit designer writes down the vendor and part number, and the electronic design is complete. This may actually be the case for a very simple product such as a calculator where complete integration of the product in a single chip is possible, and the only other components are the display, the battery, and the keypad (Fig. 1.23).

In more complex designs, complete product integration is not usually achieved. Even with SOC designs there is always some additional

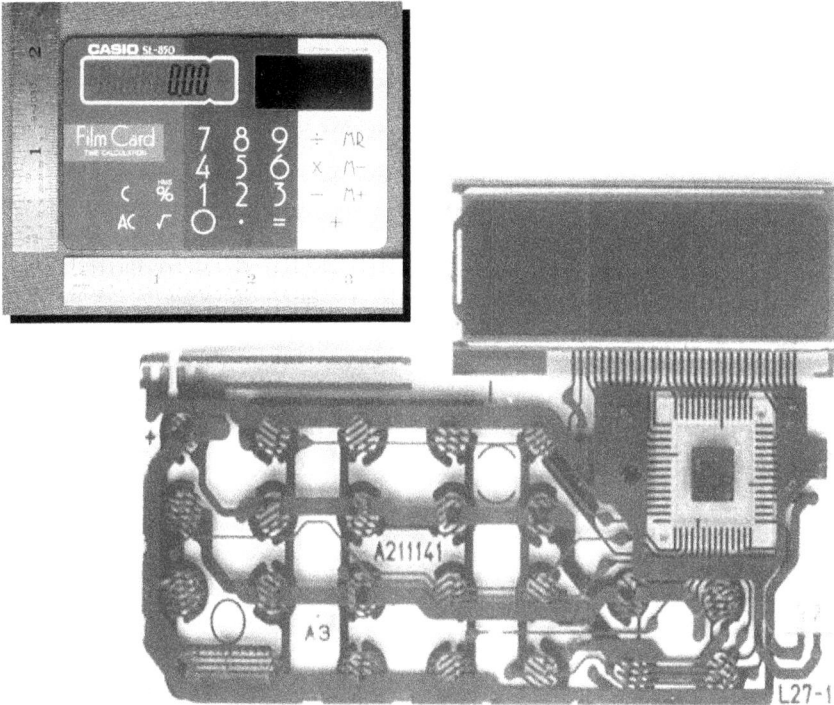

Figure 1.23 Single chip calculator product.

circuitry external to the primary silicon to support power, signal conditioning, synchronization, and other practical aspects of system implementation that are determined by the requirements of a specific product.

In some cases the system architect will have concluded that a particular functional block needs to be provided with some type of application specific integrated circuit (ASIC). In these cases, the circuit design becomes significantly more costly and time consuming. Designing custom silicon should be avoided unless the product concept is mature, and the anticipated product volumes are very high. Each custom silicon device design will typically be managed as a distinct design project.

Integrated circuit designs have become so complex that they are usually performed with an extensive array of electronic design automation (EDA) tools. The primary vendors for these tools are Mentor Graphics and Cadence. Both of these companies provide a complete suite of EDA tools that enable cell-based or full custom IC design.

The circuit design process results in the electronic schematics that integrate the key system components. All the necessary electrical components are identified including ICs, resistors, capacitors, inductors, discrete logic, filters, batteries and I/O devices such as sensors, displays, and keypads. The logical connections between all these components are specified. Electrical tolerances for each component are specified. Specific part numbers are called out when the required properties of the component are unique. The required properties for commodity parts, such as chip resistors and capacitors, are specified, but the determination of specific vendors and part numbers may be left to the purchasing organization.

The circuit designer provides all this information to the physical designers in the form of a net list. The net list identifies every component in the system and it also lists which *net* each *node* (component I/O) is connected to. A net is simply a list of nodes that are electrically connected to each other. The net list may also identify *critical nets* for which the electrical or physical characteristics of the connection must be carefully maintained. The net list may be subdivided according to the PCA subassemblies that the system architect had originally partitioned. Simulation tools and prototyping is usually conducted to ensure that the circuitry will function as intended. The net list is passed to the physical design team in electronic format.

1.3.11 Physical and mechanical design

Physical design refers to the physical implementation of the electronic schematic in the form of PCA. Mechanical design entails all other physical aspects of the design including the product housing, placement of the PCA modules, thermal management components, batteries, display modules, internal cables, etc.

The physical design of the system printed circuit boards is performed using an automated printed circuit design system. The net list is input into the system and the appropriate component mechanical dimensions are loaded from the component library. The I/O pad patterns for each component are placed on the substrate. Autoplace and manual component placements are both possible with most CAD systems.

Once the component placements on the substrate are determined, the conductive traces that connect the nodes for each net must be routed on the substrate. The designer inputs the trace routing design rules provided by the substrate manufacturer. The number of layers of routing, the routing pitch, and the size of vias used to connect between layers will impact how tightly the components can be packed onto the substrate.

Critical traces are routed first and may be routed manually. The majority of routing is then done automatically by the CAD system. If the router is unable to route the design within the designated placement area of the components, then the designer may add additional conductor layers and try again. Alternatively, the trace design rules may be reduced to smaller dimensions, but this can only be done based on process parameters supported by the substrate manufacturer. Adding layers or using more aggressive design rules will increase the substrate cost.

Some of the key vendors for PCA-automated design tools include Allegro, Mentor, PADs, Theta, and IPL. Once the PCA design is complete, the automated design system generates the necessary digital information to drive the creation of the custom tooling required for the substrate manufacturing process and the surface mount component placement programs. A completed electrical design will include the following items:

1. A complete electronics BOM

2. Circuit board (PCB) layout designs in Gerber format (for photo-tooling)

3. PCA assembly drawing and component placement data file for SMT equipment

4. Electronic test procedures for verifying product functionality

Examples of the routing patterns for each layer of a cellular phone substrate are shown in Fig. 1.24.

Mechanical design is one of the most challenging aspects of many portable electronic product designs and is often the most differentiating aspect of a successful portable product. Most mechanical design is performed on 3D CAD systems such as Pro Engineer. These systems

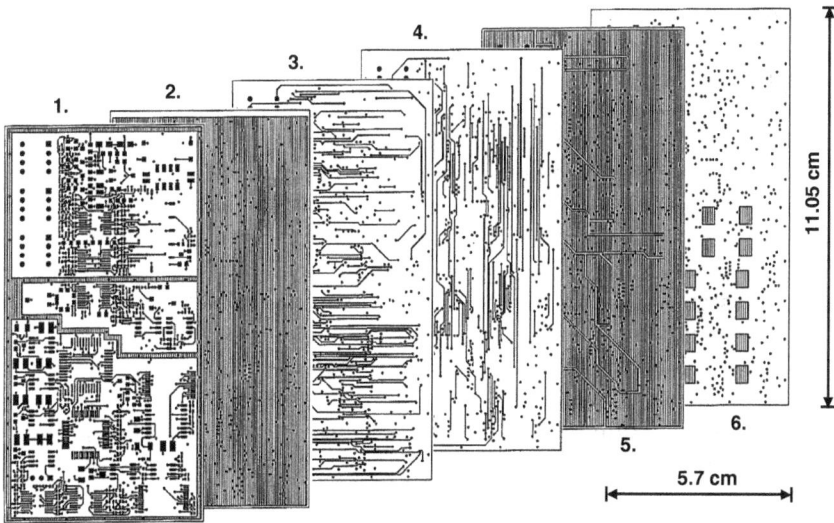

Figure 1.24 Cell phone PCB layout.

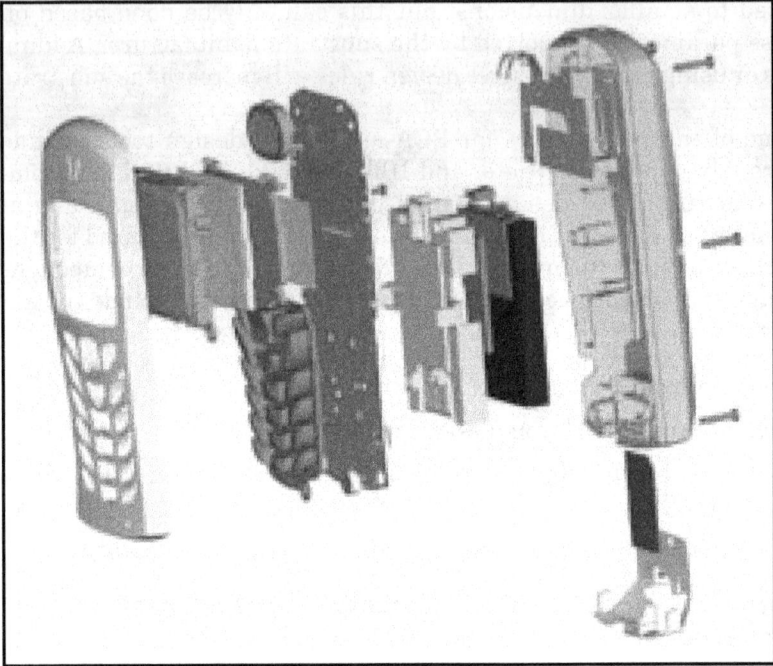

Figure 1.25   Three-dimensional rendering of mechanical assembly.

allow the designer to develop three-dimensional designs of all the custom mechanical components in the system as well as to incorporate three-dimensional representations of the various off-the-shelf components that may be used in the system.

Figure 1.25 is an example of a three-dimensional rendering of a cellular phone product. These types of renderings help the product designer verify that all the mechanical parts fit together properly and that the internal space within the product enclosure is being used efficiently.

Design tools such as Flowtherm enable the mechanical designer to conduct complex thermal analysis of the design. These systems let the designer analyze the thermal path within a model created by a 3D-design system. Effects of air movement within the product enclosure can also be simulated.

Finite element analysis (FEA) tools can assess the structural integrity and materials requirements for mechanical elements within the product. Moldflow is an example of a tool used to analyze the mold design for a plastic part, or the mechanical strength of the final product assembly.

Design for assembly (DFA) tools, such as the one produced by Boothroyd Dewherst, may also be applied to the mechanical design to ensure that the cost to assemble the product is optimized.

The output of the mechanical design process may include a number of items including

1. A complete mechanical BOM
2. A complete three-dimensional rendering of the finished product
3. A complete three-dimensional view of how all the internal system components fit together
4. Part drawings for custom mechanical parts from which machine tooling, such as plastic molds and laser machining programs, can be made

When designing a portable electronic product, it is important to put strong emphasis on the ergonomic and aesthetic aspects of the mechanical design. 3D CAD systems make it relatively easy for the designer to produce contoured designs that are both attractive in appearance and comfortable to carry and use. For a well-defined product concept, the mechanical design may be the only noticeable product differentiator from the user's perspective.

The completion of the physical and mechanical design results in the creation of the product's manufacturing and procurement documentation. The product designer must then shepherd the product into manufacturing. Many questions and potential problems will arise during prototyping and early manufacturing, which will require full knowledge of the system design in order to solve.

The manufacturing and procurement organizations will also come up with many valuable recommendations for reducing the cost of the product. The product designer must review these recommendations and ensure that they do not compromise the product's success. The product designer must take responsibility for preserving the product concept as manufacturing works through these initial implementation issues. It is only after production is up and running with a successful physical design, that the product designer may consider his task to be complete.

2

Digital and Analog Processing

The processing of electrical signals is the fundamental means by which portable electronic devices accomplish their desired functionality. System inputs are received from a variety of sources including the product user, electronic sensors, external devices, and external networks. These inputs are processed and the desired output is achieved through the appropriate output device.

The silicon *integrated circuit* (IC) is the fundamental technology that has enabled electronic processing capabilities for portable electronic products. Processing is achieved through the system electronics, which consist of integrated circuits and other electronic components mounted to a printed circuit board. The designer of a portable electronic device need not be an expert designer of integrated circuits or printed circuit boards. The portable electronic product designer must be knowledgeable in the high-level architecture of these systems and able to drive the selection of off-the-shelf components that will result in a successful system implementation. Custom integrated circuits may also be needed in many applications and the product designer should be familiar with the options available for creating unique electronic functionality.

Much of the processing that occurs in portable electronic devices takes place in the digital domain where microprocessor and digital signal processing (DSP) principles are implemented. Custom logic devices are frequently used to tie together various functions in a digital processing system. System-on-chip (SOC) devices are exploited to reduce the amount of custom design required to integrate a complex digital processing system.

The real world, however, is made up of analog signals. Sounds, images, scents, and forces must all be detected in the analog domain and converted to the digital domain for processing. Radio signals, upon which wireless communications are based, are also analog in nature and require some amount of analog processing if they are to be integrated into a digital system. Analog ICs and sensors play a critical role in enabling the portable electronic device to interface with the external environment.

Digital processing must also be supported by memory capabilities. Different types of solid-state memory devices are available to support various system memory requirements. System memory is used primarily to store instructions and data that are immediately required by the digital processing system. Mass storage is also essential in improving the usefulness of the portable electronic system. Mass storage devices, such as disk drives, are needed to make large amounts of data available to the portable electronic device.

2.1   Microprocessor

The microprocessor is the most well known of all IC types, since it is at the heart of any personal computer architecture. Microprocessors are designed to perform a wide variety of tasks, many of which may not have been envisioned when the microprocessor was designed. It is this status as a general purpose platform that has driven the microprocessor (and hence the PC) to the forefront of the technology industry.

While the design philosophy promoted in this book is one of *product focus* as opposed to creation of *general purpose* platforms, an understanding of PC architecture is extremely valuable. Notebook PCs represent one of the largest portable electronic product opportunities, and interaction with the PC environment is a major consideration in the development of any portable product.

The key elements of a microprocessor include an *arithmetic logic unit* (ALU) for performing mathematical operations, *data registers* for storing the intermediate results of programmed instructions, *instruction registers* to store the instructions being performed at a particular point in time, and a *program counter* to keep track of the memory address for the next instruction in the sequence of program steps.

A microprocessor, or central processing unit (CPU), is connected to the system memory where all the program instructions are actually stored. Input and output (I/O) data are also stored in the system memory. The CPU is connected to the system memory by a system bus. Input and output data are moved in and out of the CPU data registers and into the system memory over the system bus, as is each individual program instruction.

A simplified sequence of processor operations may be described as follows.

1. *Fetch.*   The CPU fetches the next instruction in a sequence from the program stored in memory and places it in the instruction register.

2. *Decode.*   The CPU decodes the instruction and determines the means of execution.

3. *Execute.* The CPU executes the instruction, with computational help from the ALU if needed, and stores the result in the appropriate data registers.

4. *Increment.* The CPU increments the program counter to the next memory address to find the next instruction.

A *clock* signal is also present in CPU operation to synchronize these operations. In this simplified example above, the CPU might perform each one of these operations in a single clock cycle with the result that it takes four clock cycles to complete the entire sequence.

Performance of a CPU is often measured by the clock frequency. Higher clock frequencies enable larger numbers of operations to be completed in a given time. Clock frequency alone, however, is not the only factor affecting performance. Other aspects of system architecture can have a major impact.

The processor bit size indicates the size of the registers and the resulting complexity of instructions that can be fed to the CPU in a single cycle. Thirty-two-bit CPUs are common today in notebook PCs, with 64-bit CPUs coming in the next generation of products. Larger bit sizes enable each instruction to provide more functionality.

The system memory configuration also affects performance significantly. Most system memory in a notebook PC is in the form of direct random access memory (DRAM), which is external to the CPU. Communications latency and waiting for the DRAM to refresh create delays in the ability of the CPU to carry out instructions. PC makers realized that using static random access memory (SRAM), which can be accessed more quickly than DRAM, could improve performance. SRAM is expensive compared to DRAM, but it can be used in small amounts to supplement the DRAM by storing frequently used CPU instructions. SRAM used in this fashion is called *cache*. Cache was initially implemented external to the CPU chip, but is integrated into the CPU silicon in most of the recent generation of CPU products.

In the CPU world, there are two basic architectures—CISC and RISC. CISC (complex instruction set computing) is based on the theory that by utilizing a larger library of more powerful CPU instructions, the CPU will get more work done per clock cycle. RISC (reduced instruction set computing) is based on the counter theory that a CPU will perform better with a smaller set of simpler instructions because the CPU design could be made more simple and efficient.

In practice, the RISC philosophy has proven to yield higher performance, and the X-86 architecture, although initially designed with CISC in mind, has moved in the direction of RISC.

Going back to the discussion of basic CPU operations such as fetch, decode, execute, and increment, there is another opportunity to improve performance. Rather than executing these steps from start to finish for a single instruction, it is possible to *pipeline* these operations. An example of pipelining would be for one instruction to execute, while the next is decoding, while the next is fetching, all in one clock cycle.

A similar concept for increasing performance is called superscaler execution. In superscaler execution, multiple pipelines are built into the CPU so that the same operations may be performed for different instructions simultaneously. Most advanced notebook CPUs use both pipelining and superscaler execution to enhance performance.

Since the actual sequence of instructions in a computer program can vary depending on the specific results, pipelining and superscaling operations can lead to the execution of unwanted instructions. *Branch prediction* techniques are used to reduce this effect and improve overall CPU performance. Other techniques, such as matrix processing capabilities (MMX), have also been added to CPU architectures to improve performance.

So frequency is certainly not the only indicator of CPU performance. MIPS (millions of instructions per second) is another metric that has been used to compare CPU performance. Even MIPS, however, does not take into account the efficiency of execution. Ultimately, functional testing on specific applications is the best way to compare the performance of two different CPU designs.

These fundamental principles also apply to the operation of *embedded processors*. An embedded processor is a CPU that is implemented in a system where the functionality is fixed. A notebook PC can run many different applications but a cell phone has a strictly defined feature set. Thus, the cell phone is said to have an embedded processor. The distinction between embedded processing and nonembedded processing has changed over the years and now delineates between PC processors (nonembedded) and non-PC processors (embedded). While the notebook PC and PC market are almost completely dominated by Intel's X-86 architecture, a number of different architectures are used in embedded applications. These include the ARM, MIPS, and PowerPC architectures. PowerPC has some penetration into the notebook and desktop PC markets through its use in products made by Apple.

Selecting a microprocessor or embedded processor for a portable electronic application can be a very complex decision. Traditional performance metrics, such as clock frequency or MIPS, are not as useful as they once were, since processor performance is not necessarily the key driver. Given the ample processing power of most recent generation processor designs, battery life is often a much greater concern for portable electronics.

Users probably would not perceive a difference in the execution time between a 2- and 3-GHz processor for a particular task. But those users will certainly notice if the battery life drops from 2 h to 1 h. This is the state of affairs in today's notebook computing market, where substantial but largely unnoticed performance improvements can be had with a noticeable reduction in battery life.

Given that the execution time for many functions is so much less than what a human can perceive, it is appropriate for the portable electronic product designer to look at power efficiency as an important metric for the system processor. A good question to ask is: What is the performance of this processor in terms of MIPS/watt?

2.2 Logic Devices

Logic devices include application specific integrated circuits (ASIC), programmable logic devices (PLD), and general purpose logic. ASICs have high up-front development costs but are capable of high levels of integration, and low cost if produced in high volume. PLDs have much lower up-front costs and can be developed very quickly but are not as dense as ASICs or as cost effective in high volume. General purpose logic has very low level of integration but is essential in integrating different parts of a system, hence the term *glue logic* is often applied to these devices.

There are several categories of ASICs—full custom, gate array, and cell-based ICs. Full custom ASIC devices are custom designed to the requirements of a product from the ground up. While the design cycle for these types of devices may be many months, they are appropriate for very high volume products that require high levels of integration. Full custom ASICs will often be used in new product concepts which require functional blocks that do not yet exist.

Gate arrays are devices in which the logic resources (gates) have been fabricated ahead of time, and only the final metallization layer (or layers) is (are) custom designed to connect the gates in the desired configuration. Gate arrays offer fairly high-density integration but are of lower cost to design and have a shorter design cycle than full custom ASICs.

Cell-based devices are extremely useful in creating complex portable electronic products. Cell-based ASICs leverage preexisting *core* designs to reduce the design cycle time. Thus, a microprocessor core, custom logic, and memory might be combined on a single chip instead of on separate chips. This had given rise to the concept of system-on-chip (SOC). While the tooling and fabrication may be fully custom for a cell-based ASIC, it has the advantage that a large portion of the circuit design has already been proven in other applications. The SOC approach is an increasingly popular approach in portable electronic devices due to steadily increasing product complexity. SOCs have very good electronic performance because the high level of integration minimizes signal delays that can occur if the same functions are deployed as discrete chips. The high level of integration also enables the total silicon area to be minimized, which reduces cost and potential form-factor. SOC devices also tend to consume less power than a comparable discrete implementation.

PLDs are used in portable electronic products if time-to-market and minimizing development costs are critical. Some varieties of PLD are programmed by changing the physical structure of the device. Fuse and anti-fuse schemes are used to permanently program these devices with the desired functionality. Memory-driven devices use nonvolatile memory transistors to create the desired links between logic structures. In general, fuse-type PLDs are smaller and faster than memory-driven devices. Memory-driven PLDs have the advantage of being reprogrammable.

PLDs tend to be larger and slower than ASICs but they are excellent for proving out functionality. They are also useful for developing products where there is considerable uncertainty in the final design parameters, such as a product that incorporates an emerging but yet-to-be-finalized communications standard.

2.3   Microcontroller

Microcontrollers are an important design option for many classes of portable electronic products. Microcontrollers are deployed in embedded applications and look a lot like microprocessors, but they have more system elements integrated into a single device. Microcontrollers typically have a CPU, timer, RAM, ROM, and I/O circuitry. They may also have special analog-to-digital (A/D) or digital-to-analog (D/A) circuitry to facilitate interfaces to the outside world. A microcontroller, in some cases, may be viewed as a system-on-chip (SOC) solution.

The CPU on a microcontroller may be based on an older generation microprocessor design or it may have a CPU designed specifically for microcontroller applications. Various bit sizes are available and are selected based on the complexity of the end product. Eight-bit and 16-bit processors are very common in microcontroller applications (compared with 32-bit and 64-bit in the microprocessor realm).

Microcontrollers have a small amount of RAM to facilitate the operation of the processor and have some ROM to store the basic instruction set. In newer generations of microcontrollers, this built-in memory will usually be some type of EEPROM, making it easier for the product developer to implement or change the programming instructions.

I/O ports are designed into microcontrollers to facilitate easy communications with other parts of the system. The serial or parallel nature of these data ports as well as the speed and number of ports varies depending on the particular microcontroller selected.

On board A/D and D/A converters are helpful features to design into a microcontroller because they make it easier for the designer to integrate the part into a system. Almost all microcontroller applications involve analog signals from sensors or user controls.

2.4 DSP

Digital signal processors (DSPs) are one of the most important types of devices being used in portable electronic devices today. Digital signal processing may be performed by various types of logic devices and is often performed on a microprocessor running digital signal processing code. A DSP is a logic device that is optimized to perform digital signal processing and does so much more efficiently than a general purpose microprocessor.

There are general purpose DSPs, application specific DSPs, and custom DSPs. All types of DSPs require software in order to provide the desired functionality. DSP circuitry may also be integrated into an SOC component.

Portable electronic products that involve audio or video input and output will generally need DSPs to deal efficiently with the large amount of data that is generated when audio and video signals are digitized (A/D) or synthesized (D/A). Products that use digital wireless communications may also require DSPs for the same reason.

2.5 Analog Devices

Analog devices, as the name implies, are not used for manipulating digital data but are used to process analog signals. They are important in all manner of portable electronic products ranging from cell phones to digital cameras. There are five types of analog devices (also known as *linear integrated circuits*) that are applicable to portable electronics—amplifiers, converters, regulators, comparators, and interfaces.

Amplifiers are used to amplify a signal. Common portable electronic applications include amplification of wireless signals for cell phones and WLAN cards, backlight brightness, and audio volume adjustment.

A/D (A to D) and D/A converters are used for converting analog signals from the physical world to digital signals that can be processed by logic devices. For example, an A/D converter processes an analog voice signal from a cell phone microphone and converts it into a stream of digital data so that it can be compressed and transmitted over a digital communications network. Conversely, an incoming digital signal is processed by a D/A converter and transformed into an analog signal capable of driving a cell phone speaker.

A comparator simply compares two different voltage levels and outputs one of two possible values depending on which signal is greater.

An application for this type of device might be in a battery charge detection circuit. The comparator would be used to trigger a low battery warning if the battery voltage falls below some reference voltage. A regulator device might be used to create the reference voltage. Regulators can be used to filter a noisy signal into a smoother signal or to transform a signal in some fashion including transforming it to a constant reference value.

Interface devices, such as line drivers, receivers, and transceivers, are used to transmit signals across a conductor. Interface devices condition the signal and allow the digital information to be preserved even though the transmission line may be subject to disruptive electrical noise and is fundamentally an analog transmission system.

In addition to these basic types of analog devices there are many application specific analog components that are used for a wide range of applications. In the portable electronics arena, there is a plethora of applications related to wireless and video processing that require complex custom analog devices. Likewise, analog circuits contribute to real world interfaces, the performance of which can make or break a successful product. Analog circuit design expertise is actually a rare commodity as most engineers neglected this field in the digital age. The portable electronic designer should not underestimate the challenge in finding good design support for analog-intensive circuit design.

2.6 Sensors

Sensors are devices that "sense" the physical world and convert and output an electrical signal to represent what has been sensed. Silicon-based sensors have been produced that can detect radiation, temperature, pressure, velocity, flow rate, and a variety of other physical parameters.

Many of the latest physical sensors are based on MEMS (micro electrical mechanical systems) technology. MEMS technology enables designers to fabricate three-dimensional, micromechanical structures onto a silicon chip. When combined with electronic structures on the same chip, these MEMS devices can be used to sense various aspects of the physical world. For example, a small cantilever beam may be fabricated onto a chip and the deflection due to acceleration may be measured by monitoring the change in an electric field between the beam and an electrode on the chip.

Image sensors are one of the most powerful types of sensors that have been developed because they can acquire an image that conveys a tremendous amount of information. Image sensors are used in camcorders and electronic cameras to capture images digitally. Image sensors are also being deployed in cellular phones and PDAs as product designers begin to experiment with new types of product concepts.

Image sensors present a design challenge in a number of areas. Image sensors require some type of front-end optics to create an image at the surface of the sensor device. The imaging device must be packaged such that it can be integrated into the optics chain. The optics and the image sensor must be precisely aligned during manufacturing and that alignment must be carefully maintained throughout the product life. If the product is based on a photography style usage paradigm, then additional mechanical and optical complexity must be introduced to handle zoom and focus issues.

The first digital imaging devices were based on CCD (charge-coupled device) technology. Commercial availability of CMOS sensors in the early 1990s lead to rapid price decreases for imaging devices. Along with the reduced cost of CMOS image sensors came new packaging techniques that reduced the amount of space required to package these devices. The development of low cost, injection molded, aspherical lens technology enabled expensive optics assemblies to be reduced down to a single plastic optical component.

Figure 2.1 shows a magnified view of a CCD image sensor chip. This device is wire bonded to a metal lead frame and encapsulated with an optically clear plastic material. This example is a particularly low-cost method of packaging a CCD chip. It is more common to mount these chips in a ceramic package covered with an optically clear lid.

Figure 2.1 CCD image sensor.

2.7 Wireless Communications

Recent advances in wireless communications technology have extended the horizon of possibilities for portable electronics. Advances in wireless components and network infrastructure will probably be the main driver for portable electronics for the next several years. Cellular phones are only the beginning of this proliferation of wireless technology.

The basic elements of a wireless subsystem are shown in Fig. 2.2. On the front end of the system is the antenna. In most portable products, a single antenna is used for both transmission (Tx) and reception (Rx) of signals. The antenna Tx/Rx separates the transmit and receive signals that are coupled to the antenna.

A low-noise amplifier takes the low-energy signals that have been received by the antenna and amplifies them to a level that can be processed by the receive stage. The receive stage extracts the target information from the analog carrier by subtracting the carrier frequency supplied by the frequency controller. The receive-IF (intermediate frequency) stage enables a two-stage conversion of the signal frequency to a level that can be more easily processed by the analog interface. The analog interface converts the analog signal to the digital domain (A/D) so that the baseband digital processing can be accomplished. The output of the baseband processing is a digital signal that can be further processed by the applications processing that is unique to the product, be it voice, video, data, or a combination of these.

For transmitted signals the reverse process takes place. Information from the product application undergoes baseband processing where it is manipulated into a signal that can be converted by the analog interface. The analog interface performs a digital-to-analog (D/A) conversion and

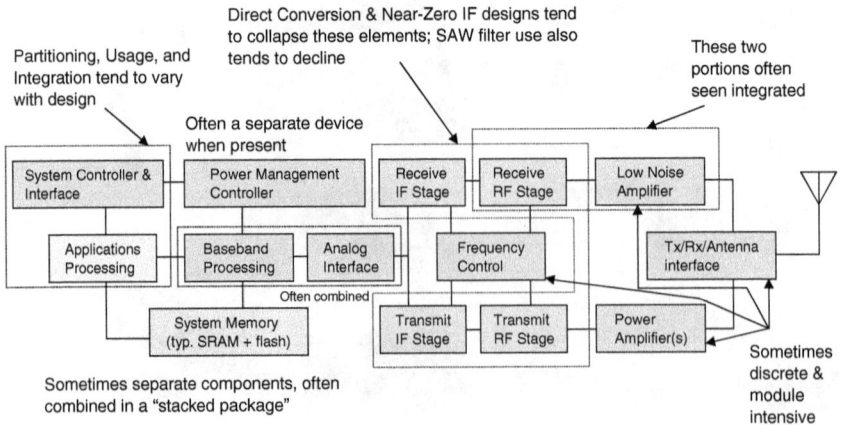

Figure 2.2 Wireless systems overview.

the signal is passed to the transmit-IF stage. The transmit-IF stage enables a two-stage conversion of the signal to an appropriate level so that it can be mixed with the transmit carrier frequency in the transmit-RF stage. Again, the precise carrier signal is supplied by the frequency control. A low-level signal of the correct frequency is then passed to the power amplifier where the signal is boosted for transmission.

A standalone RF product will also have its own system controller, power management controller, and system memory. Cell phones typically use a dedicated power management chip because of the complexity and precise timing required to reduce the effective duty cycle of the RF subsystems. If the wireless processing is integrated into a highly functional system such as a PDA, the system controller and system memory resources may be shared with other subsystems.

Recent integration trends support further miniaturization and commoditization of wireless functionality. Figure 2.3 shows how these various stages of wireless processing are being integrated into fewer components.

Zero-IF processing techniques have eliminated the need for intermediate frequency conversion and have made an integrated RF stage feasible. The RF receive, RF transmit, and frequency control can all be cost-effectively integrated into a single chip that is based on RF compatible IC manufacturing processes such as Bipolar, BiCMOS, SiGe or RF CMOS. A/D conversion, D/A conversion, Power management, and other functions can be combined on a single chip based on commodity CMOS mixed signal process technology.

The system controller, base-band processing, and some system memory in the form of SRAM can be integrated into a single chip, using leading edge CMOS processing. System memory, power amplification, and the antenna interface still require specialized processes or discrete components for a cost-effective solution. This type of integration brings the entire component set for an RF subsystem down to a few key components. Single chip solutions for the entire wireless function are beginning to appear for short-range digital radio standards such as Bluetooth.

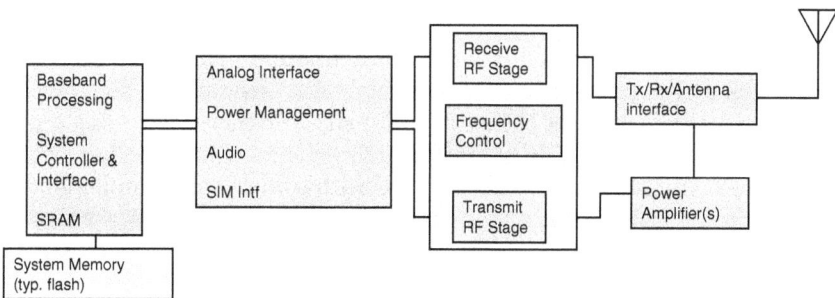

Figure 2.3 Wireless system integration.

2.8 System Memory

As discussed in the previous section on microprocessors, memory is a basic function required to support processing. There are many different types of memory available to support the design of portable electronic products. The critical factors are power consumption, cost, form-factor, and performance. Memory devices are classified as either volatile or nonvolatile. Nonvolatile memory continues to store information even after power has been removed from the device. Volatile memory must have power continuously applied to preserve the stored data.

Direct random access memory (DRAM) is the most common form of memory in notebook computer products and comprises the bulk of the system memory for these products. DRAM is a volatile memory technology and, in addition, it must be constantly *refreshed*; an operation that consumes power and saps performance. Due to its simple architecture, DRAM does provide very high memory density and has a relatively low cost per bit of data stored.

There are many varieties of DRAM that have been developed to provide modest improvements in performance or power consumption. The most important of these variations is double data rate (DDR) DRAM. DDR DRAM executes two instructions per clock cycle, improving the effective CPU performance for portable PC applications. DDR-II and DDR-III standards will be coming to market in the future to provide an additional performance boost. To improve graphics performance in notebook PCs, graphics DDR (GDDR) standards are being developed for memory that would be connected directly to the notebook PC graphics accelerator chip. These improvements will probably be of little interest to the portable electronic designer outside of the notebook PC product space. Of broader relevance to portable electronic devices is embedded DRAM. Embedded DRAM refers to a design trend toward integrating DRAM onto the same chip as the system processor. This supports the growing desire in the industry to produce system on chip (SOC) designs.

Static random access memory (SRAM) is also a form of volatile memory, but it has the advantage of not needing to be refreshed during operation. SRAM, therefore, exhibits less latency and power consumption than DRAM and enables the CPU to function more efficiently. The drawback for SRAM is that it uses multiple transistors for each bit of data stored. This reduces the storage density and increases cost per bit. In notebook PC designs, SRAM is used for cache memory and, increasingly, it is being integrated into the CPU rather than being implemented as a separate device. It is even being included on some DRAM chips to improve performance.

As SRAM becomes integrated into notebook CPU chips, the application of discrete SRAM devices in this class of products appears to be on the decline. At the same time, cellular phone products are beginning to make greater use of SRAM, as the functional complexity of these products increases.

Nonvolatile memory technologies, those which do not lose their data when the power is off, have broad application to portable electronic products. Read only memory (ROM) generally refers to memory that is programmed during the manufacturing process. The actual IC fabrication masks are configured to encode the desired information and once fabricated the information cannot be changed. This type of memory is cheap (in high volume) and fast but requires a long lead time to create, potentially affecting product time-to-market. The system BIOS for notebook PCs, cellular phones, and other high-volume portable electronic products is often stored in ROM.

Programmable ROM (PROM) devices are more flexible in that they can be encoded with data after the manufacturing process. This means that they can be mass produced and leveraged across a number of different products. PROM devices can only be programmed once.

Erasable PROM (EPROM) devices can be reprogrammed after being erased with UV light, or with an electrical signal (EEPROM). EEPROM have been used in cell phones to store the user's personal data and profiles.

Another important type of programmable, nonvolatile memory is Flash. Flash is an electrically programmable type of memory that can achieve much higher storage density than non-Flash EEPROM. There are many types of Flash but the two most common types at present are NOR Flash and NAND Flash.

NOR Flash allows data to be accessed in parallel, so it is favored in applications that require high-access speeds such as RAM supporting an embedded processor. Cellular phones use NOR Flash for storing system software. NAND Flash is higher density than NOR Flash, but it accesses data in a serial fashion and hence access times are slower. NAND Flash is currently popular for making memory cards for consumer electronics applications such as electronic still cameras and MP3 players. NAND is much less expensive than NOR and product designers are now replacing NOR Flash with a combination of NAND and DRAM.

2.9   Mass Storage

Before the days of network-connected devices a user had to input all information into a portable electronic device, or the information had to be retrieved from mass storage. For media devices, such as a cassette walkman, it is particularly critical that a standardized storage media is available with a wide selection of prerecorded content, if the product concept is to be a success. While not directly applicable to portable devices, the standard's battle in the early 1980s between VHS and Beta video tape formats is a prime example of how important standards are to consumer electronic products.

At that time, the two dominant forces in consumer electronics were Matsushita (branded Panasonic) and Sony. Matsushita backed the VHS standard and Sony backed the technically superior Betamax standard. Sony refused to license Betamax to OEMs, whereas Matsushita licensed VHS. The result was that VHS systems were made in much greater numbers and within a couple of years had conquered the market. Video rental stores stopped wasting shelf space on Betamax video tapes, resulting in the complete disappearance of the standard in the consumer market. (Interestingly, Sony continued to make Betamax products for professional recording until 2002.)

Today's portable CD players and DVD players are successful products only because definitive standards were put in place to ensure that any content will work on any appropriate player. Removable mass storage media is still very important to the portable electronics market, even though many products get their content from a network connection. Conceptually, this is due to the fact that mass storage technology can support media data rates that are 10 to 100 times what are available from network connections (particularly if the network connection is wireless).

The notion of standardized removable storage media enabling new markets for prerecorded studio content is well understood. High-bandwidth network access is beginning to impact this fundamental principle, however. Since the user can readily access content from the network and transfer that content easily to storage media that is internal to a portable device, the mass storage in this case does need to be a standardized format. MP3 players exemplify this usage paradigm.

By the same reasoning, intermediate and user-recordable storage, such as a memory stick, or a USB FLASH stick, or an SD card could be used interchangeably for the purpose of transferring network content into a portable device. There are now a number of storage formats available in the market. This seems to run counter to the lessons of the classic VHS and Betamax case study. In today's environment, it is possible for a major player such as SONY to promote a unique mass storage standard for use across its product lines, and do so for the purposes of product differentiation.

Even with the increasing emphasis on network connectivity, mass storage continues to be a central issue for portable electronic design. Network capabilities are simply not a substitute for instant access to locally stored information. Individual users are collecting large amounts of data, much of it in audio and video formats. Providing an ample mass storage solution usually enables the user to delay the decision of which data to keep and which data to preserve. With the passage of time, older versions of data can be deleted with less concern on the part of the user.

Two classes of mass storage are associated with a portable electronic device—internal and removable. Internal storage is permanently mounted inside the product and is intended to provide rapid access to frequently used data. An example of internal storage is the internal hard disk in a notebook computer. Removable storage is mounted internally to the portable device but can be easily removed by the user. It is intended to enable rapid access to large amounts of new data from an external source, to augment the internal storage capacity if necessary, to perform backups on internal storage, or to duplicate and transfer internally stored data. Examples of removable storage include a CD, a floppy disk, or a PC memory card.

Mass storage may also be classified as ROM and read-write (RW). ROM media is encoded with data once, and that data may not be changed or written over. DVD-ROM and CD-ROM are well-known examples of ROM technologies. ROM technologies are characterized by a low cost-per-bit of information stored and are used for large scale distribution of software and media content. RW technologies enable the user to record and erase data many times. Examples of RW technologies include audio cassettes, VHS tapes, magnetic hard drives, floppy disks, memory cards, and more recently CD-RW and DVD-RW.

Yet another way to classify mass storage technology is in terms of whether the storage media is tape, rotational or solid state. Tape storage is still widely used for video recording and for data backup applications. Rotational media includes all types of magnetic and optical disks. Solid-state storage includes all types of semiconductor memory chips. Solid-state storage is in turn generally divided into two categories— volatile and nonvolatile. Volatile memory must remain connected to a power source for the stored data to be preserved. Nonvolatile memory will continue to store data even in the absence of a power source.

Magnetic disk drives are currently the workhorse of the mass storage industry, being the technology of choice for computer hard disks. Notebook PCs have internal hard drives that can support 20 GB or greater of data. Typical magnetic hard disks have a diameter of over 2 in and rotational speeds of over 4000 rpm. As a result, these devices consume at least 2 W during read/write operations, which is rather high for systems that are more compact than a notebook computer. The key advantages that magnetic technology brings to the portable product are a relatively high storage density, rapid access times, and low cost-per-bit of information stored. New hard drives with smaller diameter disks are being developed that use much lower levels of power and have form-factors similar to solid-state memory cards.

Optical storage devices (CD and DVD) do not have the capacity of magnetic drives but provide a rugged and removable media for mass storage. CD format can store 650 MB and DVD format can store several GB. Writable optical disk technology (DVD-RW and CD-RW) is now being deployed in many notebook computer products which may ultimately result in the obsolescence of magnetic removable media such as floppy disks and ZIP disks.

Removable solid-state storage media is very important for portable electronic products outside of the notebook computer segment. Current rotational media, with the exception of current microdrive technology, tend to be too large and power hungry for small form-factor products. Sony has developed the memory stick media as a standard removable storage solution for a wide range of Sony products. Other formats of flash cards have also been developed to support the requirements of electronic camera products, PDAs, and other small form-factor devices.

The impact of ever denser, ever cheaper mass storage technology on portable electronic products cannot be overstated. Functionalities, such as motion video record, motion video playback, and digital music libraries, are impossible without massive amounts of compact, inexpensive memory.

Some operating systems used for notebook computer systems save entire system configurations, including the contents of all files, on a periodic basis so that the user can recover a previous system state. Such a feature is only acceptable if the cost of mass memory is low. Taken a step further, the ability to recover every email, voice mail, photograph, video, or document version ever encountered by an individual will most likely be possible if the current trends in mass-storage technology continue.

Along those lines of thinking, would not a simple electronic book product that contains every work of literature ever produced be extremely compelling?

In summary, the processing of electrical signals in the digital or analog domain is the primary mechanism by which the functionality of a portable electronic device gets accomplished. The designer should have a good understanding of the available sensor, processor, and memory technologies in order to competitively implement the functionality required for a product concept.

3

Electronic Packaging

Electronic packaging is the means by which system components are interconnected and integrated to create a product with a desired functionality. Electronic packaging must provide for electrical connections between components, physical support of each component, thermal management for the heat dissipated by each component, and connection of the system to the outside world.

The most common form of electronic packaging is the printed circuit assembly (PCA). The printed circuit assembly consists of the printed circuit board (PCB) and the various electronic components that are attached to it, including integrated circuits (ICs), discrete components, connectors, and modules. Modules are often smaller PCAs that are connected to a larger PCA (motherboard).

Through-hole and surface mount technology (SMT) are terms that describe the set of processes that comprise the assembly of PCAs. While through-hole assembly was once the predominant technology for electronic assembly, SMT is now the technology of choice, particularly for portable electronic systems. A highly simplified explanation of SMT processing is as follows:

1. The production line is *kitted* with the required PCB substrate and components for a particular PCA. Assembly instructions are downloaded to the automated assembly equipment.

2. The PCB substrate is loaded into a screen printer that deposits solder paste onto the component connection points.

3. The PCB is transferred to the *pick and place* station where robotic arms pick individual components out of a dispenser and place them onto the circuit board.

4. The PCB is transferred to a reflow oven where the solder paste is melted and then allowed to solidify, forming solder joints. These solder joints provide mechanical and electrical connection between the PCB and the components.

5. With the components attached, the PCB is now a PCA. It is transferred to a rinsing station to remove the residual solder reflow material. (A "no-clean" process has been developed in recent years and is used in many operations to eliminate this processing step.)

6. If the PCA requires components on both sides of the PCB, the PCA is transferred to another production line to repeat the process on the other side of the PCB substrate.

7. Once all necessary components are connected to the PCA, it is transferred to a test area where it is inspected and connected to test equipment. The test equipment verifies proper functionality or identifies the appropriate repair actions.

8. After testing, the PCA is sent to the product assembly operations to be assembled into the final product.

Integrated circuits provide most of the functional capabilities of a portable electronic device, but many other components, such as chip resistors and chip capacitors, are also needed to make the system function. Figure 3.1 illustrates the key elements of electronic packaging. Figure 3.2 shows a photograph of a small printed circuit assembly as a practical example of the type of electronic packaging depicted in Fig. 3.1.

Due to the expense of developing integrated circuits, these components must be leveraged across many different products to recover the development cost. Only a few products, such as a Sony or Nintendo game console, or a cellular phone, have enough production volume to justify a custom design for a semiconductor device. As a result, product designers are generally limited to off-the-shelf, semicustom, or programmable components for most of the system design. Much of the unique physical design for a portable electronic device is centered on the packaging of available components. Thus, electronic packaging technology can be a powerful differentiator in the design of a portable electronic device. Electronic packaging, particularly at the circuit board level, is where much of the physical design integration occurs for a new portable electronic device.

Of all the critical factors discussed in Chap. 2, form-factor is the most impacted by the product's electronic packaging design. A superior electronic packaging solution has often been the key to beating the competition in the portable electronics marketplace. Reliability, cost, and time-to-market are also significantly affected by the electronic packaging design. Functionality and performance are only impacted in

Cross-sectional view

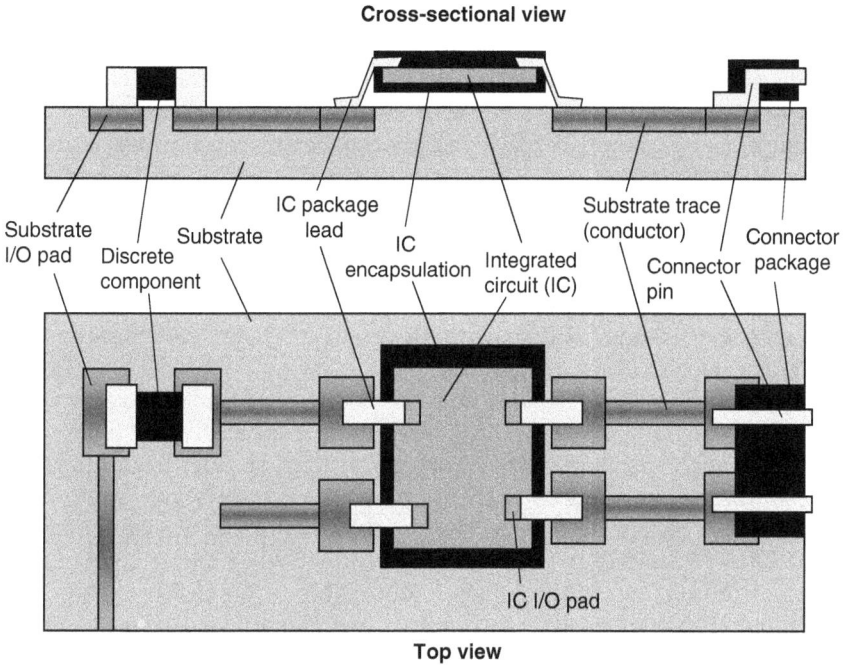

Substrate I/O pad   Discrete component   Substrate   IC package lead   IC encapsulation   Integrated circuit (IC)   Substrate trace (conductor)   Connector pin   Connector package

IC I/O pad

Top view

Figure 3.1 Electronic packaging elements.

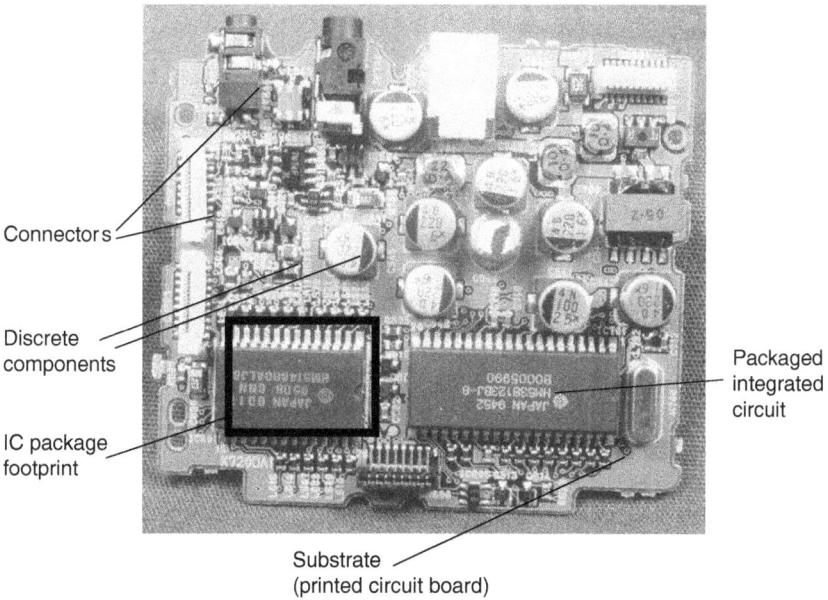

Connectors

Discrete components

IC package footprint

Packaged integrated circuit

Substrate (printed circuit board)

Figure 3.2 Electronic packaging assembly.

certain types of systems. User interface and battery life tend to be less affected by the electronic packaging solution.

The electronic packaging design of a portable electronic device is usually driven by the desire to *miniaturize* the system form-factor.

One of the most beneficial dynamics that drives rapid progress in the electronics industry is the tendency for miniaturization to result in reduced cost for a particular functionality. Smaller systems tend to use less material, require precision assembly over smaller working areas, and generally result from the use of fewer discrete components. Provided the level of miniaturization does not challenge the limits of the available processing capabilities, a smaller product should cost less to manufacture than a larger system.

It is often the case, however, that the designer of a portable electronic device will, in fact, choose to push the limits of currently available technology. This decision is made to differentiate a product from functionally similar competitors. Designers that seek aggressive miniaturization will incur a cost premium as they seek the most advanced solutions available. Presumably, the resulting product will be able to garner a price premium, or a higher market share, due to the resulting superior form-factor. If such designs do result in a very successful product, there is an additional side effect: the challenging process parameters that were implemented become more cost competitive due to the increased learning and cost absorption that results from high volume production. The advanced technology becomes more mainstream and thus accessible to a larger number of products.

Figure 3.3 is a simple representation of the key elements of an electronic packaging system.

The *device* represents a functional component such as an integrated circuit, which may have hundreds of connection pins, or it could be a simple chip resistor with two connections. The *package* is an intermediate layer or "interposer" that may be required to attach the device to the *substrate*. Various types of integrated circuit packaging and attachment

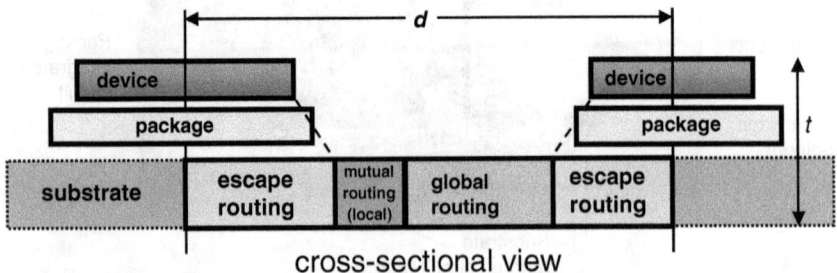

cross-sectional view

Figure 3.3 Electronic packaging and mean distance.

techniques are discussed later in this chapter. The substrate provides the interconnection between devices. (A substrate which attaches to a larger system substrate is called a *module*.) The most common form of substrate is the PCB. PCBs and other types of substrates will also be discussed later in this chapter.

Notice that Fig. 3.3 is a cross-sectional view of the electronic packaging assembly. The mean distance between devices on the board is d. The dimension d is a single dimension in the cross-sectional view, but it conceptually represents two dimensions (x,y), corresponding to the physical area occupied by the system elements. The stack-up height of the substrate, package, and device is t, corresponding to the (z) dimension or physical thickness of the system elements. Miniaturization of the system involves reducing d (area) and t (thickness) to the smallest values possible with an economically viable set of packaging technologies.

Inspection of Fig. 3.3 reveals the obstacles involved in reducing d. First, the size of the device itself is a limitation. Reducing the number of transistors required to implement a particular function is one method of reducing IC area. Another method would be to move to a more aggressive IC processing technology. Both of these solutions are in the domain of the IC supplier and may not be available to the portable electronic product designer due to economic or time-to-market constraints.

The device package is the next layer of packaging that may be miniaturized. Many devices are available in a choice of package sizes. In some cases, integrated circuits are available as *bare die*, without any package at all. These choices allow the product designer to select an option that will minimize the package *footprint* (area) if it is determined that the package size is limiting the ability to reduce d. The package footprint does not always limit d because the required *escape routing* may actually be larger than the package itself.

Escape routing requires a certain area of the substrate to attach all the electrical inputs and outputs (I/O) of the device to the substrate and to route all those connections to a perimeter location where they are accessible for routing to the I/O of other components.

Local routing requires a certain area of the substrate to connect the I/O of adjacent components. If the physical pattern of connected I/O is perfectly matched, such as with a standardized microprocessor memory bus connecting to memory devices, the local routing area requirements might be zero, since the components could be placed adjacent to each other with the escape routing zones positioned directly adjacent to each other. This efficient situation is rare in practice, and some amount of substrate area must be consumed to match I/O between adjacent components.

A particular IC device will need to be connected to many different system components and it will not be possible to place all these components adjacent to the device in question. Routing of the I/O to nonadjacent devices is referred to as *global routing* and this also requires substrate area, which can limit the minimum value of d.

Both local and global routing area can be minimized by closely locating devices that share a large percentage of mutual I/O connections. Fortunately, much of the component placement and I/O routing is performed by automated design tools that can rapidly optimize the component placements. The product designer could use such tools interactively to select the optimal device packaging and substrate technologies for a particular application, although this iterative process could be rather time consuming. In a typical product design process, the selection of appropriate packaging technologies will happen during the trade-off analysis phase of product development. (The available package types for a given component selection may be very limited in some cases.) The packaging technologies selected during the trade-off analysis will determine the extent to which d can be minimized.

Similarly, the minimization of t impacts the overall system thickness and form-factor. Electronic packaging is not always the limiting factor when it comes to minimizing the product thickness. Other system components such as hard drives or batteries may overshadow the need to minimize packaging thickness.

In situations where packaging thickness is critical, thin substrate and device packaging technologies may be selected to minimize packaging thickness. Ideally, these selections are made during the trade-off analysis phase of product development.

There is one final note relative to Fig. 3.3. While this cross section depicts a single-sided electronic packaging assembly, substrates often have components mounted on both sides. The basic concepts outlined here still apply.

Product reliability is also highly impacted by the system electronic packaging. Every physical connection in a system is a potential failure point and each type of connection has an associated failure rate. The probability of system failure due to a failed connection has an exponential relationship with the number of connections in the system. The connections from the device to the package, from the package to the substrate, from the substrate to a connector, and from a connector to its mate must all be considered. Other structures within the electronic packaging domain, such as substrate lines and vias, also have reliability factors, but their impact is a second-order contribution compared to the connections between discrete elements of the system.

Cost is clearly impacted by the electronic packaging design, although the total cost of electronic packaging would rarely be more than 15 percent of total product cost. It is important to keep this in perspective when making technology trade-offs. Doubling the cost of a substrate, for example, will have minimal impact on total system cost, but could make a significant improvement in product form-factor.

Time-to-market is a property of the specific electronic packaging technologies selected for the product design. Many suppliers specialize in quick-turn printed circuit boards that can be fabricated in a matter of one or two days, upon receipt of a design file. Other substrate suppliers may require special tooling that could require several weeks' lead time for each iteration of product design. If a highly advanced substrate technology is selected for the product design, the lead time for each new design spin should be factored into the product development schedule. Multiple iterations of the packaging design are almost always performed to correct design errors or to refine the product functionality.

Functionality and performance can be affected by the packaging design if the system components require a precise level of timing synchronization or high signal fidelity. Certain signals within a system, such as a system clock, must arrive at different parts of the system within a designated maximum period of time or else they become unsynchronized. This lack of synchronization occurs when signals must travel over significant distances to reach their destination. Even though the propagation delay in the circuit board is extremely small, the margin for timing error in a system with a multigigahertz clock is even smaller. Propagation delay is of concern on larger products that have very high-speed components, such as notebook computers. Wireless systems that have phase sensitive analog/RF signals or systems that require high-speed analog or digital signals must also maintain a high level of signal fidelity. Microwave frequency communications devices, such as cellular phones, require that certain component connections have carefully controlled impedance.

3.1 IC Packaging

The input/output (I/O) pads of an integrated circuit (IC) must be attached to a substrate in order to be electrically connected to the rest of the system. In addition, the IC package must protect the device during shipping, during the substrate assembly process, and during the life of the fully assembled product. The IC package must also facilitate alignment and testing operations required during the electronics assembly process.

The substrate provides the electrical connections between the I/O of different components in the system. The geometry of the physical interface between the substrate and package will drive the appropriate selection of substrate technology. The selection of the IC packaging and associated substrate technology will have a major impact on the form-factor of a portable electronic device.

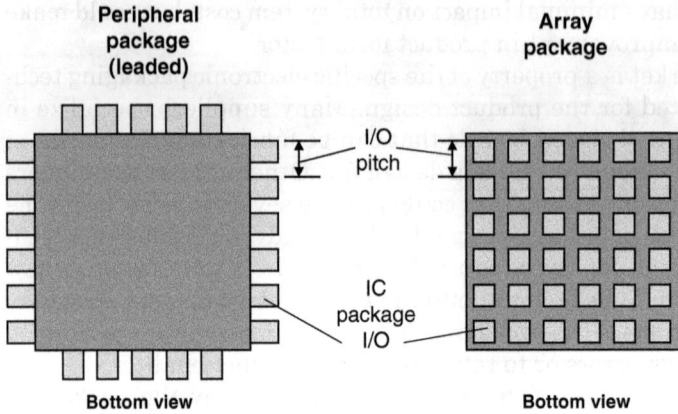

Figure 3.4 IC package connection pattern and pitch.

There are two basic types of I/O patterns that characterize an IC package—*peripheral* and *array*. Figure 3.4 shows how these two patterns differ. The peripheral pattern has a single row of I/O around the edge of the die, whereas an array pattern has multiple rows of I/O arranged in a full or partial grid.

The *pitch* is the periodic dimension that characterizes the spacing between repeating elements of a peripheral and array pattern. Thus, the I/O pattern and the pitch are generally used to characterize a package since these parameters define the physical interface to the substrate.

The *footprint* of an IC package refers to the substrate area occupied by the packaged component, as shown in Fig. 3.2. The footprint of an IC package is usually (but not always) significantly larger than the die area, and peripheral packages tend to have larger footprints than array-packaged devices.

The *thickness* of an IC package is the height that it extends above the plane of the substrate. It is not uncommon for IC packages to be less than 2 mm thick, particularly for memory devices and devices that are targeted at portable electronic applications.

Package *performance* can be an issue for some high-performance IC devices. High-contact resistance bonds, long wire bond lengths, or noise between package leads can interfere with the function of a device. Packages that minimize the length of connection to a substrate, such as ball grid array (BGA) or flip-chip (FC), are often preferred in high-performance applications.

The package connections to the IC as well as to the substrate are potential points of failure that affect reliability. Thermal stresses are one of the primary mechanisms causing failures at these connection points. Thermal stresses arise due to the fact that the substrate and the IC have different rates of thermal expansion as the temperature changes. As the electronic assembly undergoes changes in temperature (created by power dissipation during operation), stresses are created at the point of contact. While these stresses may not cause an immediate failure, the bond material may fatigue over time as the result of repeated cycling. Thermal stresses tend to be a concern in systems that involve large IC devices that operate at high power dissipations. Mechanical stresses also result from physical forces imparted on the system, such as the intended flexure of a PCB due to pressure on the keyboard, or unintended forces due to abuse of the product.

The cost of the IC package is usually low compared to the cost of the silicon IC. Obtaining IC in a package that is a nonstandard offering from a particular manufacturer can be costly, however. It is important to investigate all of the package styles available for a particular IC. If a custom package or bare die (unpackaged IC) is required, significant lead time may also be required to obtain the part. Testing and handling costs also need to be considered.

There are many types of IC packages. The Joint Electron Device Engineering Council (JEDEC) provides many of the standards that govern the physical configuration of many of the common package styles. Leaded packages are the most common, with BGA packages becoming very common for high pin-count (I/O) devices. Bare die (the semiconductor IC chip is often called a *die*) and chip scale packages (CSPs) are also becoming very common in highly miniaturized systems.

3.1.1 Leaded package

Leaded packages were developed to support the surface mounting of ICs onto a PCB substrate. There are many varieties of leaded packages including the quad flat pack (QFP), the small outline package (SOP), the thin quad flat pack (TQFP), the thin small outline package (TSOP), and a host of other standard and vendor-specific offerings. All of these packages are designed to be soldered onto a printed circuit substrate. The package sizes and lead pitches are standardized in order to facilitate a predictable design and assembly process. Smaller lead pitches reduce the package footprint, so the industry has progressed to finer and finer pitches over the years. Today, lead pitches of 0.5 mm are very common. The development of thin leaded packages has been important in enabling the miniaturization of portable electronic devices. Plastic leaded package thicknesses of 2 mm or less are available for many IC devices.

Figure 3.5 shows the basic construction of a leaded package. Notice that wire bonds are used to connect the IC I/O pads to the package leads. The wire bonds are made from a thin gold or aluminum wire, usually less than 0.002 in thick.

wire-bond   plastic encapsulation
                   (epoxy resin)

IC die

metal die pad

metal lead

substrate

solder joint

cross section

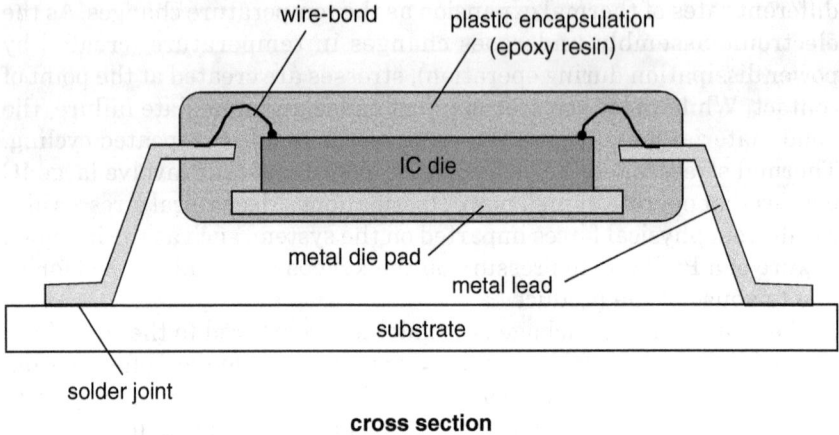

Figure 3.5   Basic construction of a plastic leaded package.

3.1.2   TAB/TCP package

Tape-automated bonding (TAB) packages were developed to achieve very thin packages with very fine pitch leads. Very fine leads are formed by photolithography of a metal layer adhering to a polyimide film. Once the leads are patterned, the polyimide film remains to support the fragile mechanical structure. TAB packages are transported and handled in a connected strip. The TAB device is generally very fine pitch and must be carefully aligned prior to bonding. Thus, TAB devices are not compatible with standard pick-and-place surface mount operations for printed circuit assembly, and a specialized assembly station must be used to attach TAB devices.

TAB was originally developed to attach high pin count, fine pitch IC packages to high density assemblies. Lead pitches of 0.3 mm or less are achievable with a TAB package. Today, however, most TAB is used to assemble LCD driver chips to the glass substrate of a flat panel display.

The tape carrier package (TCP) is a form of TAB device that is designed to be handled like a very fine pitch leaded package. TCPs are essentially a thin QFP that utilize polyimide film rather than a lead frame in their construction. The construction of a TAB package is shown in Fig. 3.6. An implementation of TAB in a PC card is depicted in Fig. 3.7.

3.1.3   COB

Chip-on-board (COB) is actually package-less technology that enables fairly high levels of miniaturization. In a COB assembly, the IC device is adhesively bonded directly to the substrate and the I/O connections are made using wire bonds. Wire-bonds must be protected with an encapsulation to prevent them from being crushed after assembly.

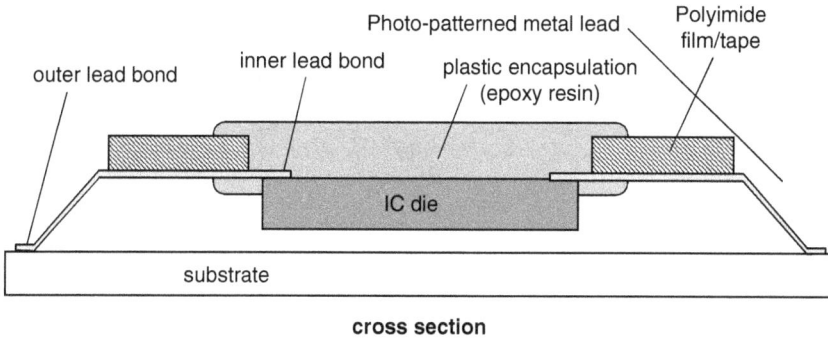

outer lead bond   inner lead bond   Photo-patterned metal lead   Polyimide film/tape

plastic encapsulation (epoxy resin)

IC die

substrate

cross section

Figure 3.6   Basic construction of a TAB package.

Peripheral I/O pitches with COB can easily match the line pitch of the highest density substrates down to 0.05 mm. Thicknesses of 1 mm or less can be achieved with COB assembly. Figure 3.8 shows a typical COB construction. Figure 3.9 shows a COB assembly that was deployed in a miniaturized 386 PC motherboard application.

3.1.4   Flip-chip

Flip-chip is another package-less technology that enables a very high level of miniaturization. The bare die is mounted directly to the substrate,

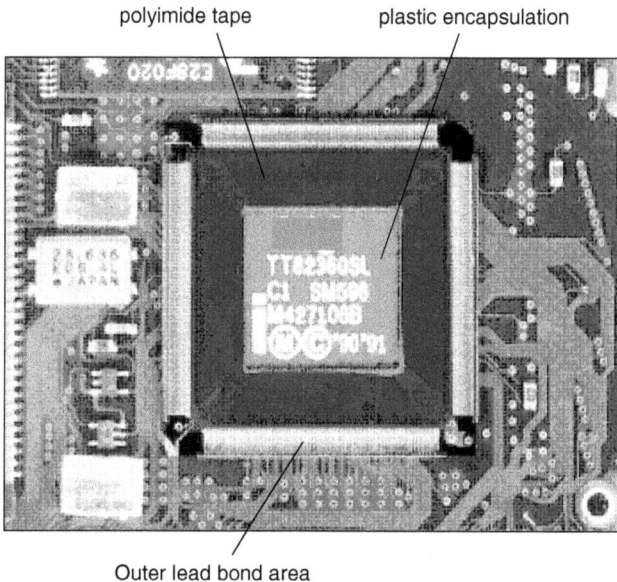

polyimide tape   plastic encapsulation

Outer lead bond area

Figure 3.7   TAB package attached to a PC card substrate.

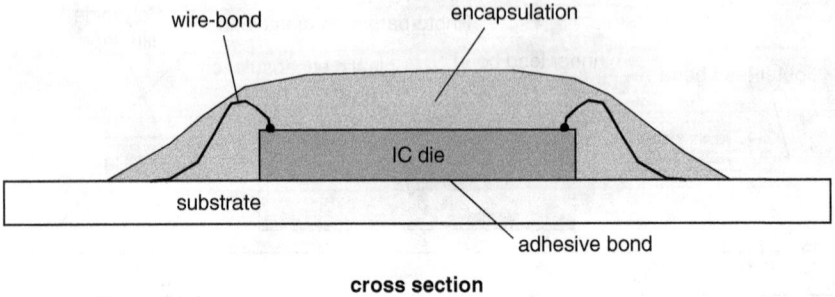

cross section

Figure 3.8   Basic COB construction.

eliminating the need for an intermediate IC package; flip-chip can be cost-effective in very high-volume applications. The electrical characteristics of a flip-chip connection are generally better than a wire-bonded device, due to the elimination of performance-sapping conductance and capacitance in leads and wire bonds.

Some drawbacks of flip-chip technology include difficulty in testing the device prior to assembly and high mechanical stresses due to thermal mismatch. Thermal stress issues are handled by underfilling the assembly with an epoxy. This adds complexity and cost to the electronic assembly process, however. Flip-chip is generally used for relatively small ICs that have low I/O counts and a high-manufacturing yield, and in applications where extreme miniaturization is of the utmost importance.

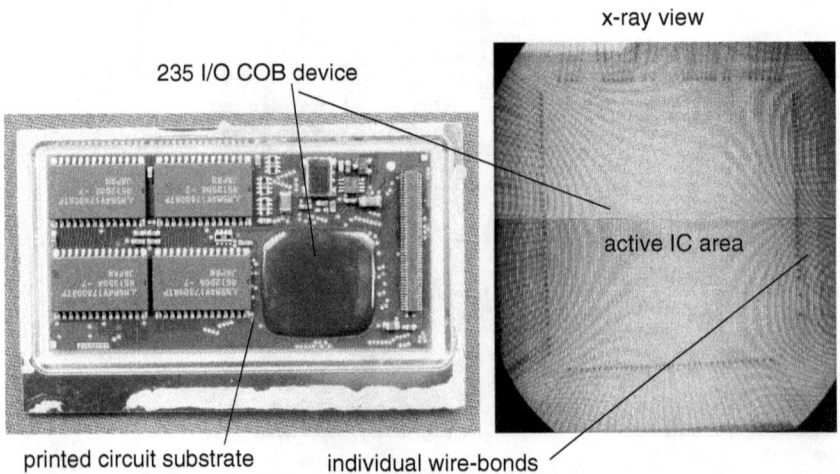

Figure 3.9   COB implementation on miniaturized PC motherboard.

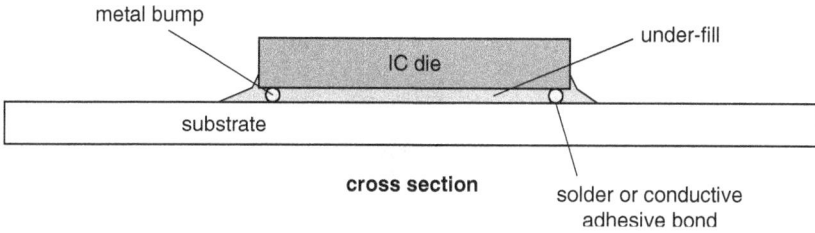

Figure 3.10 Basic flip-chip construction.

Unlike COB technology where wire bonds may be used to match any desired substrate line pitch, the flip-chip IC I/O pitch must be matched exactly on the substrate. The I/O pitch of ICs tends to be very small relative to the line pitch of available substrate technologies; thus expensive high-density substrates must often be used for flip-chip, or flip-chip simply may not be an option in some cases. Assembly thicknesses of less than 0.5 mm can be achieved with flip-chip.

Figure 3.10 illustrates the basic flip-chip construction. Figure 3.11 shows a product implementation of flip-chip technology in an electronic wristwatch application where a leaded IC package would be too large to fit in the desired product form-factor.

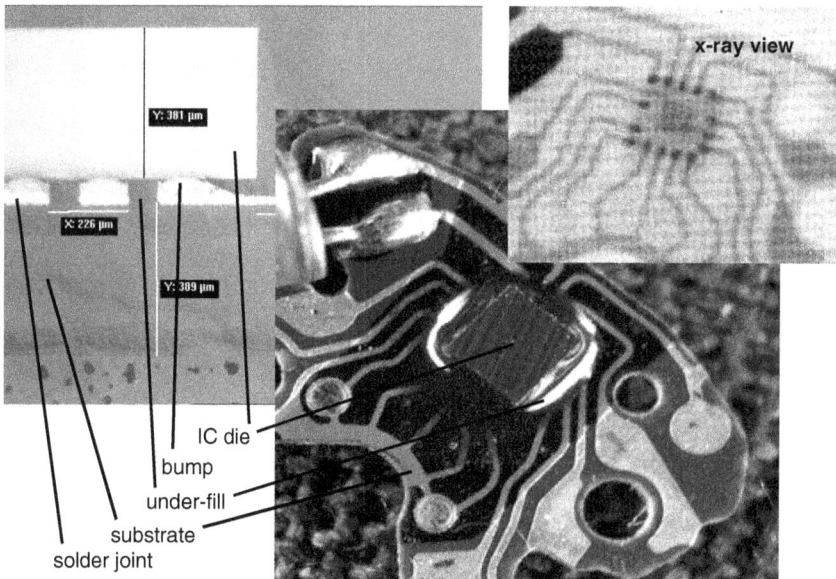

Figure 3.11 Flip-chip technology in a wristwatch.

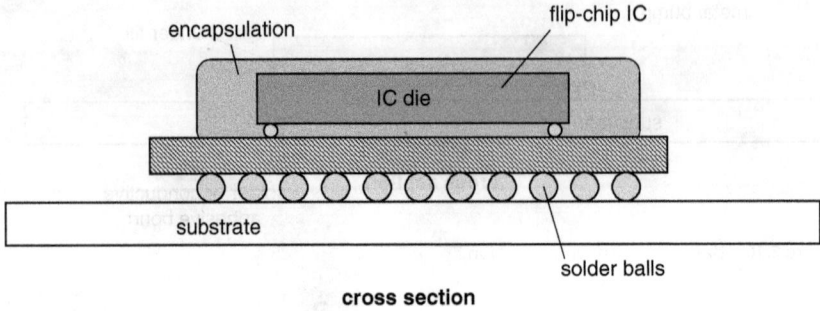

Figure 3.12   Basic BGA construction.

3.1.5   BGA

BGA (ball grid array) packages utilize a small printed circuit carrier as an interposer between the IC and the printed circuit substrate. The IC is wire-bonded or flip-chipped to the printed circuit carrier and is encapsulated in a manner similar to a leaded package. The printed circuit carrier routes the peripheral wire-bonds to an array pattern on the underside of the carrier. The array pattern is populated with solder balls that function as leads for surface mount attachment to the printed circuit substrate. Figure 3.12 shows a basic flip-chip BGA construction.

Figure 3.13   BGA package on notebook computer motherboard.

A plastic BGA package has a smaller footprint than a plastic leaded package because the array I/O pattern occupies a smaller surface area. A flip-chip package has a smaller footprint still, but it requires very high-density routing on the substrate to connect to the fine pitch I/O of the IC. The BGA package connects to this fine pitch I/O at the carrier substrate level and then fans out the I/O into a larger pitch array pattern. So, a high-density substrate is required for the carrier substrate, but not the larger system substrate (motherboard). The BGA, therefore, helps avoid the cost of a higher-density motherboard technology over large areas by applying the high density only where needed to support fan-out. BGA packages are very common in portable products requiring larger motherboards, such as notebook computers, and are comparable in thickness to leaded packages. A notebook computer application of a BGA package is shown in Fig. 3.13.

Some flip-chip style BGA packages leave the IC device exposed on the backside so that a heat-sink may be attached. This would be used for high-power devices such as microprocessor chips, where intimate contact with the chip is required for proper cooling. Figure 3.14 shows a microprocessor flip-chip BGA package, with exposed IC.

Figure 3.14 Flip-chip BGA package with exposed IC.

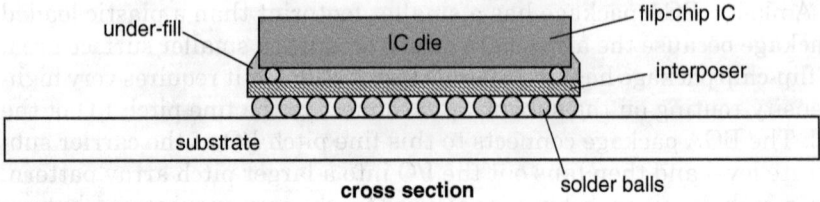

Figure 3.15 Typical CSP construction.

3.1.6 CSP

Chip scale packages (CSPs) are similar to BGA packages in concept but they have even smaller footprints and have thicknesses approaching that of flip-chip. Generally, the footprint of a CSP is only slightly larger than the footprint of the bare die. CSPs provide the miniaturization and perform-ance benefits of a flip-chip approach but are engineered to reduce mechan-ical stress and improve testability. The interposer layer in a CSP can be used to transform the I/O pattern of the device from peripheral to array.

CSP packages tend to be expensive but are becoming more affordable as their use proliferates. There are many types of constructions used to fabricate a CSP. In many of these, the interposer layer (analogous to the carrier substrate on a BGA) is fabricated directly on the semiconductor wafer before the individual IC devices are singulated. A typical CSP con-struction is shown in Fig. 3.15. As with other package types, there are many variations on the basic construction. Some variations of CSP con-struction are shown in Fig. 3.16. Some photographs of CSP devices are pictured in Fig. 3.17.

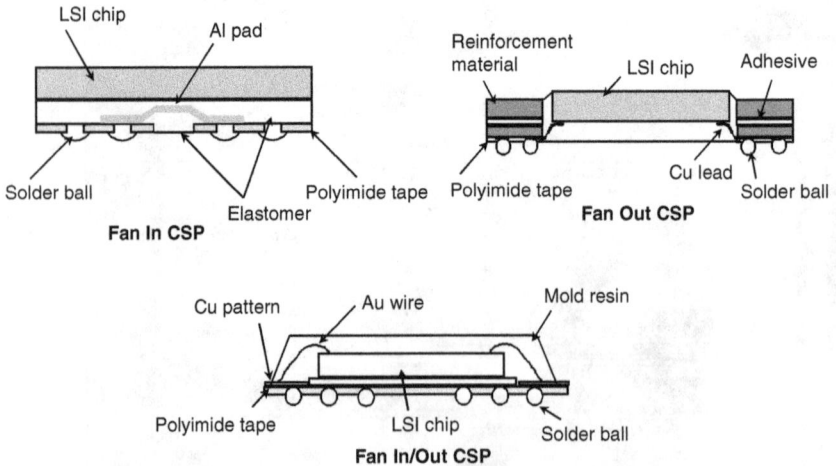

Figure 3.16 CSP variations.

Top   Cross section

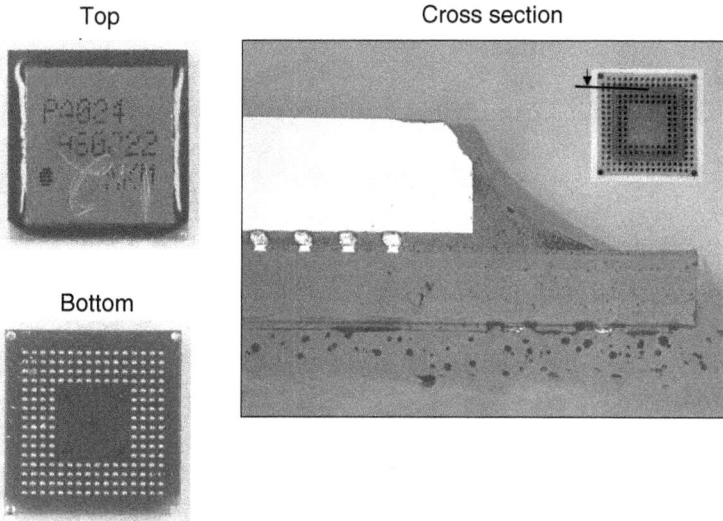

Bottom

Figure 3.17   CSP photographs.

The array pattern for a CSP will tend to be a finer pitch than for a BGA, so a higher-density motherboard is required. CSPs are not usually attached to large motherboards for this reason. They are commonly used on small modules or within small form-factor products. Figure 3.18 illustrates a prime example of how CSPs were used to achieve miniaturization in a digital camcorder design.

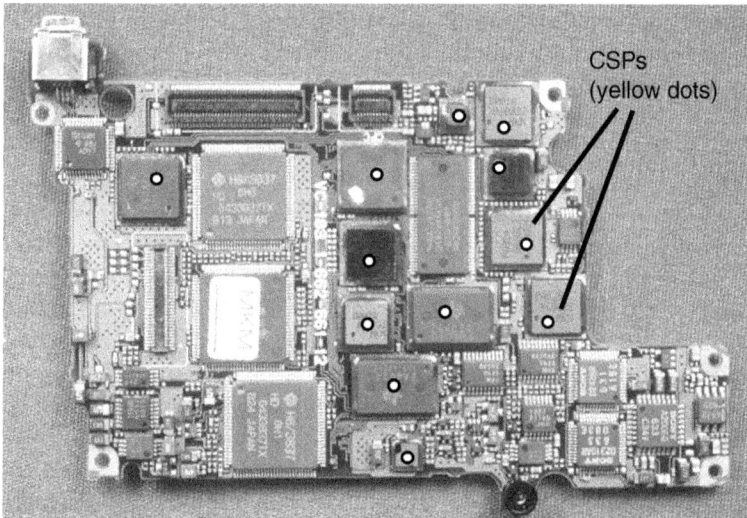

Figure 3.18   CSPs in a digital camcorder product.

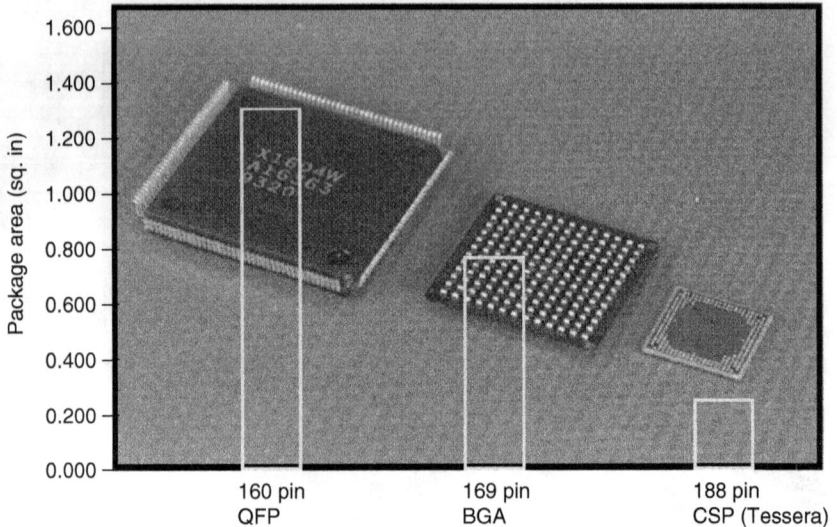

Figure 3.19   Package footprint comparisons.

Figure 3.19 shows the package footprint for a peripheral leaded package, a BGA package, and a CSP package of similar pin counts. The dramatic reduction in substrate real estate offered by array packages is apparent. The designer must be careful to match the appropriate substrate technology to the small footprint package, however, since the escape routing will not be possible without a high-density substrate technology. A complete system packaging trade-off analysis should be conducted to ensure that the use of small footprint packages is warranted, cost-effective, and of acceptable risk.

3.2   Discrete Components

Discrete components include a wide range of devices such as resistors, capacitors, inductors, diodes, transistors, oscillators, and switches. There are many sizes and shapes for discrete components, but almost all discrete components are designed to be soldered onto a PCB substrate and most have been designed to be compatible with standard surface mount pick-and-place processes. While most of the functional complexity (and cost) of a portable electronic device is embedded in the IC components, there will generally be a much larger number of discrete devices supporting the functions of the key pieces of silicon. A typical cell phone may have only three to five major ICs but over 300 discrete components.

As with integrated circuits, minimizing the footprint and thickness of the discrete component is a priority for portable electronic devices. With potentially hundreds of discrete components in a system, the aggregated footprint of all of these parts will have a significant impact on the overall product form-factor. Chip resistors and capacitors are available in a range of sizes. Standardized sizes for these devices include the 1206, 0603, and 0402 packages.

Functionality, performance, and cost are impacted by the selection of discrete devices. The cost of discrete components correlates strongly with the tolerance specified by the system designer. It is often easier to specify high tolerance, discrete, components than it is to conduct a detailed tolerance analysis. This trade-off must be weighed in light of the product cost objectives, development cost budget, and schedule constraints. In some systems, the use of high tolerance components may be avoided by tuning the system with adjustable discrete components after system assembly is completed.

Extensive reliability testing of discrete components is done by all major vendors and discrete devices are treated as a true commodity. Unfortunately, given the large number of discrete parts used in some systems, the chances that at least one part might cause a reliability or availability problem are not negligible. The circuit designer should highlight precision parts that were difficult to source during the physical design phase, as these are often the parts that will bring a production line down due to inventory shortages or vendor production problems.

Figure 3.20 shows the discrete devices on a PCA assembly taken from a digital camera product.

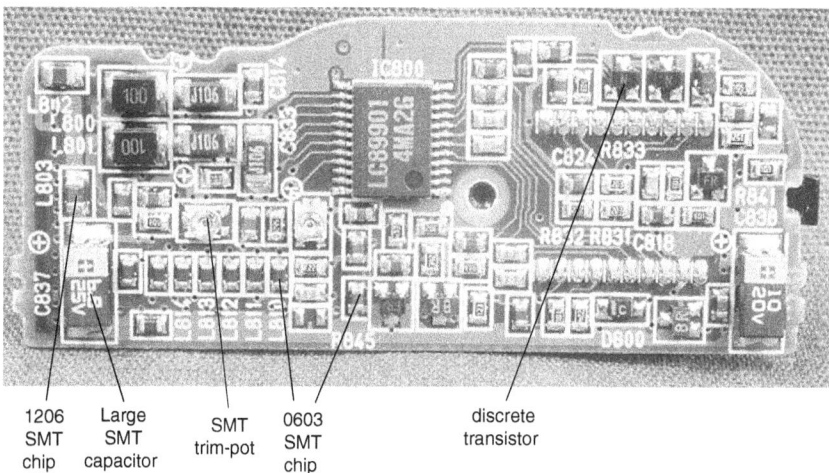

Figure 3.20 Discrete components on digital camera PCA.

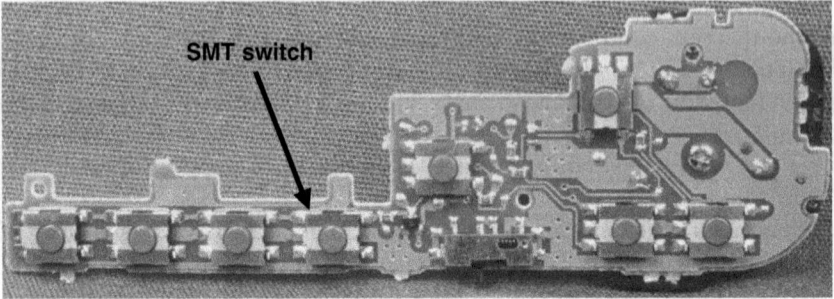

Figure 3.21 SMT switches on digital camera PCA.

The economics of making a component compatible with SMT processing is very compelling. Even devices such as user-activated switches can be mounted on a PCA inside the product housing with the button extending out through a hole in the product casing. Figure 3.21 shows a PCA that is used to provide the mechanical and electrical connections for various switches used on the product.

3.3 Board-to-Board Connectors

Connectors enable electrical connections to be routed off of the PCA, providing for electrical connections to other product subsystems, other PCAs, or to modules that are mounted on the PCA itself. Some connectors, such as network connectors, peripheral connectors, or power connectors, enable the product to connect to the outside world.

For portable electronic product applications, connectors will be of the surface mount variety, to allow cost-effective integration into the system. High-density connectors have very fine pitch connections on the substrate as well as in the mating zone where the connector connects to its counterpart.

High-performance connectors may be required, in some instances, to achieve the necessary signal fidelity. Such connectors must provide for controlled impedance of the connector and the associated cable assembly.

Since connector reliability is one of the key factors in determining overall system reliability, appropriate connectors must be selected based on the number of expected mate/demate cycles. A connector that is intended only to facilitate the original system assembly need be rated only for a small number of insertions. A connector that enables module replacement for repair or upgrade needs to be more robust since it is likely to see more insertion cycles in an unpredictable environment. Connectors that the end user sees, such as a serial port or a PC card connector on a notebook PC, must be rated for thousands of insertions. Connector manufacturers provide detailed reliability data to aid in the selection of the appropriate connector.

Connectors usually do not constitute a large percentage of the system cost. Custom connectors can be very expensive to develop, however, due to the cost of developing custom plastic molds and testing, and can usually only be justified for very high-volume products.

Most of the connectors used in portable electronic applications are surface-mounted to the PCB or interconnect cable. Board-to-board connectors allow two boards to be connected together directly without a flexible cable in between. Wire cables or flex circuits are often employed to make electrical connections between boards when they cannot be directly connected. Flex circuits are very interesting for portable electronic applications because they enable a very high density of electrical connections to be routed in a compact volume, they allow "distributed" electronics for space-constrained applications, and they enable controlled electrical characteristics and physical geometry.

When flex circuits are used, they are usually plugged into a low insertion force (LIF) or a zero insertion force (ZIF) connector mounted to the PCB. Figure 3.22 shows SMT ZIF connectors in a portable electronic application, using polyimide flex. A camcorder product is shown in Fig. 3.23, with the internal polyester flex circuits exposed.

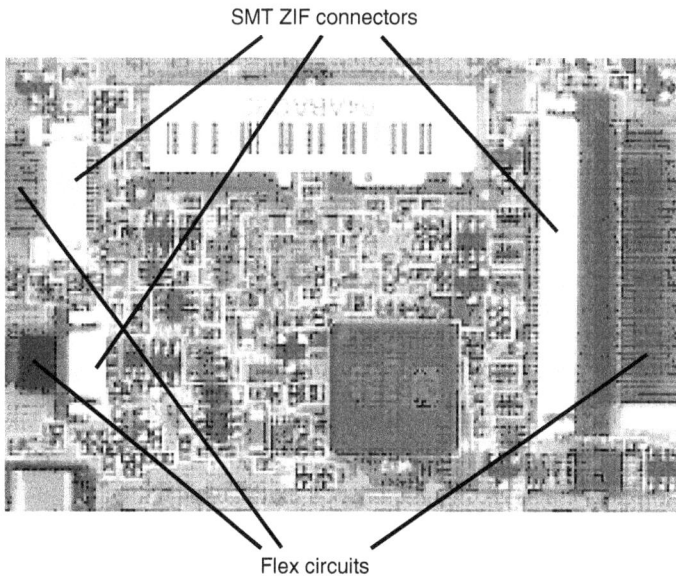

Figure 3.22 SMT ZIF connectors.

Figure 3.23 Polyester flex in camcorder product.

The polyester flex is connected to various PCB assemblies with SMT ZIF connectors.

Figure 3.24 shows a number of standard SMT pin and socket connectors. This type of connector is also used in some portable electronic applications but a smaller profile, lighter weight ZIF connector is preferred when possible.

In some products, heat seal connectors are used to achieve very fine pitch (polyimide flex) or very low cost (polyester flex). A typical heat seal connector consists of a 1-mil thick polyester flex film that is printed with traces of conductive ink. The conductive ink is loaded with carbon or silver particles. The area where the flex connector is bonded to the other components is coated with a thermoset adhesive that contains gold-plated particles. The connector is bonded to the other components (e.g., the PCB or LCD) by applying heat and pressure simultaneously for 10 to 12 s until a cured and connected interface is achieved. Figure 3.25 shows an example of a calculator product that uses a heat seal flex to connect from the system PCB to an LCD. A heat seal connection is used at both ends of the flex.

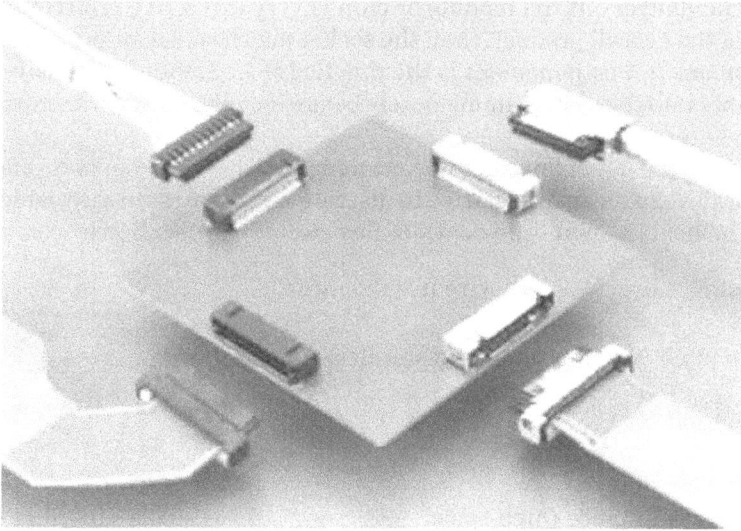

Figure 3.24   Flex connector cables with pin and socket connectors.

Microprocessor and multi-chip modules might rely on a socket connector to interface into a system. Sockets tend to be built in an array
configuration and are usually rather expensive. Sockets might be used
for a couple of reasons. If the module in question is subject to revision
or upgrade, a socket enables replacement of the module in the product

Heat-sealed flex

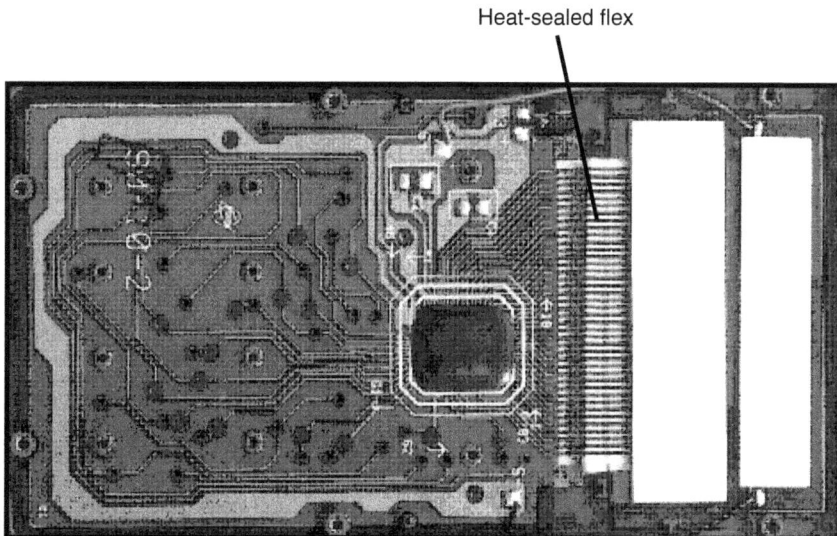

Figure 3.25   Heat sealed flex.

after initial delivery. If the module or chip is very expensive relative to the cost of the overall product, then the socket may be used for purposes of *postponement*. Postponement is the practice of assembling expensive components into the system immediately before shipping in order to minimize inventory costs.

In general, the use of flex circuitry is an important enabler in portable electronic devices as an alternative to discrete wiring or multiwire cable strips. The benefits and drawbacks of flex circuit are as follows:

1. Advantages over discrete wire interconnect
   - Size
   - Fine pitch (0.3 mm pitch and smaller)
   - SMT
   - Low profile
   - High density
   - High reliability
   - High performance (high speed, less cross talk, less delay in propagation)
   - Design flexibility
   - Easy to route
2. Disadvantages
   - More expensive than discrete wire types
   - Flex is custom made (takes long to design)

3.4 Substrates

The substrate of an electronic assembly is the system element to which electronic components are attached and their respective I/O are connected together. The substrate therefore consists of the bulk substrate material and one or more layers of conductive routing for the distribution of electrical signals or power. Each signal and power trace must also terminate at a pad, to which a single contact point, or *node*, of a component is connected. The substrate of choice for most portable electronic devices is the printed circuit board (PCB). Printed circuit boards consist of an organic bulk material such as glass epoxy and copper traces that enable electrical routing. The copper traces are often plated on the surface with an antioxidant metal or organic film coating to limit oxidation and to improve solder bonding or wire-bonding. PCBs enable high interconnect density and are relatively inexpensive due to the high volumes produced worldwide.

The design of the system PCAs is one of the primary tasks of the portable electronic product designer. PCA design is, in fact, the focal point of the physical system design and integration. Since most of the components will be off-the-shelf ICs and discretes, the PCB is likely to be one of the few custom-designed electronic elements within the system, and one where creative design can make a huge difference in the end-product characteristics. As such, a thorough understanding of PCB design is important to the portable electronic designer.

Selection of an appropriate substrate technology for each PCA in the system is an important decision that the product designer must make. Understanding the routing density potential of a particular variety of substrate is the most important factor in making this decision. Routing density describes the amount of signal routing that can occur in a given volume of bulk substrate material. Most designs utilize multilayer substrates that contain two or more layers of routing. The geometry of these layers is determined by the substrate manufacturer's *design rules* for the particular type of substrate technology.

An estimate of the maximum amount of routing that is available on a single layer is based on the minimum line pitch that is supported by the substrate technology. The minimum line pitch is equal to the minimum line width plus the minimum line spacing. The maximum routing density for a layer may be estimated by the following equation:

Maximum routing density per layer = 1/(minimum line pitch)

For a unit area of substrate, this means that the maximum routing density for one layer is equal to the total length of all of the parallel lines of minimum pitch that can be patterned into a unit area.

As an example, a substrate process technology that can form 0.08-mm lines and 0.06-mm spaces has a nominal line pitch of 0.14 mm. In 1 mm^2, 7.14 lines may be patterned. Each of these lines is 1 mm in length; therefore the maximum total length of routing resources available in 1 mm^2 is 7.14 mm. So, the maximum routing density is 7.14 mm/mm^2. Notice that the units on this figure are linear units divided by area units.

Multiple routing layers are fabricated in the substrate, and each layer will add to the wiring resource for a fixed unit area. The total substrate thickness also increases as layers are added. So a four-layer substrate would ideally have four times the routing density of a single-layer board. In the example above, the ideal maximum routing density would increase to 28.57 mm/mm^2.

But maximum routing density can never be achieved because of the irregular routing patterns that must be used in practice. In addition, *vias* are needed in the substrate to perform signal routing between layers. If the diameters of these vias are greater than the minimum line width, as they often are, then the minimum estimated pitch cannot be achieved. Vias that must route between nonadjacent layers also interfere with routing on the layers that they pass through, so routing efficiency is lost.

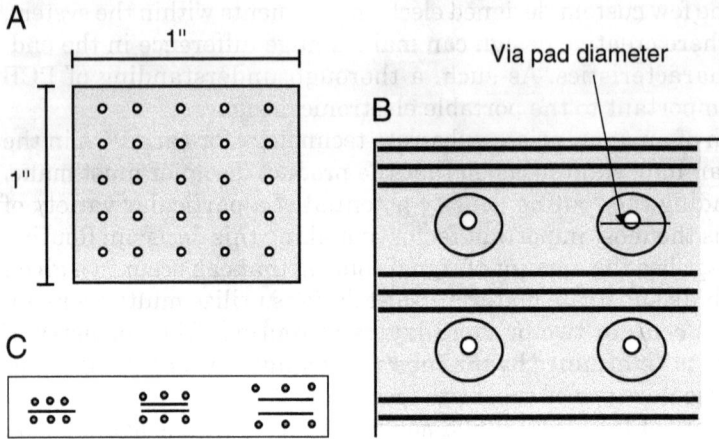

Figure 3.26 Via geometry impacts routing density.

Substrate technologies that enable very small via diameters are therefore of interest since they allow the designer to achieve much higher effective routing densities, even if well below the theoretical maximum.

Figure 3.26 sketches the physical constraints imposed by the presence of vias. The estimation of effective wiring density provided by one layer of the substrate is as follows:

1. 1 in^2 of a single substrate layer with evenly spaced via grid is used in calculating theoretical interconnection density.

2. The combined diameters of via pads (PCB design rule) deducted from available routing area.

3. The number of line-space pairs (PCB design rule) allowed per channel is calculated. The result is usually a noninteger, uneven number on a fixed even grid. Subsequent truncation results in an additional loss of interconnection density.

The actual routing density is calculated based on the actual number of lines per unit area; so,

$$\text{Density (in/in}^2) = (\text{number of line-space pairs per inch})$$
$$\times (\text{number of channels per inch})$$

Component mounting pads must also be designed on the top and bottom surface layers to attach electronic components. These pads tend to be much larger than the minimum line width and so they too have a negative impact on the maximum achievable routing density.

In practice, routing density requirements can be estimated empirically for a given set of system density and pin-count characteristics.

A heuristic estimate of the achieved routing density requirements for a PCA is discussed in Sec. 3.6.

The properties of the substrate bulk layers are also important from a miniaturization and performance standpoint. Bulk materials that can provide stable mechanical support and electrical isolation in very thin layers are necessary to keep the product thickness and weight to a minimum. Low dielectric constant materials are also important for products where line capacitance must be carefully managed for performance reasons.

In addition to the many varieties of organic PCB technology, other substrate materials are used for special purpose PCAs or modules. Ceramic substrates are often used for modules because of their thermal, mechanical, and dielectric properties. Some ceramic substrates also enable the designer to use integrated discrete devices such as capacitors and resistors, building such components directly into the internal layers. Integrated ceramic thick film resistors may even be tuned with a laser during product testing as an automated alternative to a surface mount potentiometer.

Polyimide and polyester flex circuits are not only used as connectors inside portable systems, but they often have electronic components mounted directly onto them. In some products such as digital or 35 mm cameras, flex circuits connect most of the system electronics.

Regardless of the type of substrate that is ultimately selected, all substrates require a similar design process. This design process is driven by the design rules that the manufacturer provides for the substrate technology.

PCB design process.   The design of a PCB (or any substrate) begins with a schematic for the PCA being designed. The schematic is a representation of logical connections between system components. The product designer uses schematic capture tools to generate a net list. A net is a set of nodes (component I/O) that need to be connected together electrically. It is the function of the PCB to make these connections.

The net list is loaded into the PCB design system along with all of the component information including the I/O pattern (footprint) and node identification for each component. The design rules for the substrate technology also need to be loaded into the system, including information such as minimum line widths and spacing, minimum via diameters, and other geometric-based limitations of the substrate fabrication process. SMT design rules must also be loaded into the system. These include geometric information such as minimum pad sizes, minimum component spacing, and alignment features to assist automated assembly.

The designer must consider whether dedicated power and ground planes are required for the design and define these layers before proceeding with the design.

The designer specifies the location of some of the major components (on both sides of the substrate) based on any known mechanical constraints or electrical performance issues. The design system may then be instructed to autoplace the remaining components based on an algorithm that minimizes the distance between components that are connected to the same nets.

Once component placement is completed, the designer may choose to start the routing process manually if there are signal performance issues on critical nets or if there are some obvious deficiencies that the autorouting logic may not recognize. The PCB design system is then instructed to autoroute the PCB. Each node is routed to its specified net. If the designer has selected the right substrate technology and layer count, the autorouting system should be able to complete most of the routing very quickly. The designer can work with the autorouter iteratively, to complete 100 percent of the routing. If the routing cannot be completed, the board area or the number of routing layers must be increased, or the component placement reconsidered.

A manual design check is usually performed to ensure that the automated design system has not made any obvious errors. Some systems may support various types of simulation as an additional check. Once the designer is satisfied that the design is good, it is sent to the PCB manufacturer for prototyping.

3.5   Escape Routing

Escape routing is a special topic in substrate design because escape routing requirements can drive the selection of substrate technology above and beyond the requirements of the average routing density. An escape routing study for a fine pitch BGA package is summarized below (data courtesy of Portelligent).

Figure 3.27 shows the interposer routing and footprint geometry of a CSP device. The purpose of escape routing is to get signal traces routed from a matching pattern on the substrate to the appropriate device I/O in the rest of the system. In this case, the CSP I/O pattern has the following characteristics:

Die size: 10.2 mm × 10.2 mm

Number I/Os: 250

Bond pad pitch: 0.163 mm

Bump pitch: 0.5 mm

Bump capture pads: 0.2 mm

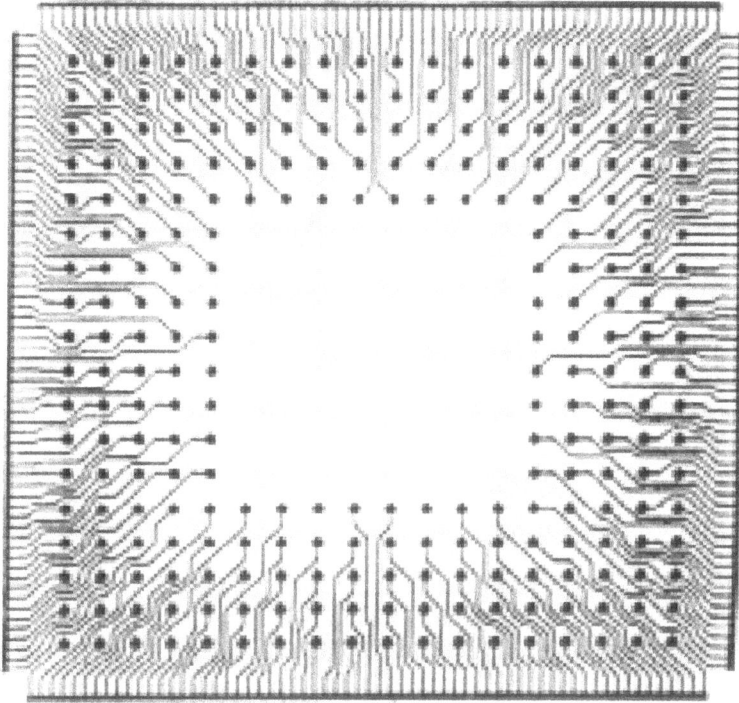

Figure 3.27   CSP footprint for escape routing.

The escape routing was performed with a number of high-density substrate technologies to compare their relative effectiveness. Each figure in this study shows the CSP mounted on a particular substrate technology solution.

Figure 3.28 shows the cross-sectional substrate design characteristics for a surface laminar circuitry (SLC, IBM trade name) substrate with the following parameters:

- Two layers of photosensitive dielectric
- Fine-line metal is used (FC2 & FC3)
- FC2 used as a ground plane
- FC1 is the top layer of the PCB
- Photo vias are approximately 10-mils in diameter
- Line width and spacing is about 3 to 4 mils
- Photo vias over Thru-Hole via not allowed

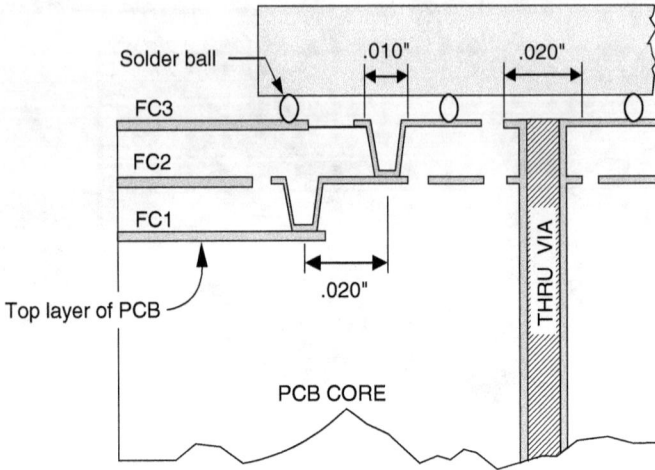

Figure 3.28 Surface laminar circuitry.

Figure 3.29 shows the cross-sectional design parameters for a film redistribution layer design. Film redistribution is a high-density substrate option offered by several manufacturers. The parameters for the design study are as follows:

- Only one layer of photosensitive dielectric
- Fine-line metal is used (FC2)
- FC1 is again used as a ground plane

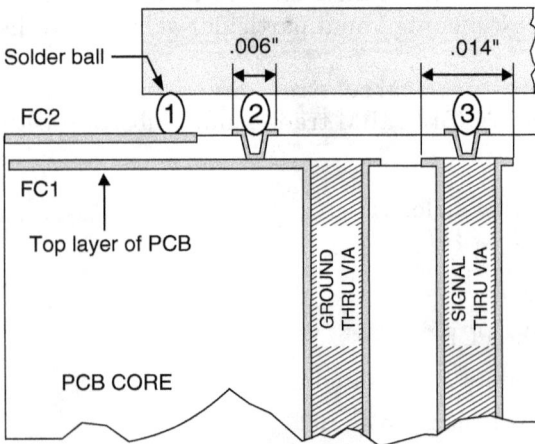

Figure 3.29 Film redistribution layer.

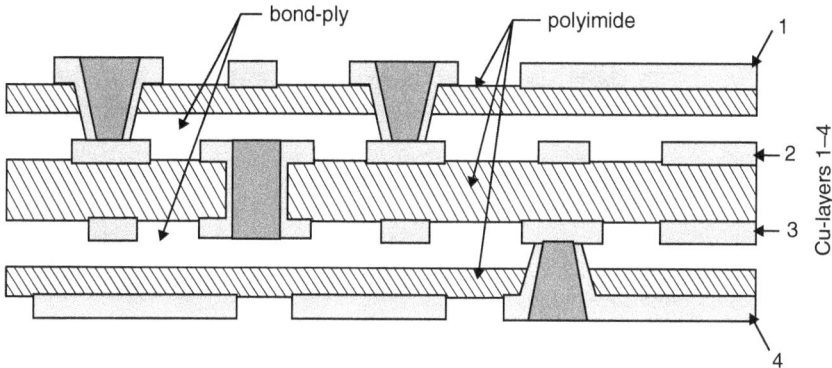

Figure 3.30 Dycostrate.

- Photo via: 4 mils. Photo via land: 6 mils
- TH vias: 8 mils. TH via land: 14 mils
- Line width: 2 to 3 mils. Spacing: 3 mils
- Photo vias over filled TH vias are allowed

Figure 3.30 shows the dycostrate (Dyconex, trade name) substrate with the following cross-sectional design parameters:

- Uses epoxy-based plasma-etched redistribution developed by Dyconex
- Single- and double-sided boards are possible using a variety of dielectric materials
- Variation shown in which FR4 core is replaced by a polyimide core
- Trace width and spacing: 70 to 125 μm
- Plasma etched via holes: 75 μm. Via lands: 300 to 350 μm
- Cu trace thickness: 20 to 25 μm. Plated Cu thickness: 15 μm
- Core material thickness (polyimide): 25 μm outer layer; 50 μm inner layer

Figure 3.31 shows the Tessera-laminated substrate (Tessera, trade name) with the following cross-sectional design parameters:

- Two-sided sublaminates (typically four to six are used) are drilled and metalized in volume and fabricated on either an advanced PCB or flex circuit line
- Premanufactured sublaminates are then personalized
- Typically power/ground is used on one side and signal on the other

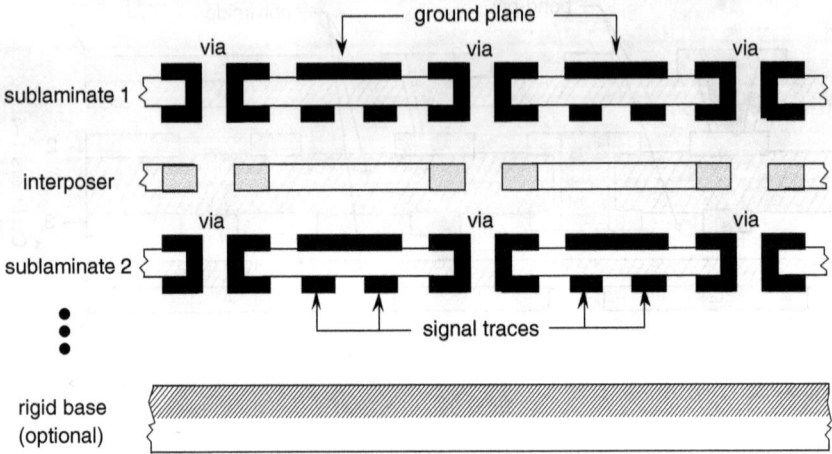

Figure 3.31 Tessera laminated substrate.

- Sublaminates are stacked, aligned, and laminated, using a proprietary interposer layer that serves to connect sublaminate layers together
- Via grid: 500 μm. Bump capture pad: 225 μm
- Min line width and spacing: 50 μm
- Line thickness: 15 μm. Interposer thickness: 75 μm
- Dielectric thickness: 50 μm

Figure 3.32 shows the z-link (Sheldahl, trade name) substrate with the following cross-sectional design parameters:

- Double-sided plated-through-hole circuit boards are either flexible or rigid
- Final lamination process uses anisotropic adhesive known as z-axis adhesive
- Adhesive has suspended tin-lead alloy spheres about 1.5 mils in diameter
- Heat and pressure applied during final lamination process forms fusion bonds between conductive particle in the z-direction
- z-link pad sizes: 0.5 mm, 0.25 mm, and 0.125 mm. Via land: 0.2 mm
- Line width and spacing: 2 to 3 mils

For each of these substrate technologies the substrate layout designer used a variety of escape routing techniques to enable the signal traces to exit the area in the immediate vicinity of the CSP footprint. Figure 3.33

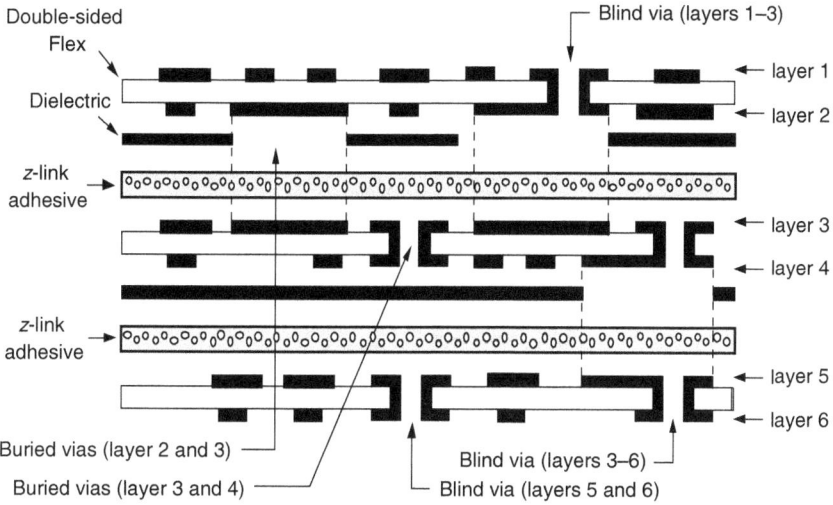

Figure 3.32 Sheldahl z-link.

gives examples of these various exit routing geometries. The results of the study indicate how many layers are required with each technology to achieve the escape routing and are given in Table 3.1. Ideally, the number of layers is less than or equal to the number of layers required to provide the necessary global average routing density for the system.

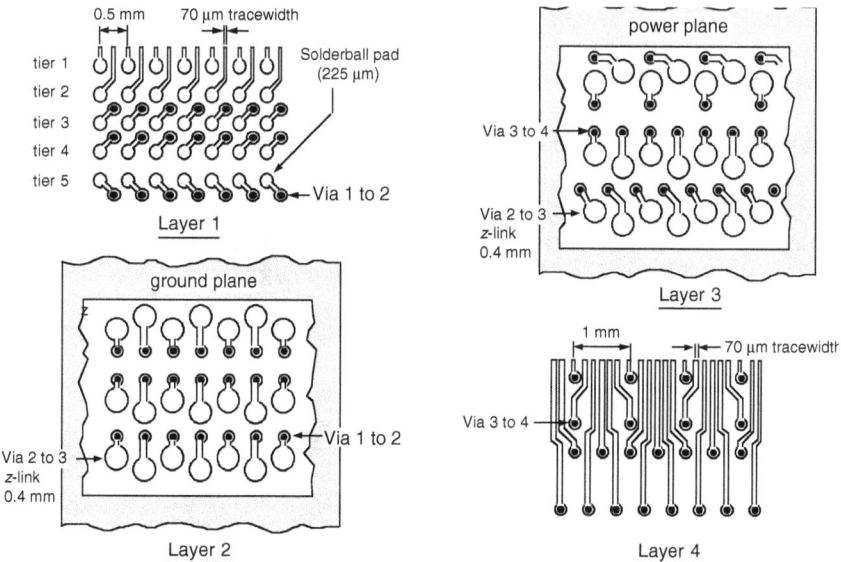

Figure 3.33 Escape routing.

TABLE 3-1   Escape Routing Layer Requirements

Surface laminar circuitry	2 film layers
	1 core layer
	+ P/G layers
Film redistribution layer	0.5 mm bump pitch
	1 film layer
	3 core layers
	+ P/G layers
	0.53 mm bump pitch
	1 film layer
	1 core layers
	+ P/G layers
DYCOstrate	4 layers
	including P/G
Tessera laminated substrate	Requires uniform bump grid
	3 layers
	+ P/G layers
z-link	4 layers
	includes P/G

If the escape routing requirements are significantly higher than the global routing density requirements, then the designer should consider partitioning the CSP along with some of the connected components into a higher density module, or reconsidering the merits of a CSP.

3.6   PCA/Module Design Metrics

Designing PCA modules is one of the primary functions of the portable electronic designer. Completing a successful design can take many weeks or months depending on the product complexity. During the trade-off analysis phase of product design, it is not feasible to do a PCA design each time different trade-offs are being explored. For this reason, it is important that the designer be able to characterize and assess the PCA requirements in terms of readily available metrics that would be available during the trade-off analysis process. These metrics can also be used to benchmark the effectiveness of an actual design once it is completed.

3.7   Electronic Packaging Metrics

A set of electronic packaging metrics was developed at the Microelectronics and Computer Technology Corporation (MCC) throughout the 1990s. These metrics were developed specifically for use with portable electronic devices and are now the property of Portelligent, Inc. (Austin, TX). Portelligent conducts cost modeling of portable electronic devices as a part of the information services that they provide to the electronics industry.

Below are Portelligent's current definitions of key metrics for portable electronic systems. It is included here because it represents a definitive set
of metrics for conducting an electronic systems packaging analysis.

Electronic Packaging Metrics—Overview and Discussion

In our product teardowns, we gather a series of metrics for product profil-
ing and comparison. Some metrics focus on system characteristics such as
total silicon area, total system semiconductor storage capacity, and total
connection count. Other metrics reflect more subtle aspects of electronics
assembly such as connection density, average component I/O count, and
silicon-tiling density.

Taken as a whole, the metrics allow deeper comparison and benchmarking across multiple disciplines and multiple products. Key metrics we
gather on products are described below along with their definitions and
what they tend to say about the system under study. Most metrics can be
used both in comparing similar products for benchmarking purposes or for
quantifying differences in levels of complexity between dissimilar product
types. Data fall into two categories—either "raw" measured data or ratios
of these measured data sets.

Total silicon area.   This metric describes the total area of silicon as
measured from x ray or direct measurement of ICs. The area is an expression of the enclosed bare die area and excludes packaging area. The aggregate silicon area is a good benchmark to show how integrated a design
might be when making comparisons to similar systems. Total silicon area
also reflects the major cost driver for most systems we examine.

Silicon-tiling density.   Ratio of total silicon area to total printed circuit
board "projected" area (i.e., the simple board area and not the cumulative
surface area of both sides of the board). This metric directly reflects the level
of efficiency and aggressiveness in integrated circuit packing and placement. Single digit silicon-tiling density is typical but silicon coverage of 10
to 20 percent has been seen in some of the most advanced products we have
examined. Higher-tiling densities often correspond with the use of chip
scale packaging (CSPs) or other small form-factor IC packaging technologies. High-density circuit boards are also often a supporting technology.

Number of parts.   Total component count including ICs, passives, modules, connectors, etc., each separated out in our reporting.

Number of connections.   The total number of connections corresponds
to the total number of interconnects introduced by the aggregate component set and reflects any electrical connection observed (solder joints, adhesive interconnect, or connector terminal interfaces).

Opportunity count.   This is the total number of parts plus the total
number of connections; the name indicates that each of these constituent
elements represents an opportunity for failure. A high opportunity count
means more complex and riskier electronics assembly.

Average pin count (APC).   Ratio of total number of component terminals
to total number of parts, at the system level. This metric reflects the "average" terminal complexity of the components and often provides a signature

of integration level and/or "digital-ness" of the overall product. Low APCs reflect a high number of discretes or other low-pin count devices often characteristic of analog circuitry. Conversely, high APCs are characteristic of highly integrated, high-pin count assemblies, often those composed largely of digital integrated circuits.

Connection density.   This metric is a ratio of the total number of connections to total printed circuit board assembly area, in units of connections per cm^2. The metric provides data related to the silicon-tiling density above, but with an emphasis on complexity of I/O interconnect. For example, with a fixed connection density, high-tiling density of low-pin-count memory chips is more readily achieved than comparable silicon tiling of high pin-count logic.

Part density.   This metric is a ratio of the total number of parts to total printed circuit board assembly area, in units of components per cm^2. The metric provides data related to the silicon-tiling density and connection density as described above, but with an emphasis on density and complexity of component packing efficiency. For example, low part density of high-pin count devices can pose an equal challenge in connection density to high part density of low-pin count devices. High part density does reflect challenges in surface mount assembly in terms of (typically) precision of placement, number of placements and engineering of part clearances.

Routing density (heuristic estimate) = 3 × (average pin count) × √part density. The routing density metric is an empirically derived relationship that characterizes the wiring density of the interconnect used to support the interconnection of components in a planar electronic assembly (i.e., the circuit board). Architectural issues such as bussing or other factors affecting the regularity of wiring impact the actual routing density needed to support a given application, but the metric provides a ready measure of wiring complexity.

The routing density formula is particularly useful because it enables the designer to determine the required wiring resources needed in a substrate based on the average pin count and required part density of the product. As mentioned above, the routing-density metric is calculated as follows:

$$\text{Routing density} = 3 \times \text{average pin count} \times (\text{part density})^{1/2}$$

Dimensional analysis yields:

$$(\text{cm/cm}^2) = (\text{parts}^{1/2}/\text{pins}) \times (\text{pins/part}) \times (\text{parts/cm}^2)^{1/2}$$

Notice that the units for the routing constant "3" are $\text{parts}^{1/2}/\text{pins}$.

Conceptually, $parts^{1/2}$ may be viewed as the linear pitch between parts spread across a two-dimensional space. A value of $3(\text{parts}^{1/2}/\text{pins})$ then is a multiplier that estimates the length of wiring for each pin to be three part pitches. This has proven empirically to be a useful method of analyzing wiring density although increased accuracy could certainly

TABLE 3-2   Comparison of Packaging Metrics for Different Product Types

Product category	Cell phone	Digtal camera	PDA	Notebook PC
Routing density (cm/cm^2)	33	73	40	38
Part density (parts/cm^2)	12	29	7	5.9
Average pin count	3.2	4.5	5.2	5.3
Connection density (conn./cm^2)	38	132	34	30
Total silicon area (mm^2)	170	130	473	774
Silicon-tiling density	4.0%	14.1%	7.3%	2.8%
Number of parts	500	270	420	1650
Number of connections	1,600	1,200	2,200	8,700
Opportunity count	2,100	1,505	2,600	10,000

be obtained by developing a more complex function for the routing constant.

Table 3.2 gives some typical values for several key electronic packaging metrics. Digital cameras tend to have the highest routing density among portable electronic products. This is driven by the intense competition among electronic camera makers to produce high functionality in the smallest possible form-factor. Digital cameras have a high number of discrete components and the component-packing density is double or triple that found in most cell phone products. To achieve these high routing densities the most advanced printed circuit board technologies are used. The extremely high silicon-tiling density achieved within these products is also evidence of the extreme pressure to miniaturize.

Notebook computers, in contrast, have relatively low component densities. These products have a few very high pin count IC devices and a considerable number of discrete components, but the form-factor requirements tend to be held at bay by the need to accommodate a large flat panel display. Without this pressure to reduce the product area, the main PC board tends to occupy much of the available system area and lower density, lower-cost substrates are used. Silicon-tiling density is relatively low for notebook computers when compared to other portable products such as cell phones and digital cameras. The latest generation of ultraportables may mark a change in this trend, as some manufacturers are beginning to use smaller high-density boards to make room for hard drives, optical drives, and batteries, within a more constrained volume.

PDA products are similar to notebooks in terms of average pin count and part density. PDA products tend to be built around a more integrated architecture and do not need to support a lot of legacy system I/O infrastructure. So, even though these products are driven by form-factor more than notebook computers, they achieve miniaturization more through silicon integration than through electronic packaging.

Cell phones have historically required very high routing densities when compared to other products. Over the past few years, however, cell phone functionality has become well defined in the marketplace, paving the way for greater levels of silicon integration. As a result, cellular phones still require fairly high component densities but have just a few high pin count parts, which result in a lower average pin count.

3.8   I/O Hardware

Various components that enable the input and output of information to and from the system are also integral to the electronic packaging design. Buttons, switches, and dials are needed so that the user can control the device. Speakers and microphones are needed for audio input and output. Antennae are needed to make wireless connections to external data sources, as are external connectors.

3.8.1   Buttons, switches, dials, and touch screens

Portable electronic devices are controlled "at the push of a button," or increasingly "at the stroke of a pen." Buttons, dials, keypads, and other forms of tactile input devices are the primary mechanism by which a portable electronic product interfaces with the user. These devices may also be thought of as sensors that are intended to detect the desires of the user; they also serve an output requirement, by helping users calibrate their input and verify that the input was successful.

Buttons and switches are the most common user interface device. They are simple devices that can be put into one of two states—off or on. Buttons and switches can have a single function or their function can change depending on the state of the system. The function of a button is determined, of course, by the system design, not the button itself. The primary characteristics of a button that matter, therefore, are the electrical, mechanical, and tactile properties of the device. A common type of SMT switch is shown in Fig. 3.34 along with a picture of the corresponding buttons on the product.

There is a wide range of options used to implement button and switch devices. The button has to be able to handle the voltage and current of the signal being switched. The physical configuration must be able to be packaged within the physical constraints of the product. The tactile feedback must match the function that the button provides. The tactile properties of a button or switch can have a big impact on the user's perception of product quality. Buttons that have a smooth z-axis travel and a crisp "click" allow the user to verify through the sense of touch that the button has been activated. Buttons that can be activated by casual contact are very annoying, as are buttons that require significant pressure and provide no tactile feedback.

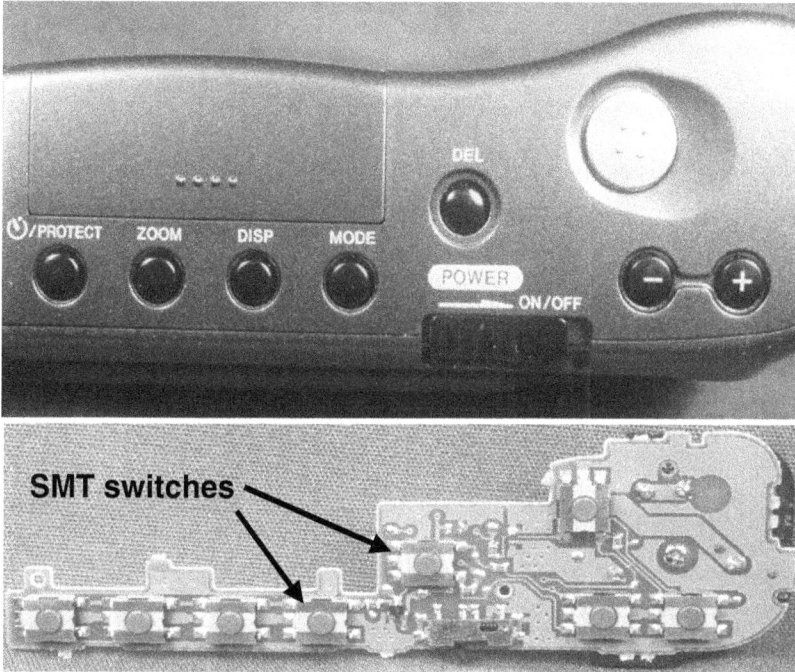

Figure 3.34 SMT switches.

Computer keypads are a collection of buttons and they are subject to the same issues. Users have varying preferences on keyboard feel, and the ability to adjust button characteristics in software can be important for notebook computers. One of the challenges of selecting a notebook keyboard is to achieve a good tactile response in a thin package. Membrane switch technology can be used to get very thin keypads but they usually sacrifice z-axis travel—an undesirable effect for systems that require good tactile feedback.

Dials are very useful for control functions that require variable rather than binary input. Volume control for a portable CD player is a prime example of this type of application. It is currently a common practice for product designers to avoid dials and use buttons instead. A continuous adjustment with two buttons (one up and one down) is used in conjunction with visual feedback on the display to show the current level. Present trends aside, however, dials should still be considered for this type of system control. Sony has implemented a dial scan function on a number of products to allow scrolling through a list of displayed options with a thumb dial (Fig. 3.35). Selection of an item is achieved by pressing the dial.

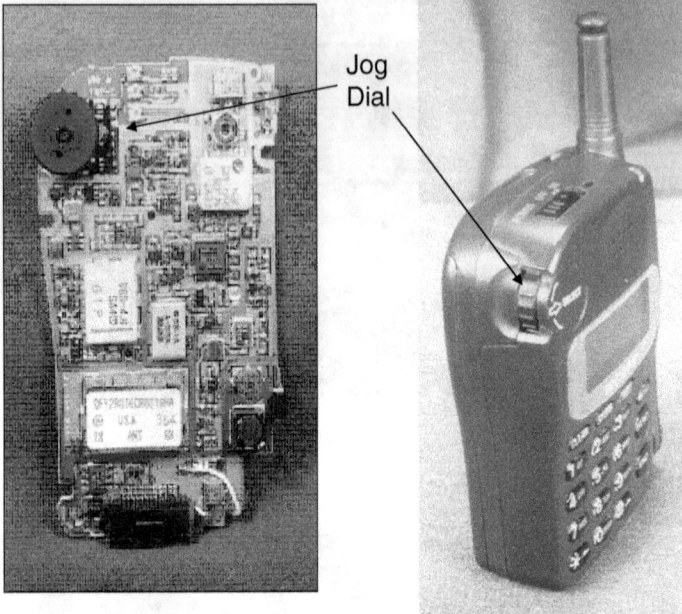

Figure 3.35 Jog dial.

Touch screens are a powerful user interface mechanism for portable devices, because they enable the user interface to be software-based rather than hardware-based. Often the visual representation of a touch panel control will simulate a button, dial, or keypad. What the touch screen lacks in tactile response, it makes up for in product flexibility and the level of interactivity that may be achieved with graphical content. Touch panel mechanisms are generally integrated into the display module, and this is described in more detail in the display section of this chapter.

Reliability is a major concern with buttons, switches, and dials. Since they are mechanical in nature, these devices are one of the weakest links in system reliability. Inexpensive membrane switches are thin and low cost but tend to have high failure rates as the flexible contact surfaces wear down through repeated usage. The product designer should thoroughly review the reliability data provided by the manufacturer for these types of devices.

Buttons, switches, and dials do not represent a large part of total system cost, but they play a critical role in the user's satisfaction with the product. It is not uncommon for a user to select from a choice of portable electronic products based on the way the control devices feel. This is all the more reason that the product designer should invest in high quality devices.

3.8.2 Speakers and microphones

Speakers and microphones facilitate a system interface with the user and are critical to voice communications products such as cell phones. Built-in speakers and microphones are used in such products although a connector for an external headset is often included. In these audio-centric products, the product housing is generally designed to accommodate or enhance the acoustical requirements of the system, as well.

Portable products that require high-quality sound will need to utilize an external stereo headset since the small form-factor of a portable product does not enable high-performance acoustic design.

3.8.3 Antennas

Interfacing to a network is an increasingly common requirement for all types of electronic devices. For portable electronics, there is the growing expectation that network links should be wireless. High-bandwidth WLAN network connections operating at gigahertz plus frequencies are very sensitive to antenna and circuit board geometries. Integrating the antenna design with the product shielding strategy also needs to be accomplished.

Increasingly, portable electronics users are being drawn to products that have internal antennas. External antennas create an awkward form-factor and represent a reliability problem, as external antennas are subject to breakage. On the plus side, as frequencies increase, the length of the antenna decreases, making integration into a small form-factor much easier.

Antennas are not particularly expensive components, but good RF design is an expensive skill set and many companies fail to recognize the importance of accessing good RF expertise when trying to field a quality product.

3.8.4 External connectors

External connectors are physical connectors that are used to interface the product to a cable or docking station. They are often associated with some type of communications port. These communications ports may be custom to the product or, more often, they are based on a communication standard.

Selection of the communications port capabilities is part of the system requirements creation, and selection of the specific external connector and its placement is part of the physical design process.

Most external connectors inside of portable electronic products are surface mounted to the system PCA and mounted flush to the product housing. Physical connection is then made through a window in the product housing.

External connectors have an impact on system form-factor, reliability, and cost. Their reliability is an issue because of the unpredictable nature of the mechanical forces involved when the user is plugging and unplugging a cable. A strain relief mechanism should be included for external connectors to minimize the amount of force transferred directly to the connector and PCB interface. Smaller form-factor connectors and highly durable connectors will generally command a premium. While off-the-shelf connectors do not typically comprise a large percentage of system cost, a custom connector or newly introduced connector product can be surprisingly expensive, due to the manufacturer's need to quickly recoup the development costs.

There is growing pressure in the industry to eliminate physical external connectors altogether and to replace them with wireless connections. One other type of communication port that is often used with portable devices is the IR (infra-red) communications port. This consists of an IR diode and sensor pair that can communicate with a similar pair in another system. IR ports can be a compact and cost-effective alternative to physical connectors in many applications. More recently, the Bluetooth standard has been developed to replace external connectors with a digital RF connection. Besides eliminating cables and making the product easier to use, these wireless connection options should yield more reliable systems.

This chapter has only touched on the basic issues of electronic packaging. The serious electronic product designer will take the time to become familiar with this particular area because of its relevance to the actual implementation of product design. Electronic packaging is one of the most powerful tools that the designer can use to differentiate the product form-factor.

Chapter

4

Displays

In terms of the critical factors that affect portable electronic design, the user interface is most affected by the selection of the display. Almost by definition, the display represents the physical embodiment of the user interface for many products. The appropriate selection of display technology determines whether the user will be satisfied with the look and feel of the product, as it performs its primary function. Matching the visual quality of the display to the user interface requirements is of paramount concern to the portable electronic system designer, and displays have a variety of properties which must be considered and matched against the user interface requirements.

Pixel resolution determines the level of graphic detail that can be displayed in a single image frame. Resolutions vary from low-resolution, monochrome, segmented, character-only displays to high-resolution, color displays capable of displaying millions of pixels. Figure 4.1 shows a close-up view of the pixels in a color LCD. The pixel pitch is 117 μm in this example.

The color range of a display determines the number of variations of color that can be displayed. The color range is usually described in terms of the number of bits that are available to specify the intensity of the color subpixels. Displays range from single color monochrome displays to full color displays capable of delivering a near infinite palette of colors.

The brightness indicates how much light is emitted from the surface of the display. Reflective displays have no brightness and rely on reflected ambient light to reveal the contrast and color created on the display surface. Emissive displays generate light at each pixel. This light creates the colors and contrasts to form the intended image. Transmissive displays use a backlight which is filtered by the display surface to create color and contrast. Transflective displays reflect ambient

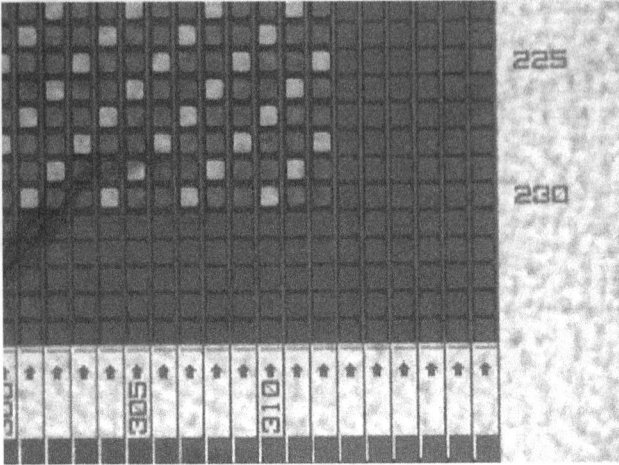

Figure 4.1   Close-up of LCD pixels.

light back through the filtering display surface to create an image. Transflective displays consume very little power, but their brightness and contrast ratio is inferior to other display types.

Contrast is a measure of the difference in brightness that is achievable within a local region of the display panel. Contrast can actually have more of an impact on the display quality than brightness because the human eye will adapt to the average brightness level; however, the recognition of graphic details requires contrast to establish detailed features. A plot of contrast ratio versus viewing angle is shown in Fig. 4.2.

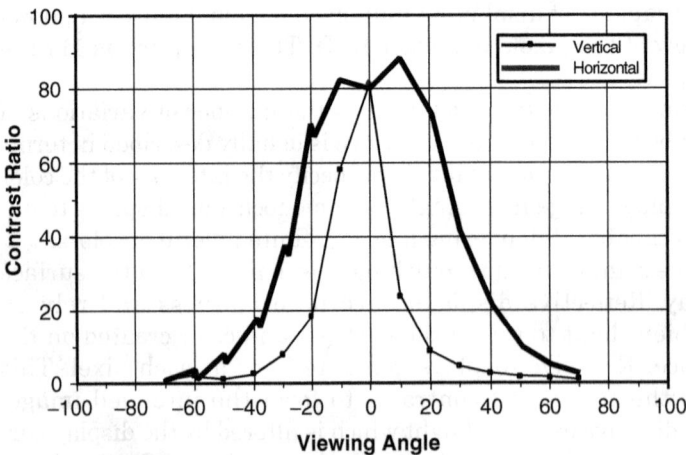

Figure 4.2   Contrast ratio vs. viewing angle.

The interplay between brightness and contrast should be thoroughly understood relative to the particular product application. Viewing a display in broad daylight, for example, can cause glare, which will washout the contrast of the display. In the case of a cellular phone, a user who makes a lot of outdoor calls may prefer a monochrome, high brightness, and high-contrast LED character display to a high-resolution color TFT display. Lack of resolution would put limits, however, on the ability to use remote web browsing services or other more graphic intensive functionality.

Pixel refresh rate is also an important visual property of the display if the user interface involves motion video or the use of a floating cursor. While there is a great deal of technical complexity involved in characterizing the visual properties of a flat panel display, the product designer should stay focused on the resolution, color range, brightness, and contrast as perceived by the end user in the expected usage environment. A precise definition of key terms is presented in Sec. 4.6.

With respect to the critical factor of functionality, the display is a key enabler for many product concepts. They must be capable of delivering the primary functionality of the device to the end user. In some cases, the display only facilitates the user interface, as with a traditional cell phone, where the display shows the phone number being dialed. The primary function is, of course, the phone call, in which the display plays little or no part. In a portable DVD player the display may facilitate control of the device, but it is also critical to the primary function, namely, displaying a video from DVD. The product designer should make a clear distinction between the display requirements that are driven by user interface control issues and those driven by delivery of the product functionality. This exercise will lend clarity to the trade-offs that must be made in selecting the best display technology.

Product performance is not driven directly by the choice of display. Systems with graphics-intensive functionality may require significant processing power and will also tend to have a high-resolution display in order to deliver that functionality.

Thus, high-performance systems will tend to have high-performance displays. Matching product form-factor to the display technology is another key decision for the product designer. Larger displays are capable of displaying a larger number of pixels (for a given level of display technology), but larger displays also increase the product form-factor. Often display size is used as the product differentiator, with a larger display being positioned as a premium product, such as in the notebook computer market. It is important to compare the actual pixel density (pixels per unit area) when comparing larger flat panel displays. If display size is increased and the pixel density is held constant or increased, then the display will have an improved appearance to the end user. If, on the other hand, the pixel density decreases, as would be the case when moving from a 14.1-in VGA screen to a 15-in VGA screen, some images may look slightly less sharp on the larger display when viewed from the same distance. (Since both displays have the same number of pixels, the larger display has a lower pixel density.)

It is well worth the designer's time to explore this dynamic between form-factor, anticipated viewing distance, and perceived resolution for the particular product usage scenario.

The selection of display components can have a major impact on battery life. Reflective LCDs require very little power but rely on ambient light for viewing. Most products that need only character or numeric displays rely on reflective LCDs. A front light may be added to allow occasional viewing in low ambient light settings and this does not have a huge impact on battery life. Backlit displays tend to be the most power hungry devices in the products in which they appear. Power management schemes to shut off the backlight when not in use are essential to achieving an acceptable battery life. Emissive displays tend to be more efficient at producing a luminous display because light is only generated at the pixel where it is required. Emissive displays must also be shut down, however, when not in use, to achieve a reasonable battery life.

In terms of cost, display modules tend to be a fairly substantial portion of the system cost. This is true because designers generally want to present the user with the highest quality display that makes economic sense for the product concept. In notebook computers and DVD players, the display module alone can comprise over one-third of the total system cost, and is the single most expensive component in the system. This proportion is usually lower in less display-centric products.

Display modules are a highly strategic component in the notebook computer industry. Ensuring access to the latest generation of flat panels is a top priority to notebook computer OEMs. Since the display is one of the key selling features of the product, being first to market with a larger or higher resolution display can result in a rapid increase in market share.

Balancing improved display capabilities with reliability is a challenge, however. New display technology is heavily dependent on materials science, and as such, it is exposed to extensive reliability issues. New types of organic light emitting diode (OLED) materials show great promise in enabling a new generation of displays, but they are very sensitive to moisture and contamination.

Basing a product on a new materials system can be a risky proposition for a product designer. One approach that has been employed in the past is to design a product to accommodate multiple *pin-compatible* display modules that use different technology. This approach allows a designer to introduce a product with the option of using a stable display technology or a riskier, but potentially superior, emerging technology. Small numbers of the version that uses the new technology can be introduced into the market and carefully tracked for reliability concerns. As confidence in the new technology improves, the entire product line can transition to the newer technology. If major reliability problems are discovered, the customers with the advance product can be refunded or outfitted with the more stable version, and the technology experiment can be phased out. In this scenario, the users of the new technology should be aware of the technology risk and provided with the product only as an informed *beta* customer.

Leveraging pin compatible display modules also enables product risk reduction through *multisourcing* of the required display components from different vendors.

There are a number of established and emerging display technologies. The advantages of the most interesting technologies relative to portable electronics are discussed below.

Most portable electronic devices that have a display use a flat panel display (FPD) and, in most cases, that FPD is based on LCD technology.

4.1 Display Technology Overview

There are several types of flat panel displays (FPDs) that are important to portable electronic design, as summarized in Fig. 4.3. Liquid crystal displays are the market leader by a wide margin, but various emissive display technologies have been under development for a number of years

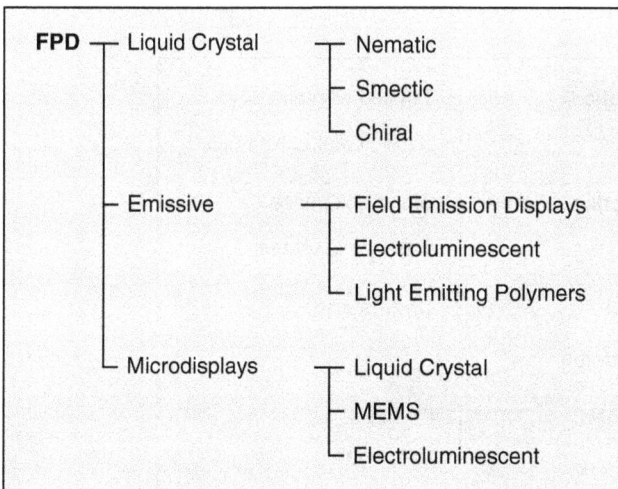

Figure 4.3 Major types of FPD.

and are likely to revolutionize the parameters of portable electronic design. Microdisplays offer a compelling display alternative for portable products, but a successful product implementation is still needed.

4.2   LCD

While there are many variations of LCD, they tend to leverage the same infrastructure, making it more difficult for new technologies to replace them. The major types of liquid crystal are outlined in Fig. 4.4. Nematic LC is the most common type and is used in almost all familiar products, including notebook computers and PDAs. Smectic and chiral LC materials are currently used in small volumes.

LCDs function as a light filter by blocking the backlight (or reflected light) with perpendicular polarizers. The LC material rotates the polarization of the light passing through it when the LC material is under an electrical bias. By rotating the polarization of the light, the light which passes through the first polarizer is then able to pass through the second polarizer as well. Twisted nematic (TN) and super-twisted nematic (STN) LC materials both allow light to pass in this fashion. The TN material rotates the polarization 90°, while the STN material rotates the polarization 270° (Fig. 4.5). STN provides better overall optical performance than TN.

Passive matrix LCDs (Fig. 4.6) are the easiest type of display to produce because of the relative simplicity of the structures patterned on the substrate (usually glass). Passive matrix LCDs do have the disadvantage of poor contrast ratio because of the cell discharge during refresh. Reflective passive matrix displays are particularly cost-effective,

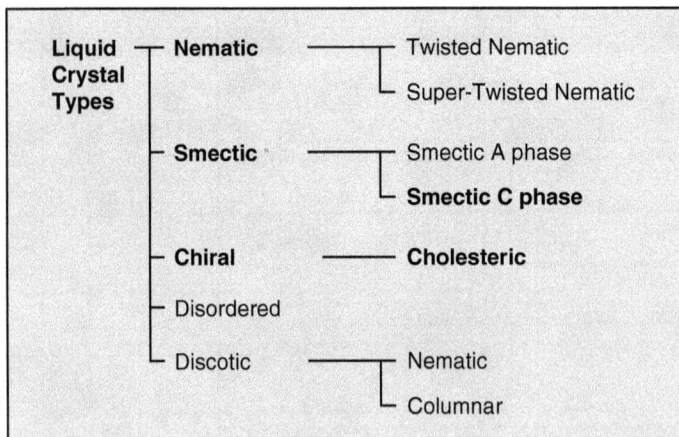

Figure 4.4   Liquid crystal types.

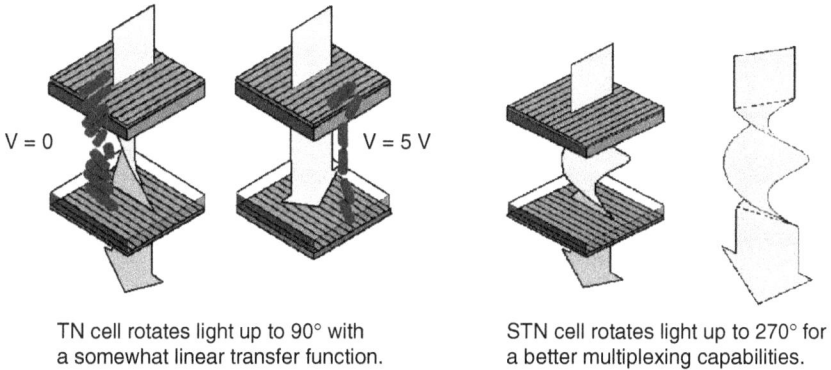

TN cell rotates light up to 90° with a somewhat linear transfer function.

STN cell rotates light up to 270° for a better multiplexing capabilities.

Figure 4.5   TN and STN LCDs.

however, and should be considered whenever the product usage paradigm does not require high display performance or a backlight.

Active matrix LCDs (Fig. 4.7) provide the best optical quality available to most product designs. TFT or MiM (diode) structures are used to drive the bias on each individual pixel, reducing cross talk and increasing the uniformity of bias across the panel.

LCDs are fabricated primarily on glass, but silicon and plastic are also used, as shown in Fig. 4.8. Glass is more dimensionally stable than plastic and more cost-effective than silicon. Silicon-based displays have the advantage that the driver electronics can be readily fabricated on the same substrate as the display pixels. Plastic substrates have the potential for low cost roll-to-roll processing. The flexibility and light

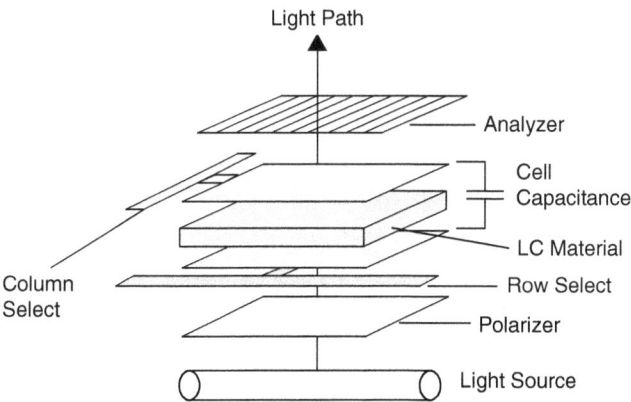

Figure 4.6   Passive matrix LCD.

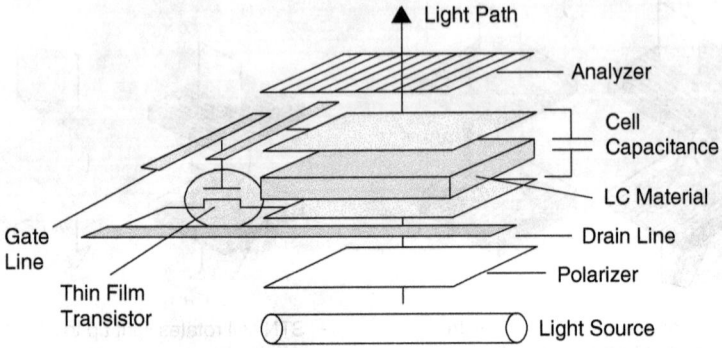

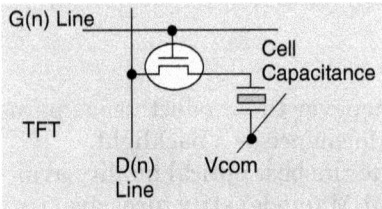

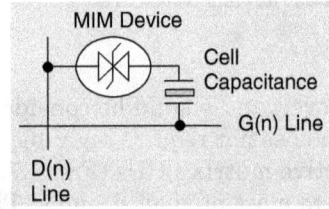

Figure 4.7 Active matrix LCD.

weight of plastic could also be exploited to make a notebook computer, or similar product, that is much less susceptible to breakage.

The substrates are coated with an alignment layer material (Fig. 4.8). The alignment layer is subjected to a rubbing process prior to the application of the liquid crystal material. The rubbing produces directionally

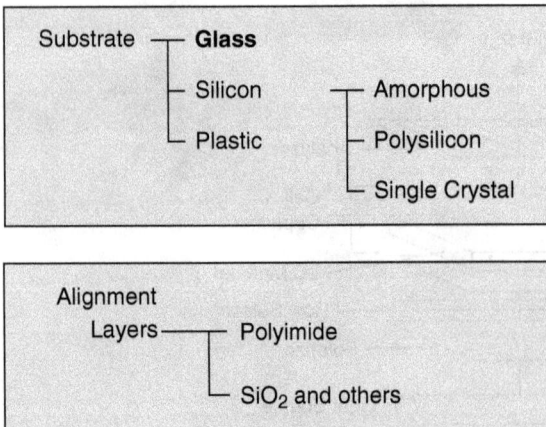

Figure 4.8 LCD substrate materials.

controlled roughness on the substrate surface, which provides physical orientation for the liquid crystal material.

Backlighting technologies for LCDs include cold cathode florescent (CCFL), electroluminescent (EL), light-emitting diode (LED), and organic light-emitting diode (OLED). CCFL is the predominant technology for notebook computers because it is the most cost-effective way to produce the bright white light needed for a full color display; however, CCFL backlight solutions are thicker than EL, LED, or OLED. EL backlights are very energy efficient and thin but are monochrome and have relatively low brightness. LED backlights are compact and bright but are inefficient at producing white light and are relatively expensive. OLED backlights are still under development. A typical inverter board used to drive a CCFL backlight is shown in Fig. 4.9.

Various types of optical films are used in LCDs, and other types of displays, to improve the visual characteristics. Polarizers are key components in LCD fabrication. Most films are based on stretched polymers that enable a compromise between polarization rejection ratios versus optical throughput. Compensation films are used for viewing angle enhancement and color compensation in STN applications. Some films are temperature compensated to exactly offset effects of display cell. Antireflection films are used to enhance contrast ratios by minimizing the front reflection from the display (which competes with the desired optical information). Brightness enhancement films are used to couple light more efficiently from backlight diffusers. Reflection films (holographic) are used to reflect light off-axis from the front surface reflection, which allows the display to offer better contrast ratios. Figure 4.10 shows some of the optical elements that are mounted in a display module along with the flat panel.

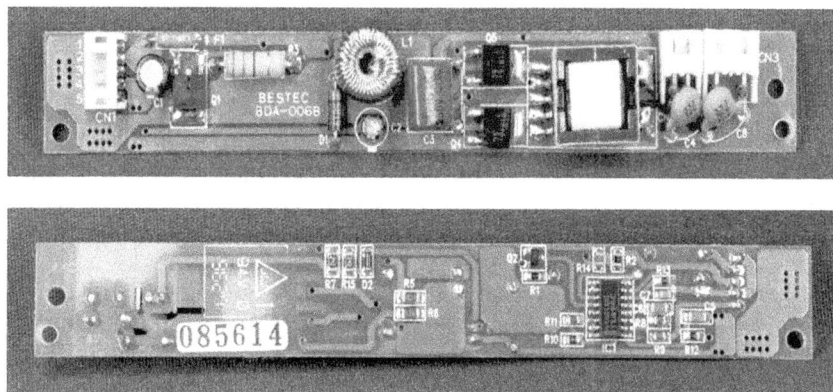

Figure 4.9 Backlight inverter board.

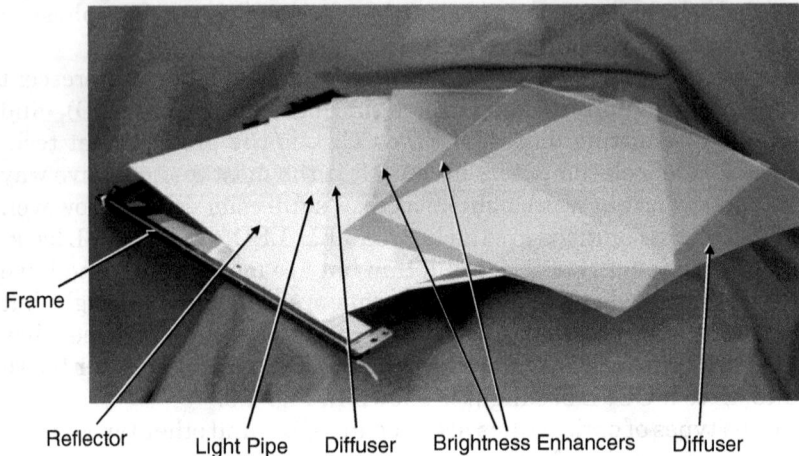

Frame

Reflector   Light Pipe   Diffuser   Brightness Enhancers   Diffuser

Figure 4.10 Optical films.

An LCD requires specialized ICs to support its functionality. Row and column drivers are needed to bias individual pixels. Standard display interface circuitry is also needed to support communication with the rest of the system. Power control circuitry is also needed both for the LCD drivers and the backlight. In many cases these various display support electronics can account for over half the cost of the display module.

One of the reasons that new technologies have difficulty in displacing LCDs is that the driver ICs have become so cost-effective because of the tremendous volumes produced. A competing flat panel technology must not only beat the price of the LCD panel, but must also utilize less expensive driver electronics if it is to compete. This is almost impossible unless the contender can leverage LCD drivers or has an incredibly simple drive scheme.

Display row and column drivers presented a difficult packaging problem to LCD manufacturers in the early days of the notebook computer industry. The IO from the driver chips must be attached to the display rows and columns at significantly finer pitches than previously encountered. Since the display row and column traces are made of indium tin oxide (ITO) on glass, soldering is not a viable method of connection. Two packaging solutions emerged for high-resolution displays, namely, TAB and flip-chip. Lower resolution passive matrix displays often use heat seal or Zebra strip connectors.

One of the first applications of flip-chip mounted display drivers was in a Sega handheld video game manufactured in 1994. Figure 4.11 shows the underside of the mounted flip-chip device through the display module glass.

Figure 4.11   Flip-chip mounted display driver chip.

Figure 4.12 shows a TAB-based notebook computer display module. A close-up of the TAB devices is shown in Fig. 4.13. This photo shows two of the 10 column driver ICs used in the product. The spacing between TABs is 1 mm. Due to the thermal expansion mismatch between the glass and the PCB, anisotropic conductive adhesive (ACF) is used to connect the TAB to glass and PCB. The open areas in the TAB polyimide allow the TAB to bend easily and also allow the copper conductors to absorb the mismatch stress. If solder were used at the PCB connection, the copper leads would prove stiffer than the solder at elevated temperature, so the solder would undergo plastic deformation and eventually fatigue. The ACF maintains its high elastic modulus at elevated temperatures, forcing the copper leads to deform. They too will eventually fatigue, but only after delivering acceptable field life.

The use of low-temperature poly-silicon technology enables the display manufacturer to produce the LCD drivers, using the same fabrication process as the active matrix transistors used to drive each pixel of the active matrix display. This represents the ultimate integration of display driver chips, since they are fabricated as part of the display panel and discrete driver ICs do not need to be mounted to the LCD glass.

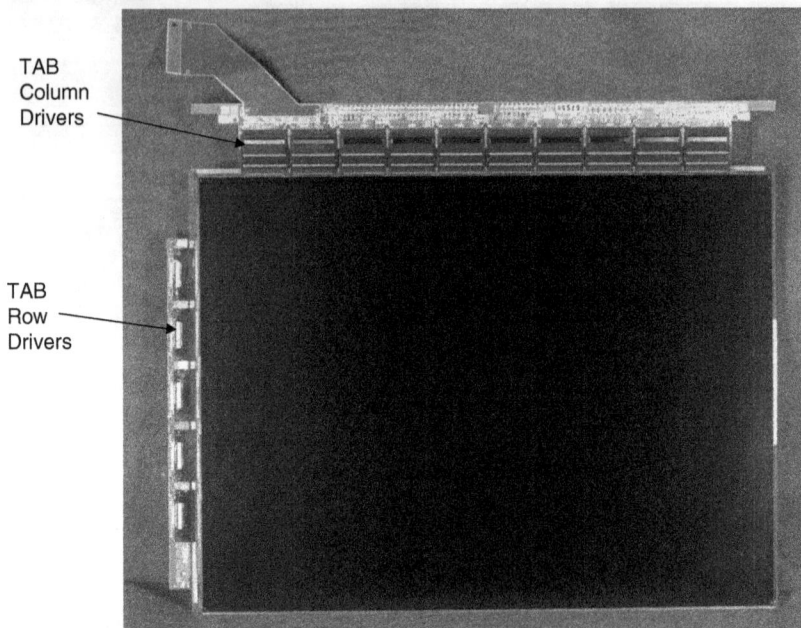

Figure 4.12 TAB-based notebook display module.

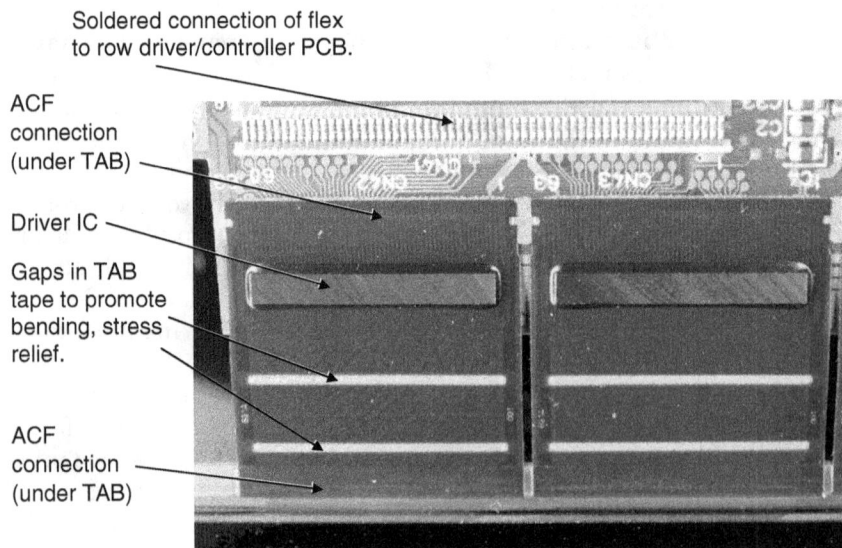

Figure 4.13 TAB driver close-up.

4.3   Other Display Technologies

While LCD still dominates the portable electronics market, a number of other technologies have potential applicability in portable electronic applications. Polymer OLED displays continue to receive a lot of attention because they are an emissive display technology. Emissive display technologies are promising for portable electronics applications because the theoretical energy efficiency is higher than LCD technology. Since LCDs function as a light valve by blocking the backlight of an OFF pixel, an LCD is always producing the maximum amount of light when in use. Emissive technologies, such as polymer OLED, generate light at the pixel source; thus, light is only produced where needed to create an image. If each pixel is capable of generating three primary colors, then a color display can be produced without the need for color light filters, increasing energy efficiency even further. Polymer OLED technology has been under development for many years and is finally beginning to appear in portable electronic products with impressive results.

Other emissive display technologies include electroluminescent (EL) display technology and field emission display (FED) technology. EL displays have been adopted into some low-resolution applications, but have mounted no significant challenge to LCDs in the overall portable electronics market. FEDs initially appeared to offer an energy efficient, high-brightness display technology, but huge investments have failed to produce a viable technology.

4.4   Microdisplays

Microdisplays are an important concept to portable electronics because they enable alternative form-factors and usage paradigms. Head-mounted display devices and heads-up displays that project a reflected image onto a pair of glasses require some type of microdisplay device to create an image. Small LCD and polymer OLED-type displays can be fabricated for this type of application. Figure 4.14 shows a small polysilcon LCD module used in a camcorder application.

Another technology, particularly well suited to microdisplays, is microelectromechanical systems (MEMS) technology. The use of MEMS has recently increased in high-definition projection TV applications. The small size of these devices (Fig. 4.15) makes them ideal for microdisplay applications.

Figure 4.16 shows a photograph of a head-mounted display system manufactured by Sony in the mid 1990s. This product allows the user to see the surrounding environment while superimposing the displayed information over the user's field of view. Figure 4.17 is a simplified representation of the physical design of this system. Two AMLCDs are mounted horizontally in the viewing section of the headset. The LCD

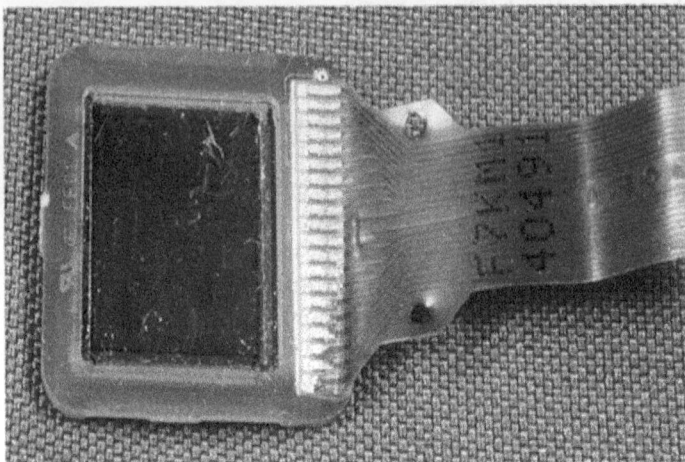

Figure 4.14 Polysilicon LCD microdisplay—viewfinder.

Rows and columns driven
from alternate sides of array

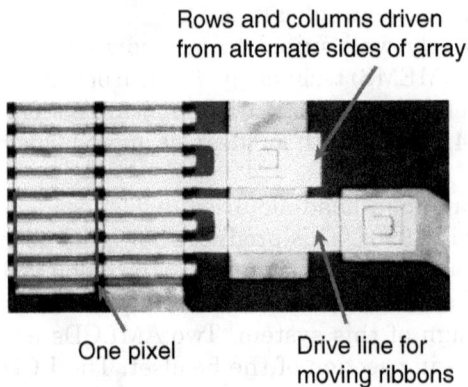

One pixel

Drive line for
moving ribbons

Figure 4.15 MEMS display.

Figure 4.16 Head-mounted display system.

images are viewed through the inclined flat mirror and the front-curved mirror. Both mirrors are partially silvered. Partial mirroring of the curved mirror is solely for "see-through" capability. The degree of see-through is further controlled by the "shutter" liquid crystal device, which is basically a three-character plastic-film LCD. The curved front mirror ensures that all points on the displayed image are at the same focal distance from the user's eye. A single CCFD tube spans the width of the

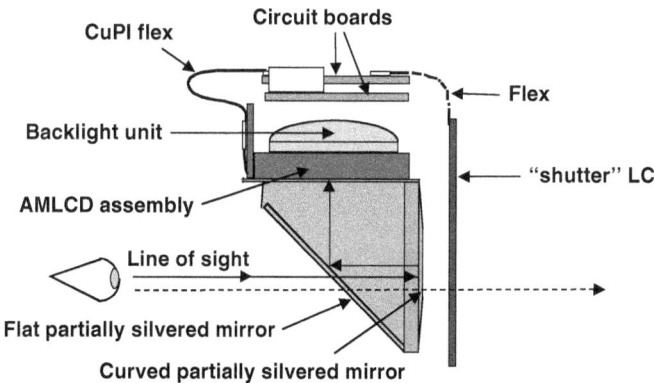

Figure 4.17 Sony Glasstron PLM-50.

TABLE 4.1   Display Technology Comparisons

	LC			EL	Plasma	OLED	FED
	TN	DSTN	TFT				
Matrix	Passive	Passive	Active	Passive	Passive	Passive	Passive
Resolution	80°	10°	140°	160°	160°	160°	160°
Contrast ratio	Poor	Better	Best	Best	Best	Poor	Best
Brightness	n/a	n/a	n/a	Good	Best	Fair	Best
Viewing angle	Poor	Poor	Better	Best	Best	Best	Best
Switching speed	Slow	2x TN	Fast	Very	Very	Fast	Very fast
Linearity	Good	Better	Fair	Best	Best	Too early	Best
Refresh	Slow	2x TN	Fast	Fast	Medium	Fast	Medium
Drive voltage	Low	Low	Medium	High	Very high	Medium	Very high
Temperature range	OK	OK	Better	Best	Best	Too early	Best
Lifetime	Long	Long	Long	Long	Long	Too early	Medium
Maturity	Very	Very	Very	Very	New	Evelopmen	New

headset, backlighting both LCDs. The LCD controller and backlight power supply PCBs are mounted above the backlight.

Table 4.1 compares the properties of various display technologies. When developing system architecture for a portable electronic product, estimating the display bandwidth is necessary in determining the system processing requirements and the interface requirements for the display module. Table 4.2 shows the uncompressed data bandwidth requirements for several different resolutions of display. In practice, the display bandwidth problem is handled through a combination of high-speed signaling design and data compression.

4.5   Pen Input

Pen input is an increasingly popular capability for many types of portable electronic devices that utilize a display. Many pen input displays are also touch sensitive, allowing the user to select an on-screen button by touching

TABLE 4.2   Display Bandwidth Requirements

Graphics std. acronym	Resolution pixels (H × V)	Bandwidth(mbytes/second)		
		256 colors	64K colors	16M colors
VGA	640 × 480	25	50	75
SVGA	800 × 600	40	80	120
XGA	1024 × 768	65	130	195
SXGA	1280 × 1024	108	216	324
UXGA	1600 × 1200	162	324	486

the display with a finger. The pen input capabilities are generally integrated into the display module supplied by the display manufacturer.

The most common type of pen input technology is the resistive overlay. With this approach, two very thin transparent sheets of Mylar or a similar material are laminated together with a small gap in between, maintained by some type of spacer material. The opposing surfaces of these sheets are coated with a transparent, electrically conductive material such as indium tin oxide (ITO). When pressure is applied to a point on the sheets, they are forced into contact, creating an electrical short at the point of contact. Circuitry connected to the edge of these sheets measures the resistance and interpolates the point of contact based on a calculation of the electrical resistance (resistance of the conductive film) measured from each sheet.

This is the simplest and, usually, the most cost-effective method of creating a pen input capability, but the optical quality of the display generally suffers with the additional layers of optically clear material. The optical degradation may include reduced brightness, reduced contrast, reduced viewing angle, and increased glare. This optical degradation may be mitigated by introducing an index-matching layer between the resistive overlay and the LCD glass. One of the transparent sheets may also be eliminated by using the top surface of the LCD glass as the lower member of the resistive overlay structure. Other types of pen input methods use capacitive coupling or various electrical field detection methods to determine the location of the pen contact. Most of these methods require a special pen with some built-in electronics.

This chapter has provided an overview of display technologies as they apply to portable electronic applications. Since the display module can be one of the most visible differentiators to the end user, and since the display also has a major impact on battery life, the product designer should spend ample time in viewing and understanding the attributes of various display solutions.

4.6   Definition of Key Terms

Aperture ratio.   This is the ratio of active, contrast-producing area to the overall area of a display element. Aperture ratio is reduced by bus lines, masks, or switching devices on the display substrate that are not optically active. Aperture ratio is also called *fill factor*.

Aspect ratio.   This is the ratio of screen width to height. Popular ratios in use today are 4:3 (NTSC standard) and 16:9 (HDTV).

Brightness.   Brightness or luminance is a measurement of luminous flux or the total power of light emitted from a single source. An equivalent definition is the flux through a unit-sized solid angle (steradian). A steradian is the solid angle that closes an r^2 sized area on a sphere of radius r. Because the steradian is dimensionless, luminous flux has the same units as luminous intensity. Units of luminous flux are lumens (lm), which are defined as one candela per steradian. The following is the luminous output from common sources:

Light source	Lumens
Tungsten lamp (25 W)	235
Tungsten lamp (100 W)	1,750
Fluorescent tube (40 W)	3,250
Mercury vapor lamp (400 W)	20,000
Mercury vapor lamp (1000 W)	60,000
Metal halide lamp (400 W)	34,000
Metal halide lamp (1000 W)	100,000
High pressure sodium vapor (1000 W)	140,000

Candela.   The fundamental unit of luminous intensity. One candela is the luminous intensity of 1/60 of a square centimeter of projected area of a black-body radiator maintained at the freezing point of platinum (2042 K).

CCFL.   Cold cathode fluorescent light, a fluorescent light that does not use a heating element to start electron emission, but instead creates a field emission initially through use of a high-voltage arc (starting voltage). An alternative to hot cathode fluorescent (HCFL) lights.

Color temperature.   The color temperature of a white light source is equivalent to the temperature at which a black-body radiator (such as xenon or carbon) emits light most closely matching it. Higher color temperatures have a bluish cast, while lower color temperatures have a reddish cast. Well-designed displays provide for adjustable color temperature. Many color TVs are set up with higher color temperatures than the accepted NTSC standard of 6500 K in order to appeal to customers in the retail environment. Common color temperatures are as follows:

Source	Temperature (K)
Sky (north)	7,500
Daylight (average)	6,500
Carbon arc	5,000
Flash bulb	3,780
Fluorescent lamp	3,500
Tungsten halogen	3,300
Tungsten lamp	2,900
Sunset light	2,000
Candle flame	1,900

Contrast ratio.   This is defined as the luminance of an ON pixel divided by the luminance of an OFF pixel. In some cases there is a distinction made between "small area" contrast ratios versus "large area" contrast. The "small area" measurement may be for only a pixel (monochrome) or subpixel (color), while the "large area" may be for up to half of the entire display area. Example contrast ratios are as follow:

Item:	Contrast
CRT next to window	200:1
Newspaper	50:1
Photocopy	500:1

Dielectric constant.   A dielectric is a material that increases the capacitance of parallel plate capacitor structure. A perfect vacuum has a dielectric constant of 1. Air is very close (within 0.01 percent) of this value. Commonly available dielectric materials go up to 50. Some specialty materials (PLZT) have dielectric constants up to 34,000. Dielectrics are important in LCD cell construction where holding charge in a pixel element is very critical for device performance.

Dithering.   This is similar to half toning where the number of ON pixels per unit area are changed to produce gray scale or to hide systematic errors.

Dot inversion.   This is a drive scheme where every other pixel (or dot) is written with an alternating net positive and negative bias with each successive frame write cycle. Inversion is necessary so that there is no net DC bias build-up on the liquid crystal materials. See Field inversion for an alternative implementation.

Ferroelectric liquid crystal (FLC).   These liquid crystal materials demonstrate behavior similar to ferroelectric materials in that they exhibit bistable behavior.

Field emission display.   A field emission display is a type of flat CRT that uses strong electric fields to generate electron emission from a metal surface rather than thermoionic emissions as are used in filament-based CRTs. Common field emission cathode structures are Spindt-type structures.

Field inversion.   This is a drive scheme where the entire display is written with an alternating net positive and negative bias in each successive frame write cycle. Inversion is necessary so that there is no net DC bias build-up on the liquid crystal materials. See Dot inversion for an alternative implementation.

Fill factor. See Aperture ratio.

Frit seal. A glassy material that is used to seal two pieces of glass together. A frit is made by mixing powdered glass with organic binders and solvents. This material is screen-printed onto one of the glass surfaces and heated at a high-enough temperature to decompose all of the organic material, leaving behind a uniform bead of soft glass. The opposing piece of glass is then placed on top of the fritted glass plate and the sandwich is heated at a high temperature to remelt the glass bead, thereby forming a bond between the two glass plates. This is the basic process used to seal the plates of a plasma or FED display panel.

Illuminance. This is a measure of light reflected off of a passive surface, as opposed to light radiated from an active emitter. Because illuminance is defined in terms of reflected light, it is dependent upon both the light source and the reflectivity of the illuminated material. Illuminance is defined as luminous intensity times the material's luminance factor (LF). Luminance factors for common objects are in the range of 20 percent. Shiny metal has a much higher factor while black felt has a much lower factor. A footcandle is the illumination on a surface, which is 1 ft^2 in area, over which there is a uniformly distributed flux of 1 lm. Units commonly used to measure illuminance are as follows:

Name	Abbreviation	Derivation
lux	lx	lumens/square meter
foot-candle	fc	lumens/square foot

Some references for illuminance and luminance, with a 20 percent luminance factor are listed below:

Item	Illuminance (lux)	Luminance (nits)
Bright Sun	50,000–100,000	3,000–6,000
Hazy Sun	25,000–50,000	1,500–3,000
Cloudy, bright	10,000–25,000	600–1,500
Cloudy, dull	100–2,000	6–120
Office	200–300	12–18
Living room	50–200	3–12
Sunset	1–100	0.06–6
Full moon	0.01–0.1	0.0006–0.006
Star light	10^{-4}–10^{-3}	10^{-7}–10^{-6}

Luminous efficiency.   This is a term used to measure the ratio of light output to power required to generate that light. Luminous efficiency is measured in terms of lumens per watt. Various light sources are better light generators than others. Some produce high light output along with high heat. The theoretical maximum is 680 lm/W at a wavelength of 555 nm. A high-pressure sodium vapor lamp approaches 25 percent of the theoretical maximum, while a tungsten lamp barely reaches 3 percent.

Luminous emittance.   This is the flux density of the power per unit area. The SI units are lumens per square meter (lm/m^2).

Luminous intensity.   This, sometimes called candlepower, is the power per unit solid angle. It is also the flux density of light through a solid angle emanating from a source. The fundamental unit of luminous intensity is the candela, which is rigidly defined in terms of physical constants.

Luminance.   The luminance of a display is the amplitude of the light coming from its surface. Luminance is the photometric quantity that most closely approximates the intuitive concept of brightness, which is a psychological interpretation of the intensity of light energy. Luminance, like sound, is measured on a logarithmic scale, since that is the way it is perceived. The SI unit for luminance is candela per square meter (cd/m^2) or nit. Direct view large screens are often described in terms of nits. Units commonly used to measure luminance are as follows:

Name	Abbreviation	Derivation
nit	nit	$candela/m^2$
foot-lambert	ft-l	$candela/ft^2$
lambert	L	$candela/cm^2$

MicroLCD.   A small liquid crystal display, typically created as a liquid crystal on silicon device. These devices are made with single crystal silicon backplanes that are typically designed as a static or dynamic RAM and processed as a standard IC device. These devices benefit from the chemical-mechanical polishing planarization advances of recent years.

Nit.   A candela per square meter. See the definition of luminance.

Resolution.   Resolutions are quoted as the number of horizontal pixels followed by the number of vertical pixels. Commonly used standard resolutions are as follows:

Acronym	Resolution
QVGA	320 × 240
HVGA	640 × 240
VGA	640 × 480
SVGA	800 × 600
XGA	1024 × 768
SXGA	1280 × 1024
UXGA	1600 × 1200

Response time.   This actually consists of two figures, T_{on} and T_{off}. A display may have excellent contrast and brightness, but if the response time is slow, the amount of time for the screen to show the intended image will be noticeable. In general the human eyes cannot detect a time-lag of less than 75 ms. Response time is important for applications where an LCD is being used with a mouse pointer. In these applications, the display cannot keep up with the rapid movement of the mouse, and on quick movements, the mouse's pointer disappears and reappears some moments later after the mouse has stopped moving.

Subpixel.   A subpixel is a primary color *building-block* element used to compose a full-color pixel on a display. The most commonly used color primaries are red, blue, and green for displays where colors add together and yellow, cyan, and magenta where colors subtract from each other. Most LCDs used in portable computers have RGB subpixels, while some *stacked-element* reflective displays use CYM subpixels. Generally there are three times as many subpixels as pixels for a given display (e.g., an 800 × 600 display has 480 K pixels and 1.44 M subpixels).

Thin film transistor.   A thin film transistor (TFT) is a switching element fabricated on thin, uniform films of various materials in layers that range in thickness from 100 Å to 20 μm. TFT devices used in LCDs are typically fabricated on amorphous silicon, which offers device performance that is several thousand times slower than that of regular transistors fabricated on single-crystal silicon.

Power Sources

Power sources refer to all types of technology that can provide power to a system including AC outlets, fuel cells, photovoltaic cells, and batteries. Batteries are the most common type of power source for portable electronic devices although fuel cells are showing great promise for the future.

Many portable electronic devices are capable of utilizing power from an AC power outlet, but almost all use some type of portable power storage technology. If AC power is used, it is generally for the purpose of extended use or for recharging the system battery.

Battery technology has been evolving very rapidly in recent years to meet the demands of the portable electronics industry. Advances in battery technology have contributed greatly to the extreme miniaturization that has taken place.

There are many types of batteries that use a wide variety of wet and dry chemistries. We will only discuss batteries that are of interest for portable electronic applications.

Batteries are usually divided into two types: *primary* and *secondary*. Primary cells are intended to be charged once, used until discharged, and then discarded. Secondary cells have the ability to be recharged and discharged many times.

Energy density is a measure of how much energy a battery can store in a given volume or mass. Volumetric energy density is usually measured in watthours per liter (Wh/L). Gravimetric energy density is generally measured in watthours per kilogram (Wh/kg).

Some secondary batteries exhibit a behavior known as *memory effect*. When these batteries are fully discharged, they may be recharged to their original energy capacity. If these batteries are only partially discharged before recharging, however, they will begin to exhibit a reduction in charge capacity. Over the course of many partial discharge and recharge cycles, such batteries will degrade to the point of being useless.

Cycle life is the number of charge and discharge cycles that a battery can withstand before it is no longer able to hold a useful charge.

The battery *working voltage* is the voltage available from a single cell as determined by the battery chemistry.

The *self-discharge rate* is the rate at which a battery will discharge while sitting unused.

5.1   Battery Technologies

5.1.1   Ni-Cd

Nickel-cadmium (Ni-Cd) batteries are the most common form of household battery, appearing in a wide range of products. Ni-Cd cells can provide high current density and are often used to power devices that have small motors. The memory effects of Ni-Cd cells, the high self-discharge rate, and the low-energy density that they provide make them a poor choice for many types of portable electronic devices such as cellular phones and notebook computers.

5.1.2   Alkaline

Alkaline batteries have energy density that is slightly better than Ni-Cd and are popular with consumers as a single use battery. Rechargeable (secondary) alkaline batteries are available but the energy density of these batteries degrades quickly with multiple recharges.

5.1.3   Ni-MH

Nickel-metal hydride (Ni-MH) batteries were commonly used in cellular phones and notebook computers because of their reasonable cost and relatively high energy density. They do have a self-discharge rate that is high enough to make them unsuitable for some applications. This battery technology bridged the gap between Ni-Cd and lithium ion but has become obsolete as lithium-based technologies have become more cost-effective.

5.1.4   Lithium ion

Lithium-ion batteries have high energy density and are very common in cellular phones and notebook computer products. Lithium-ion cells can be fabricated in thicknesses as low as 0.5 mm. Lithium-ion has become increasingly cost-effective in recent years.

5.1.5 Lithium polymer

Lithium polymer batteries have high energy density and can be formed into a variety of shapes to conform to various product form-factors.

5.1.6 Photovoltaic cells

Photovoltaic cells convert ambient light into electrical power and can be used in very low power devices such as calculators.

5.1.7 Fuel cells

Fuel cells convert hydrocarbons into electrical energy and have a very high energy density. Recharging a small fuel cell would be similar to refilling a lighter with lighter fluid. Fuel cells offer the potential of providing three to five times the energy density of lithium-ion batteries, but fuel cells are impractical for portable electronic device applications today.

Table 5.1 gives the critical metrics for the most important power sources for portable electronic devices.

Dead batteries are the most common reason that portable electronic products fail to perform when needed. Designing a battery capacity and recharging strategy that is consistent with the use-case scenario is a critical aspect of portable electronic product design. Batteries are also limited in the temperature extremes in which they can operate. While most of the batteries we are concerned with will perform very well in moderate environments, environmental constraints should be checked if the product application involves extreme temperatures. Battery safety issues should also be taken into consideration. The battery manufacturer will provide a detailed description of safety issues and charging conditions that may cause battery failure or explosions. Lithium batteries have always posed a safety and reliability concern because of the unstable nature of lithium compounds. Lithium battery design has constantly improved, however, so that lithium batteries are generally considered safe and reliable. Final disposal of lithium batteries must also

TABLE 5.1   Critical Metrics for Battery Technology

	Ni-Cd	Ni-MH	Li-ion	Li-polymer
Working voltage (V)	1.2	1.2	3.6	3
Energy density (Wh/L)	120	240	260	264
Energy density (Wh/kg)	50	60	115	250
Cycle life	300–800	300–800	1200	1200
Memory effect	Yes	Yes	No	No
Cost ($/Wh)	1	1.3	2.5	2

be considered by the product manufacturer. Given the volatile nature of the constituent components, some countries require the electronic industry to make provisions for systematic recycling and/or reclamation of battery cells.

5.2   Product Implementation

The best battery technology for a portable electronic device is determined during the trade-off analysis process. The designer must strike the right balance of high energy capacity, small form-factor, and system cost impact to enable a successful product concept. To make the battery solution a reality, the product designer must consider the available form-factor, recharge/replacement requirements, mechanical mounting, connectors, and power management electronics.

There is an endless variety of battery form-factors available to the product designer. Common form-factors include the familiar AA, AAA, C and D cells, and lithium button cell batteries that are purchased at the grocery store checkout. These types of batteries are desirable if the product concept requires that the user be able to easily obtain replacement batteries rather than be responsible for recharging the system.

Lithium-ion and Ni-MH batteries are available in a wide variety of prismatic (rectangular, noncylindrical) sizes and a standard product can usually be located for any product application. Custom form-factors can also be provided but will usually require significant nonrecurring engineering (NRE) costs.

The recharge and replacement requirements for the product will greatly affect the battery solution. It is important to understand whether the battery will be permanently mounted within the product, or whether the user will be removing the battery, and if so, how frequently. Some usage paradigms may allow the user to recharge the battery without ever having to gain direct access to the battery. Other usage paradigms may require that the user carry a spare battery and be able to swap out the used battery when it is depleted. In yet other cases, the user may simply buy new batteries and discard the used ones.

The mechanical mounting design must reflect the frequency of battery removal. For permanently installed batteries, the battery may be placed virtually anywhere in the system.

For batteries that are occasionally removed, a battery access port should be designed into the product housing that can be opened manually or with a screwdriver. Battery orientation markings or nonambiguous battery connectors should be designed-in to avoid incorrect installation. The battery itself should have an appropriate surface finish to be handled by the user. For example, exposed electrodes, easily punctured surfaces, adhesive residues, and sharp edges should all be avoided.

For batteries that are frequently removed, the battery access port should be able to be opened manually and the batteries should be smoothly finished and ergonomically comfortable from a visual and tactile point of view. AA, AAA, etc., type cells meet the ergonomic requirement quite sufficiently. Again, battery orientation should be addressed.

For some product concepts, battery design is a major aspect of the product design. Cellular phones, camcorders, and notebook computers are good examples of products that may require extensive battery swapping. In these types of applications, custom battery packs are often designed. These battery packs are designed to snap into the product quickly. It is common for cell phone battery packs to be attached to the outside of the product, thus becoming an extension of the product housing. Custom battery packs are typically composed of custom plastics encasing standard form-factor prismatic cells.

In all cases, since the battery tends to be a sizeable part of the system volume and has a major effect on product weight distribution, the location of the battery should be selected to effectively balance the feel of the product from an ergonomic standpoint. Battery location can also impact system thermal management, both as a source of heat as the battery charges and discharges, and as a major structural element in the thermal cooling path for the overall system.

The battery connector is also affected by battery design issues. Permanently mounted batteries are usually connected to the rest of the system by a connector that is designed to optimize the product assembly process during manufacture. Such connectors may be very compact and may require special tools to release the connector. User-removable batteries require an easily demated connector, or better yet a snap-in system that allows the user to quickly fit the battery securely into the product with the correct battery orientation. The designer should also make sure that the connector is designed to carry enough current to meet the product power requirements. Surface metallurgies of the battery and the connector should be reviewed to ensure that the surface contact will not degrade over time. Connector failure is a leading cause of product failure in portable electronic devices, so the manufacturer's specification for the number of mate/demate cycles needs to match the product lifetime requirements.

The following product examples serve to illustrate various product battery design implementations.

Figure 5.1 shows a digital minidisc music player that uses a lithium-ion battery cartridge but also has an attachable battery module that uses two AA batteries. This approach enables the user to enjoy the long battery life of a lithium-based battery, but if the charge is depleted unexpectedly, the player can operate off batteries that can be readily purchased near any urban location.

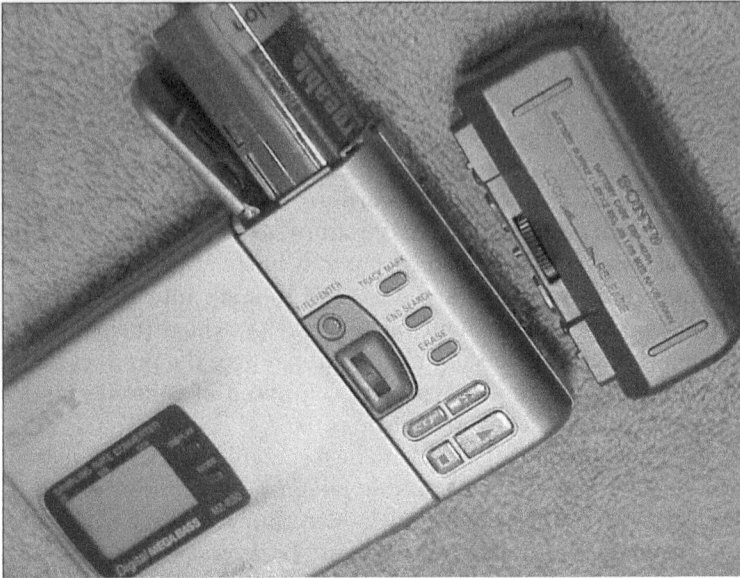

Figure 5.1 Mini-disc music player with attachable AA battery pack.

Figure 5.2 shows a similar concept. This product uses a Ni-MH battery pack (3.6 V, 430 mAh), but can also use three AAA batteries if the user depletes the battery and is not able to conveniently recharge the battery pack.

Figure 5.3 illustrates a Ni-MH cellular phone battery (3.6 V, 750 mAh), which has a wired connection to the phone electronics. This type of connection is suitable for products that do not require the user to replace the battery.

Figure 5.4 pictures a large Ni-MH battery that latches onto the back of a cell phone and serves as the rear housing of the product. The battery pack is actually shaped to accommodate a comfortable grip for the user. This is a Ni-MH battery (6 V, 1200 mAh) weighing 138.2 g, which is nearly half of the total product weight.

Figure 5.5 depicts a *flip-phone* style design and a typical lithium-ion battery pack. Notice the recessed electrode contacts on the battery pack. These electrodes make contact with spring-loaded connector pads when the battery pack is clipped into place.

Figure 5.6 is a picture of a very compact cell phone design with a very thin lithium-ion battery pack (7.2 V, 600 mAh). With this battery pack, the electrodes are recessed on the inward-mounted surface of the pack. When the battery is snapped into place, it compresses four pogo pins that

Figure 5.2 Cell phone battery pack interchangeable with AA cells.

maintain good contact with the battery electrodes. Notice also that the battery pack has a manual release latch molded into the casing.

Figure 5.7 depicts a Ni-MH battery, (3.6 V, 370 mAh) which is clipped into the plastic housing of a cellular phone. The plastic housing is firmly attached to the battery pack but is separate from the battery pack housing. This approach allows for later changes in the battery pack, which will not require retooling of the entire product plastic housing. Only the part that is attached to the battery pack would need to be modified.

Figure 5.8 shows a rather compact Ni-MH battery. The size is approximately 58 mm × 48 mm × 13 mm, and the weight is approximately 58.5 g. Since this is a 4.8-V, 600-mAh battery, the energy density is 79.6 Wh/L or 49.7 Wh/kg. It is interesting to compare these energy density metrics to the values given in Table 5.1. The actual energy density is significantly less than the numbers given in Table 5.1. This will often be the

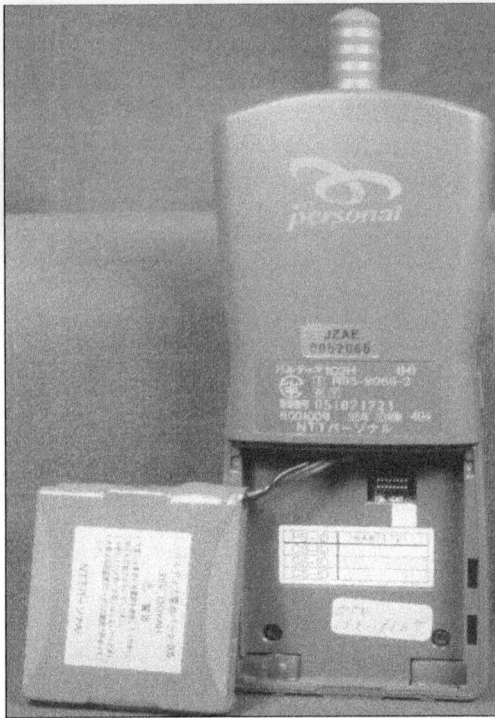

Figure 5.3 Cell phone battery with wired connection.

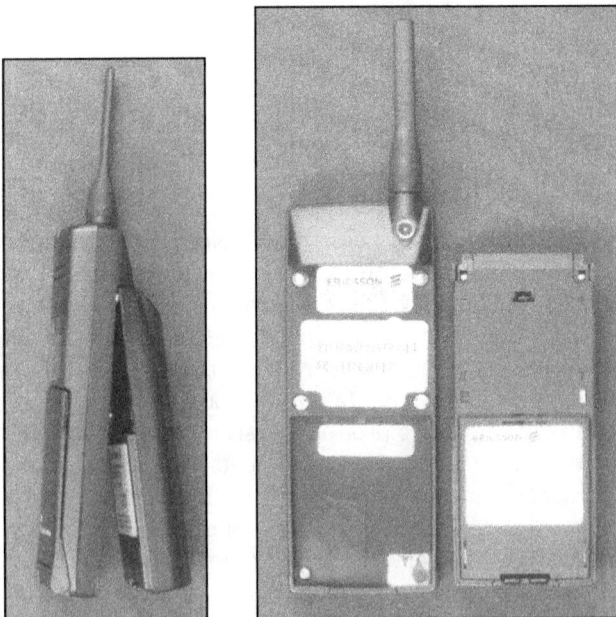

Figure 5.4 Large cell phone battery pack.

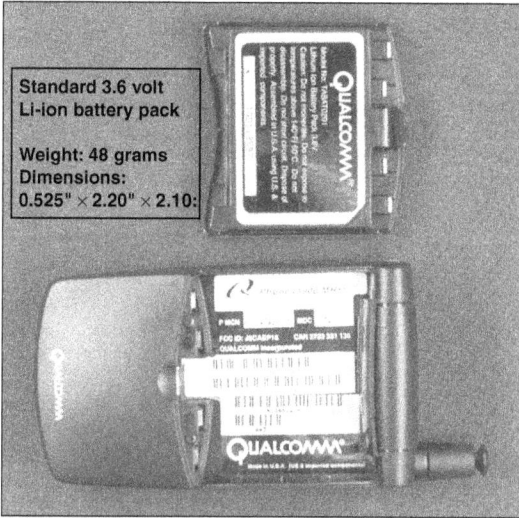

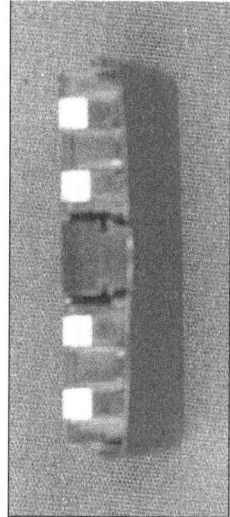

Figure 5.5   Flip phone style battery pack design.

case since the physical implementation of a particular battery cell will always fall short of the ideal performance. In this particular example, the mass power density compares more favorably with the ideal than does the volumetric density. We might therefore conclude that there is unused space in the battery pack.

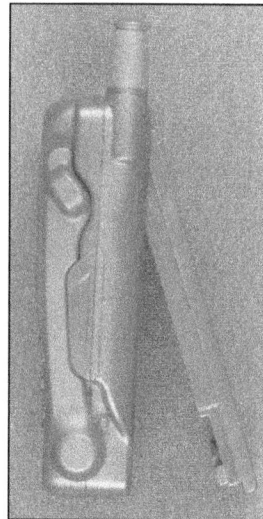

Figure 5.6   Compact lithium-ion battery pack.

Figure 5.7 Ni-MH battery with attached plastic housing.

Figure 5.8 Compact Ni-MH battery pack for a cell phone.

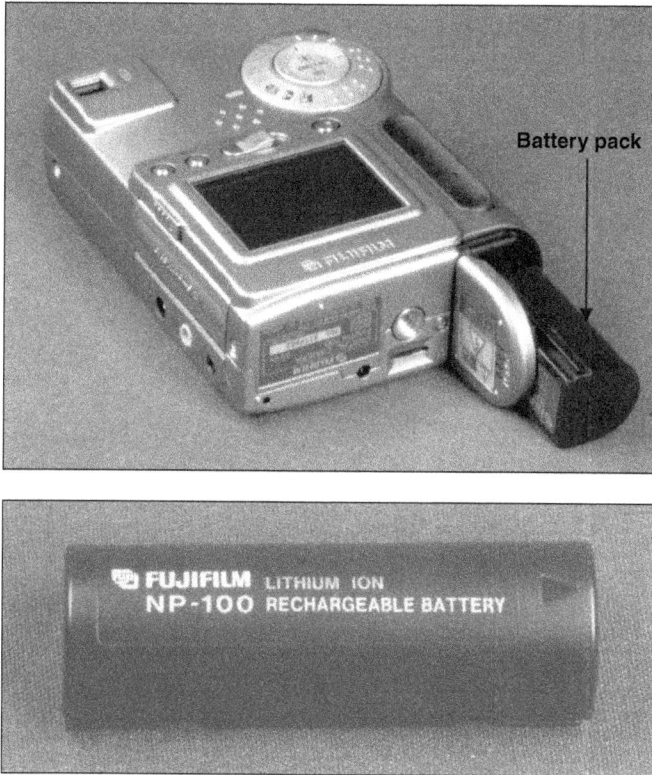

Figure 5.9 Digital camera with lithium-ion battery pack.

Digital cameras represent another product category where high energy density is required. Figure 5.9 depicts a digital camera product that uses a lithium-ion battery cartridge (3.6 V, 1350 mAh). At 47.2 g, this battery has an energy density of 103 Wh/kg.

The digital camera in Fig. 5.9 is designed around a usage paradigm that requires the user to recharge the battery. Figure 5.10 shows a digital camera that is powered by four replaceable AA battery cells. Battery orientation marks can be seen on the inside of the plastic door that covers the battery compartment.

The digital camera design pictured in Fig. 5.11 achieves the best of both worlds. This product can use two AA cell batteries or an optional lithium-ion battery in the same battery compartment.

Notebook computers provide a particularly tough challenge when it comes to battery design. Notebook batteries tend to be large, due to the high power consumption and the desire for long battery life to support the extended working sessions of business users. Many notebook

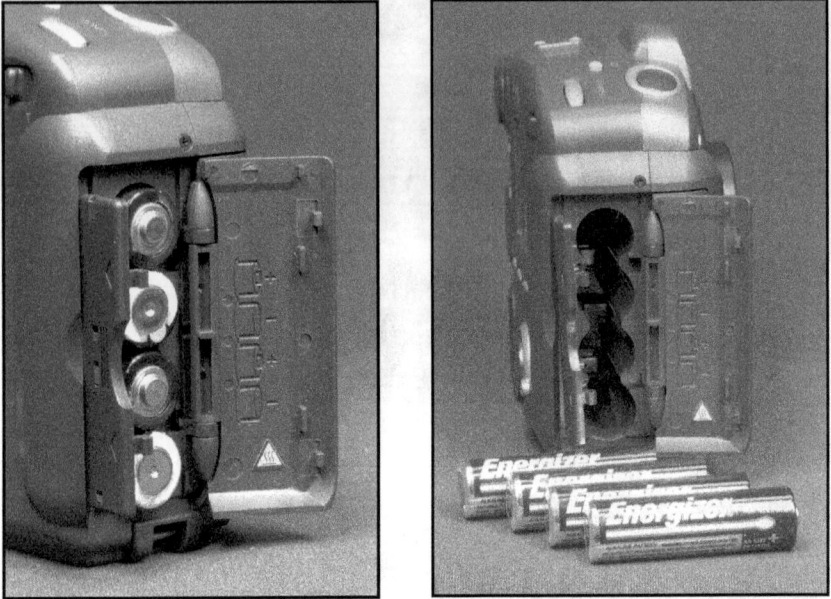

Figure 5.10   Digital camera using four AA cells.

Optional 3.6 v., 3.6 Wh Li-Ion battery ⟶

Figure 5.11   Digital camera using AA cells or lithium-ion cell.

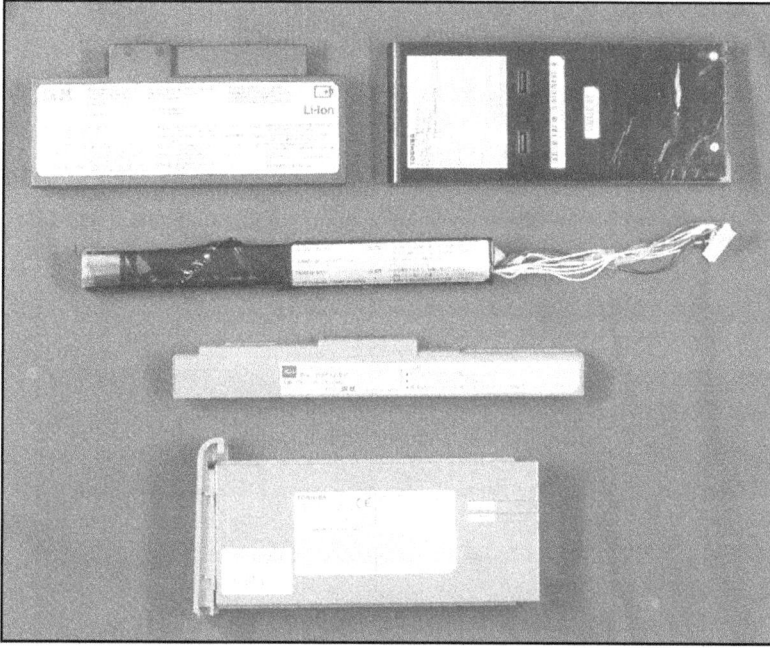

Figure 5.12   Assorted notebook computer battery packs.

products use a custom-designed battery. This is required because of the difficulty in efficiently using every bit of space inside the notebook computer housing. (This tendency is reinforced by the fact that the purchase of spare batteries by the customer, after the original sale of the computer, has high profit margins.) Figure 5.12 shows a number of battery packs from different notebook computer systems.

Figure 5.13 shows a closer view of a specific notebook battery pack and the bay into which it is plugged. Notice that the battery connections in the product are male connectors. The electrodes on the battery are female connectors to keep them from being damaged or exposed to physical contact. With high-energy-capacity batteries such as these, special provisions should be taken to ensure that the electrodes cannot be shorted by incidental contact with a conductor.

Lower power consumption devices do not need to use high-capacity batteries. Devices such as IR remote controls for a TV set can operate for many months on a single AAA-cell battery. Button-cell batteries are also a popular alternative for products that do not exhibit high power consumption. Figure 5.14 depicts a miniature voice recorder that operates using four Ni-Cd button cells and a lithium-button cell as a backup to prevent loss of stored data.

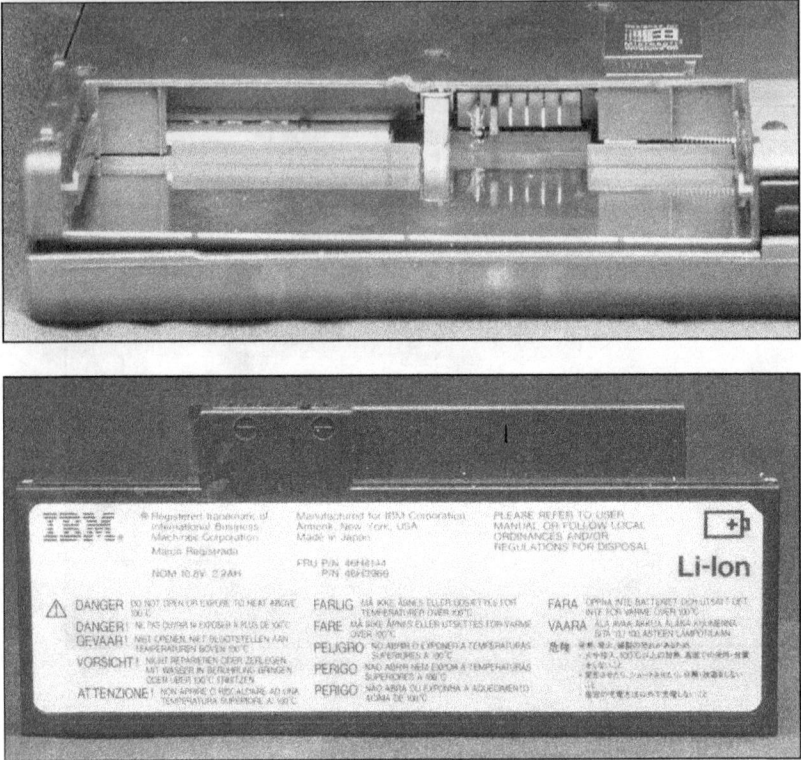

Figure 5.13 Notebook battery and battery bay.

For products that try to reach a very low price point, the philosophy is often: "Batteries Sold Separately." Figure 5.15 depicts a very low-cost digital camera that uses a 9-V battery (sold separately).

Custom battery packs are commonplace for higher-end products that require high energy density, as observed with notebook computers. While the product designer will not generally need to get involved in the details of a custom battery design, he should check the power-density metrics and other specifications provided by the manufacturer to ensure that a competitive design is being implemented. Observation of the internal cell configuration can shed light on the special efficiency of a custom battery pack. Figure 5.16 shows the internal cell configuration for a number of notebook battery packs.

The technology, voltage, capacity, weight, and energy density for each of these cells is shown in Table 5.2. Battery number 1 is an example of one of the first lithium-polymer batteries deployed in a notebook computer. While the theoretical energy density of lithium-polymer is higher

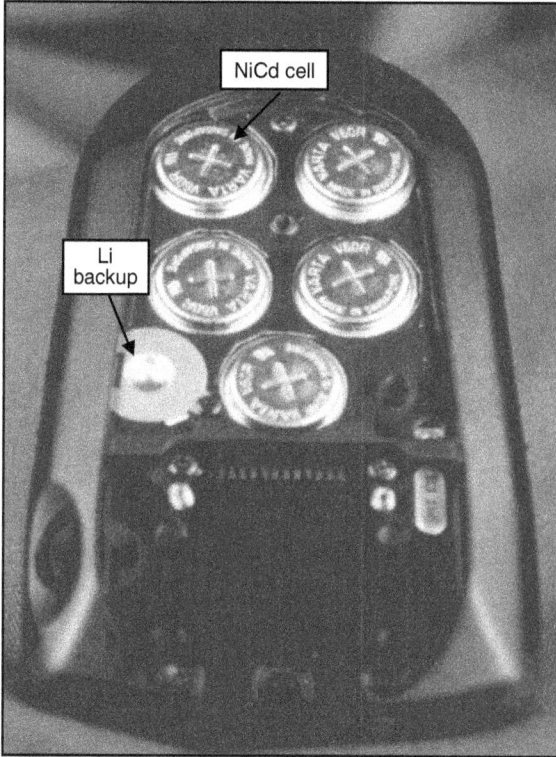

Figure 5.14 Voice recorder with button cell batteries.

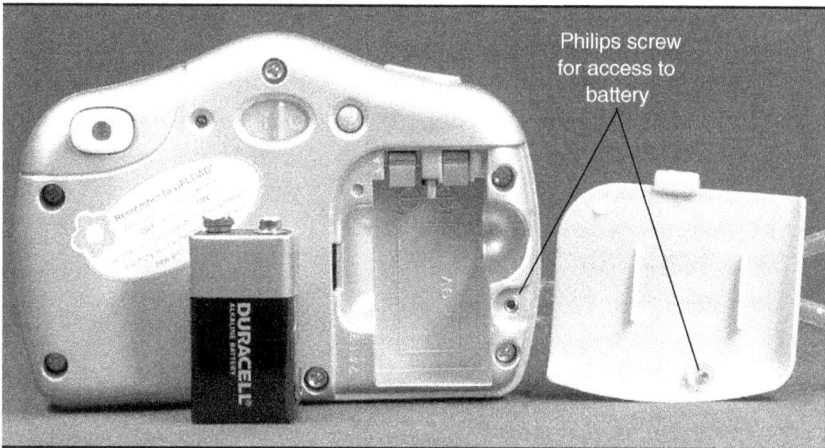

Figure 5.15 Batteries sold separately.

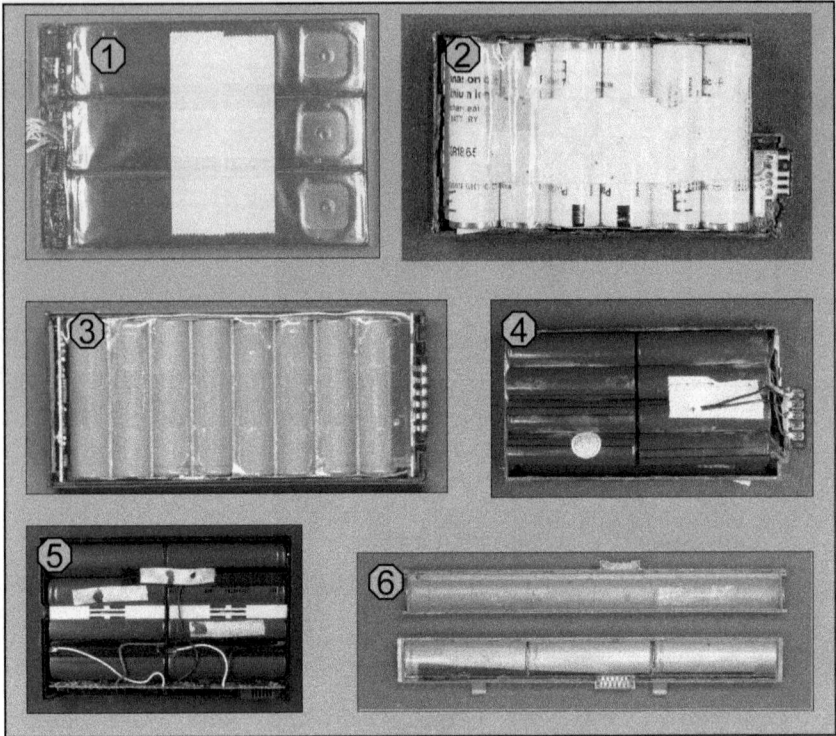

Figure 5.16   Internal cell configuration of notebook battery packs.

than lithium-ion technology, the higher cost and failure to achieve significantly higher energy densities in practice have slowed the adoption of lithium polymer. The lithium polymer cells have a consistently higher energy density than Ni-MH (battery number 4). Lithium-ion is the technology of choice for current notebook computer applications.

TABLE 5.2   Notebook Battery Metric Comparisons

No.	Technology	Voltage (V)	Capacity (Ah)	Mass (g)	Energy density (Wh/kg)
1	Li-polymer	11.1	1.7	194	97
2	Li-ion	10.8	3.2	314	110
3	Li-ion	14.4	2.7	408	95
4	Ni-MH	9.6	3.5	488	69
5	Li-ion	14.4	2.7	402	97
6	Li-ion	11.1	1.55	159	108

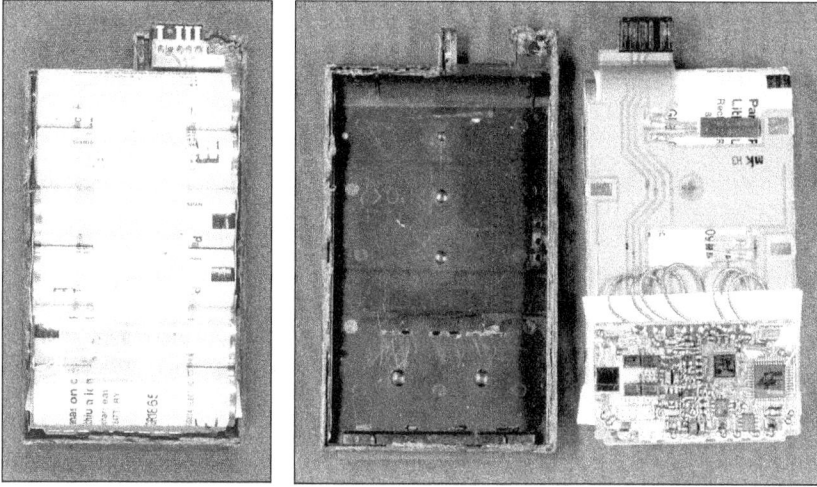

The 10.8 V Li-ion battery pack is rated at 3200 mAh(34.56 Wh) and weighs 313.7 g. The six 3.6 V cells (Panasonic CGR18650) are connected in a series-parallel combination.

Figure 5.17 Notebook battery pack and internal electronics.

Battery number 2 is pictured again in Fig. 5.17. Here we also see the electronics that are packaged inside the battery pack. These electronics enable features such as charge capacity detection, rapid recharging, and safe recharging.

Power management electronics can pose a significant obstacle to successfully achieving the product concept. Recharging circuitry and docking stations are often designed as external accessories to the core device. If these accessories must be carried around by the user, however, then they still contribute to the encumbrance associated with the use of the product. (Likewise, the need to carry additional batteries can be a drawback if constant battery swapping is required.) For a product concept that relies on an extremely small form-factor, it may be equally important to design a very compact charger and cradle.

Power management circuitry internal to the product is usually required to adjust the voltage levels of the source battery to those required to operate various subsystems within the product. Small motors, displays, integrated circuits, and display backlights have very diverse voltage and current requirements. In addition, small button-cell batteries may be required to maintain the contents of volatile memory components. Minimizing the number of disparate power requirements of the product early in the design process will result in reduced cost, board real estate, and complexity of the power management circuitry. Even an efficient power management module can be a significant source of heat and power consumption in products that have relatively high power consumption, such as notebook computers.

Battery life requirements do not increase in a smooth continuum. Understanding this is key to setting the correct target for a product battery life. Battery life requirements tend to be centered on cyclical break points that correspond to the individual's daily routine.

For example, a cellular phone that has a battery life of just 1 h, as many early cell phones did, requires the user to keep a recharger handy and to charge the phone at several points throughout the day when not in use. To keep from forgetting to recharge the phone, the user might have three chargers—one located at home, one in the car, and one to keep in the office or carry on business trips. The process of recharging can thus be integrated into the daily routine, albeit somewhat inconveniently.

Extending the battery life to 2 h adds some value, but for a heavy phone user, a recharge strategy still needs to be in place. If the battery life is extended to say 10 h, however, a significant change in the charging strategy takes place. The user can keep a single charger and plug the phone into the charger each evening on a routine basis. Without having to be concerned about the charge level during the day, the product has become much more user friendly.

Interestingly, extending the battery life beyond 10 h, to say 14 h, may seem of little benefit to the users of the product, as they will continue to recharge the product on a daily basis. Even extending the battery life to 20 h (two work days) may go unnoticed, since it is easier for many users to plug in the charger every night than to figure out whether the charging is really needed.

If the battery life is increased to a week, however, then the user can get into a routine of plugging in the phone over the weekend, and a convenient routine is established. Thus, achieving cyclical break points in battery life such as a day, a week, a month, a season, or a year is more valuable in the eyes of the user than intermediate battery life improvements.

5.3   High Level Power Analysis

The average sustained power consumption in a portable electronic product is the sum of the sustained power consumption for each of the system components. Consider the following symbols for a product with a number of subsytems:

P_n = Power consumption for subsystem n (watts)

D_n = Duty cycle for subsystem n (ratio of 1 or less)

T = Total number of subsystems

P_A = Average system power consumption = $\sum_{(\text{for } n\,=\,1 \text{ to } T)} P_n D_n$

The duty cycle is the fraction of time that the subsystem is on over an arbitrary time interval. From a thermal design standpoint, it may be necessary to assume a worst-case duty cycle of one for every component when assessing worst-case thermal conditions. From a battery life perspective, it is desirable to minimize the duty cycle of each subsystem based on the specifics of the usage paradigm.

Notebook computer designs require very sophisticated power management in order to achieve a competitive battery life. The power budget for the subsystems in a contemporary notebook product is as follows:

Subsystem	Watts ($P_n D_n$)
Processor	2.0
Chipset and memory	4.0
GFX (integrated in chipset)	0
LCD 14.1" SXGA+ and backlight	6.0
HDD	1.2
DVD	0.5
LAN (wired or wireless)	0.5
Power supply loss	2.0
Fan	0.5
Clock generation	0.7
Rest of platform	1.5
Platform total (P_A)	18.9

The numbers shown above would be the average power consumption over a typical user's session. With a 74.0-Wh eight-cell lithium-ion battery, this notebook would be expected to deliver a 3.9-h (=74.0 Wh/13.0 W) battery life.

Battery life in a notebook computer system is a valuable product differentiator, so notebook manufacturers have adopted very elaborate power management techniques to extend battery life. The first step in power management is to classify the mode in which the system is operating. For example, notebook systems can determine whether they are operating from an external power source or from the battery. If operating only on battery power, the notebook might execute the following actions to conserve power:

1. Dim the display backlight

2. Stop the hard drive from spinning unless needed

3. Reduce the voltage and/or frequency of the CPU (reducing performance)

4. Shut down other parts of the system that are not being utilized

5. Go into hibernation mode if the system has not been active for some period of time

In practice, there are a number of different power management states that the system can enter into, based on user preferences and activity. The Advanced Configuration and Power Interface (ACPI) standard was developed to enable very sophisticated control of the power consumption within a notebook system. This control of power consumption is determined by the operating system, which communicates to the ACPI in the system BIOS.

Within the ACPI parlance there are a number of states including **G** (system states), **D** (device states), and **C** states (CPU states). The progression of states C1, C2, C3,..., for example, denotes a progressive reduction in CPU functionality and subsequent reduction in power consumption.

In the G0 state, the system is considered to be *on*. While in the G0 state, the CPU may operate in a number of states ranging from C0 (full power), C1, C2, C3, or C4. C3 and C4 are entered into when the system is idle. C4 is the lowest power state. Other devices such as disk drives and displays can have power states ranging from D0 (full power) to D2, D3, or D4.

In the G1 state, the system is considered to be *sleeping*. Within G1 are the designations for the sleep states. S0 is considered *full on*. S1, S2, and S3 are "suspend" states. S4 is the "hibernate" state. In hibernation mode, the entire system is stopped and the states of various devices are stored to disk.

Because notebook PCs are so complex, the ACPI standards for power management were developed by the notebook computer industry to make it easier to design systems with a longer battery life, while using components and software from a variety of vendors.

Other types of portable electronic devices will tend to have custom power management schemes. Cellular phones, for example, are even more sophisticated in their power management design. Power management is controlled in a deterministic hardware and software environment that allows power optimization throughout the system at microsecond intervals.

In summary, power sources and power management are among the most critical issues for portable electronic devices. For highly functional products that dissipate significant power, such as notebook computers, or for smaller form-factor imaging and RF products, managing the trade-off between battery life and form-factor is essential to the design of a competitive portable electronic product.

Mechanical Design

The mechanical design of a portable electronic product is an interesting and challenging job for the product designer and for the mechanical engineers that support the product design. From a mechanical perspective portable electronics require the simultaneous consideration of the product housing aesthetic and structural attributes; the system's electromagnetic shielding requirements; the thermal management of the electronic components; the three-dimensional integration of mechanical, electronic, and optical parts; and the manufacturability of the entire product assembly.

6.1 Housings

The housing design of a portable electronic device impacts all aspects of the product. First and foremost, the housing provides mechanical support for the key system components. Internal components such as PCAs, disk drives, batteries, and displays must be mounted to a stable chassis. The chassis must be rigid enough to maintain the relative position of these various components. The housing must protect the components from being damaged by contact with outside objects and from shock due to impact. The housing must also protect the product from undesirable electromagnetic radiation, incidental moisture, and, in some cases, from pressurized immersion in water.

At the same time, the product housing must stay within the size and weight limits specified for the product form-factor and must meet the cost objectives of the overall product concept. In some products the housing must also provide a thermal path for heat to be removed from the system.

While providing all these functions, the housing must also have an aesthetic, or even stylish, appearance and be ergonomically formed to provide a comfortable user experience. Even the surface finish and texture of the housing can have an important effect on the success of the product.

Housing design is generally performed with a CAD system such as ProEngineer. These systems enable a full three-dimensional representation of the product housing, allowing the designer to visualize the appearance of the final product. These systems are also critical in ensuring that internal components fit into the housing in the most efficient manner possible, minimizing the product form-factor.

CAD systems let the designer create a data file from which mechanical tooling such as plastic molds can be made. Stereo lithographic modeling may be used to create prototype parts without committing funds to a production mold. Stereo lithographic modeling uses a laser polymerization process to fabricate complex plastic parts on a low-volume basis. This process is also useful for testing the fit of nonplastic parts that comprise the system.

Acrylonitrile-butadiene-styrene (ABS) plastic is probably the most commonly used material for portable electronic product housings. Other rubbery plastics are often used in handheld devices to cushion impact and create a more "grip-able" feel for the product user. Metal housings are also common and are sometimes used to differentiate high-end products from lower-cost versions. Metal housings have obvious thermal, EMI (electromagnetic interference) shielding, and ruggedness advantages, but carry penalties in weight and cost of fabrication.

Magnesium metal has found its way into a number of product designs ranging from cellular phones to notebook computers. Magnesium has the structural, EMI, and thermal benefits of other metals, but offers an impressive strength-to-weight ratio that the user can readily perceive when holding the product. For these applications the magnesium is actually molded rather than machined. Unfortunately, the magnesium molding process tends to wear out the mold tool rather rapidly, so the fabrication and material costs are both high.

The housing design can have a major impact on the overall product reliability. Portable electronic products tend to get dropped. The housing design determines where the impact force of dropping is transferred within the system. The goal of basic chassis design is to minimize the exposure of functional elements to impact. Portable electronic housings often have internal ribbing to provide the necessary support, but may also allow flexure in strategic locations to absorb shock without transferring the force of impact to the functional elements.

Surface texture and color are also key properties of the product housing design. Painted finishes can be used to create a variety of surfaces ranging from metallic to velvet. Painted finishes have the drawback that they wear off over time, leaving a perception of poor quality with the user. Painted finishes are very common in low-cost products.

A more durable approach is to apply color and texture to the bulk housing material. Colored plastic resins can be molded with a variety of svelte finishes that will give the end user a feeling of quality and aesthetics. Likewise, metal finishes, such as polished chrome or brushed stainless steel, make a product very attractive to the end user. Bulk metals can also be colored or oxidized for a lasting color finish.

Integrating pass-throughs for buttons, I/O connectors, indicator lights, air cooling, latches, hinges, or battery compartment covers can greatly complicate the housing design. These features complicate the mold design and compromise both the structural integrity and the EMI-shielding effectiveness of the system.

Hinges are also an important element in the design of the product housing. Hinges are used in portable electronic products for two main reasons. First, they enable the carrying form-factor and the usage form-factor to change. The flip-phone design, for example, is very compact while being carried, but the phone expands by means of a hinge when in use, to cover the distance from the ear to the mouth. Similarly, note-book computers are closed while being carried, but open up to expose more surfaces for a display and a keyboard.

The latest generation of *transformable* tablet computers uses sophisticated hinge designs to provide a choice of usage form-factors.

The design of the product housing is perhaps the most important aspect in differentiating a portable product from products with similar functionality. The challenge of the product designer is to create a housing that meets thermal, EMI, and structural requirements while providing a highly aesthetic and ergonomic experience for the user.

A few final concepts to keep in mind when designing product housing:

- Contoured edges and rounded corners, if properly shaped, make the appearance of a product more aesthetic and can enhance user comfort.

- Metal housings instill a feeling of quality and have excellent thermal, EMI, and structural benefits. Magnesium is an ideal, but expensive, material for portable electronic products.

- Plastic materials are lighter and cheaper than metal, but have a lower-quality appearance and scratch easily.

- Plastic housings can be made to look like metal by using various finishing techniques such as plating or the application of metallic paint.

- Plastics can be molded with more expensive colored resins and a textured finish or coated similarly, to produce a very high-quality appearance.

- Edges and corners can be trimmed with a rubber molding to create a ruggedized appearance and create a more comfortable, "grip-able" product.

- Never underestimate the extent to which the attributes of the product housing, such as styling and color, will differentiate the product in the eyes of the end user, and thus influence the purchasing decision.

6.2   EMI Shielding

EMI stands for electromagnetic interference. EMI shielding is used to suppress unwanted emissions (usually RF) that are created by the system electronics. Shielding is also used to protect the system circuitry from the effects of external interference. Improper shielding designs are often responsible for delays in product introduction due to an inability to meet government emissions regulations.

To design an effective shielding system for a portable electronic product, the designer must understand the potential sources of emission and their respective frequencies. Generally speaking, as the frequency of signal sources in a product increases, the need to provide shielding also increases. Fortunately, the ability to shield also improves as the frequency increases.

Solid sheet metal or foil is commonly used for shielding and is very effective. Metallic-painted or plated plastics are also frequently used in portable electronic devices because they are lighter and more integrated than metal shields. The disadvantage of integrated plastic shielding is that it is more expensive to design and produce.

Shielding can be positioned very close to the source of emission or it can encase the entire product, depending on the situation of various sources within the product. Highly localized shielding tends to be used when a particular source has the potential to interfere with other circuitry within the system, or when the source is very high frequency and needs to be completely sealed from an EMI perspective.

Quantifying shielding effectiveness is the first step in comprehending shielding design. EMI radiation has both an electrical component and a magnetic component. Figure 6.1 shows the equations describing shielding effectiveness.

EMI emissions decrease in intensity as the distance from the product increases. At a distance $>2\pi/\lambda$ from the source (defined as far field) electric and magnetic field shielding differences may be neglected.

Containment   **Immunity**

E_1, H_1

E_0, H_0

Noise
source

E_0, H_0

Noise
source

E_1, H_1

Shield   Shield

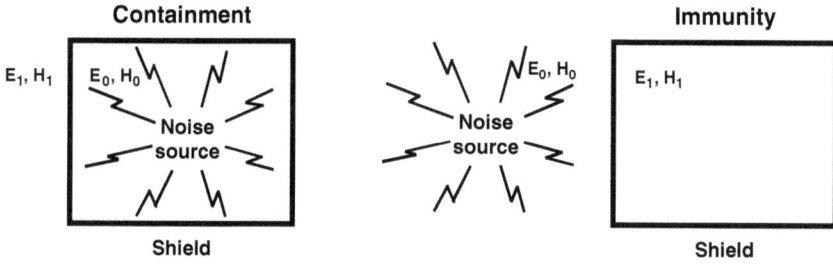

Shielding effectiveness (SE) = 20 log $\dfrac{E_0}{E_1}$ dB (electric fields)

Shielding effectiveness (SE) = 20 log $\dfrac{H_0}{H_1}$ dB (magnetic fields)

Figure 6.1 Shielding effectiveness of electrical and magnetic components of EMI.

(λ is the wavelength of the emission.) Figure 6.2 shows whether the magnetic or electrical field is predominant depending on the wave impedance.

There are several modes of loss that help to attenuate the emissions occurring within a system. Absorption loss is simply a loss of signal energy as it passes through the shielding material. Reflection loss is the redirection of signal energy by the shielding surface interfaces back into the system. The multiple reflection correlation factor deals with the subsequent absorption and reflection of previously reflected signals.

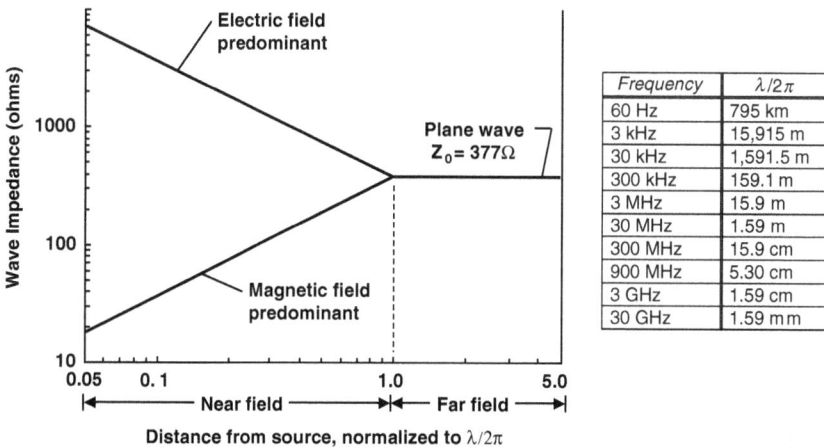

Frequency	$\lambda/2\pi$
60 Hz	795 km
3 kHz	15,915 m
30 kHz	1,591.5 m
300 kHz	159.1 m
3 MHz	15.9 m
30 MHz	1.59 m
300 MHz	15.9 cm
900 MHz	5.30 cm
3 GHz	1.59 cm
30 GHz	1.59 mm

Wave Impedance (ohms)

Electric field predominant

Plane wave
$Z_0 = 377\Omega$

Magnetic field predominant

0.05   0.1   1.0   5.0

|← Near field →|← Far field →|

Distance from source, normalized to $\lambda/2\pi$

Figure 6.2 Near field vs. far field.

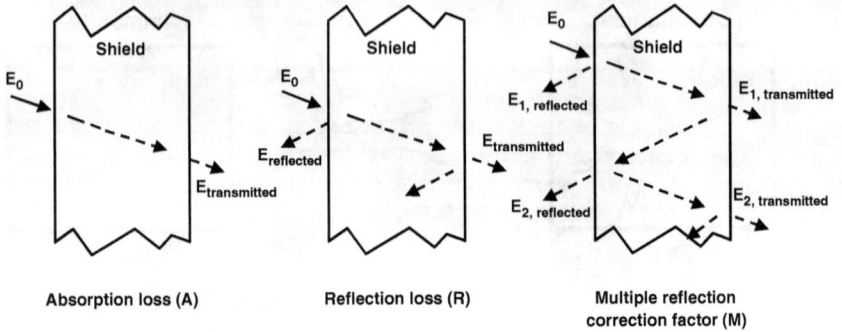

Shielding effectiveness (SE) = A + R + M dB
Electric and magnetic fields, near fields and far fields

Figure 6.3   Shielding loss modes.

Figure 6.3 depicts each of these modes and their cumulative effect on overall shielding effectiveness. Figure 6.4 charts the methodology for characterizing absorption loss and provides some typical properties of common shielding materials. Notice that the dB loss is proportional to the shielding thickness. Figure 6.5 illustrates a reflection loss calculation example for a 0.020-in copper shield.

Figure 6.6 demonstrates the impact of the multiple reflection correction factor, which subtracts from the overall shielding effectiveness. The impact on electrical fields and plane waves is negligible.

Figure 6.7 demonstrates how these different loss modes combine to create a total loss effect in terms of dB versus frequency. At high frequencies, the absorption loss dominates. The magnetic field component is the most difficult to shield. At frequencies above 100 MHz, however, any solid shielding should be sufficient.

A comparison of shielding effectiveness for different metals is shown in Fig. 6.8. Steel is a very effective shielding material at all frequencies and is also very cost effective. Figure 6.9 shows the shielding improvement of metal plating on plastic. This is due to the increased number of available reflection surfaces. While plated metals will cost more, the shielding effectiveness is increased dramatically. Figure 6.10 shows the shielding effectiveness of conductive coated plastics. The shielding performance for these coatings is poor compared to discrete metal shielding, but coated plastics may be sufficient for many applications.

Solid shields will generally provide excellent shielding (>100 dB) in almost all cases except for low-frequency magnetic fields. It is very difficult to produce a truly solid, or sealed, shielding enclosure.

- **Applies to electric fields, magnetic fields, and plane waves** $A = 3.34\,t\sqrt{f\mu\sigma_r} = 8.69\,t/\delta$ dB

- **Thin materials provide effective absorption losses at high frequencies**

- **Skin depth (δ):** $\delta = 1/\sqrt{\pi f\mu\sigma}$ in
  - Distance needed for wave to be attenuated to 37% of its original strength
  - Varies with material and frequency

Frequency	δ, copper	δ, aluminum	δ, steel	δ, mumetal
60 Hz	0.335	0.429	0.034	0.019
100 Hz	0.260	0.333	0.026	0.011
1 kHz	0.082	0.105	0.008	0.003
10 kHz	0.026	0.033	0.003	—
100 kHz	0.008	0.011	0.0008	—
1 MHz	0.003	0.003	0.0003	—
10 MHz	0.0008	0.001	0.0001	—
100 MHz	0.00026	0.0003	0.00008	—
1 GHz	0.00008	0.0001	0.00004	—

Thicknesses in inches

Shielding material that is one skin depth thick ($t/\delta = 1$) provides approximately 9 dB of absorption loss; doubling the thickness doubles the dB loss

f = frequency (Hz), μ= permeability (H/in), σ = conductivity (mho/in), t = thickness (in), μ_r = relative permeability (free space), σ_r = relative conductivity (copper)

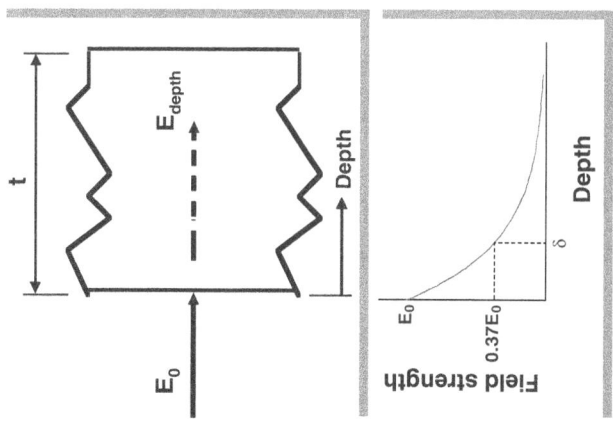

Figure 6.4 Absorbtion loss.

$$R = C + \log\left(\frac{\sigma_r}{\mu_r}\right)\left(\frac{1}{f^n r_r^m}\right) \text{ dB}$$

Field	C	n	m
Electric	322	3	2
Plane wave	168	1	0
Magnetic	14.6	-1	-2

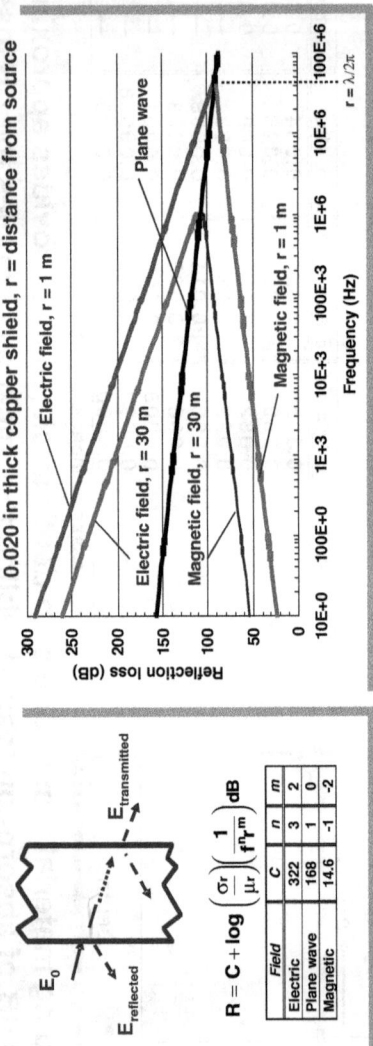

0.020 in thick copper shield, r = distance from source

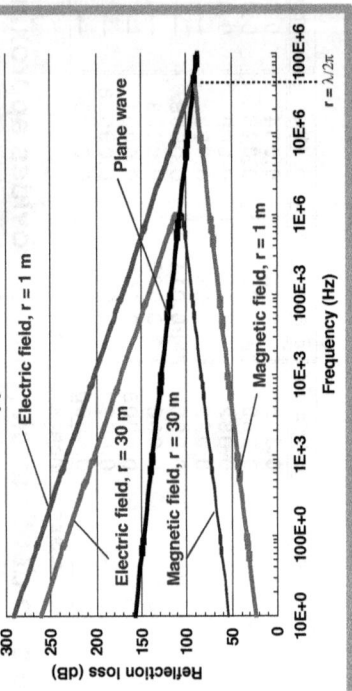

Reflection loss (dB): 0, 50, 100, 150, 200, 250, 300

Frequency (Hz): 10E+0, 100E+0, 1E+3, 10E+3, 100E+3, 1E+6, 10E+6, 100E+6

Electric field, r = 1 m
Plane wave
Electric field, r = 30 m
Magnetic field, r = 30 m
Magnetic field, r = 1 m
$r = \lambda/2\pi$

Electric fields:
- Reflection loss decreases with frequency
- Largest reflection is at first shield boundary
- Thin materials provide good reflection loss

Magnetic fields:
- Reflection loss increases with frequency
- Largest reflection is at second shield boundary
- Multiple reflections must be accounted for

μ_r = relative permeability (free space), σ_r = relative conductivity (copper), f = frequency (Hz), r = distance from source (m)

Figure 6.5 Reflection loss.

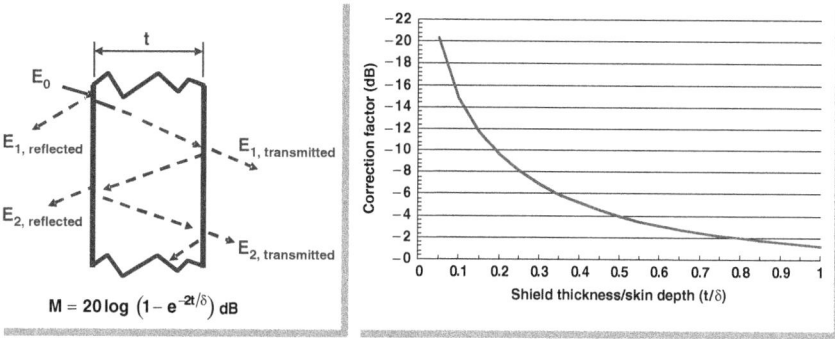

- **Correction factor subtracts from overall shielding effectiveness**
- **Neglect for electric fields and plane waves**
- **Neglect for thick shields ($t \gg \delta$)**

For magnetic fields, thin shields reduce shielding effectiveness

Figure 6.6 Multiple reflection correction factor.

In practice, shielding effectiveness is compromised by the presence of slots, openings, and seams in the shielding assembly, through which emissions are able to leak. The maximum linear dimension of these apertures is the determining factor in the extent of the leakage (not the area). Figure 6.11 shows the shielding effectiveness of different slot lengths. The results in Fig. 6.11 are for single slots only. Multiple openings further reduce the shielding effectiveness, as shown in Fig. 6.12. Openings on different shielding surfaces radiate in different directions, so their impact on shielding effectiveness may not be cumulative.

Figure 6.13 provides a comparative example for several different shielding situations. Considering a point noise source on a PCB, the first situation is to shield the source with a solid-plated steel can. This provides very effective shielding but it adds weight to the product, particularly if there are a number of such sources in the system. In the second instance, the can has holes punched in it. Provided the holes are not too large, a certain level of shielding effectiveness is maintained, while eliminating much of the weight of the can. In the third instance, conductive paint is added to the product housing to boost the overall shielding effectiveness. This might make economic sense if there are a number of sources in the system that require additional EMI attenuation.

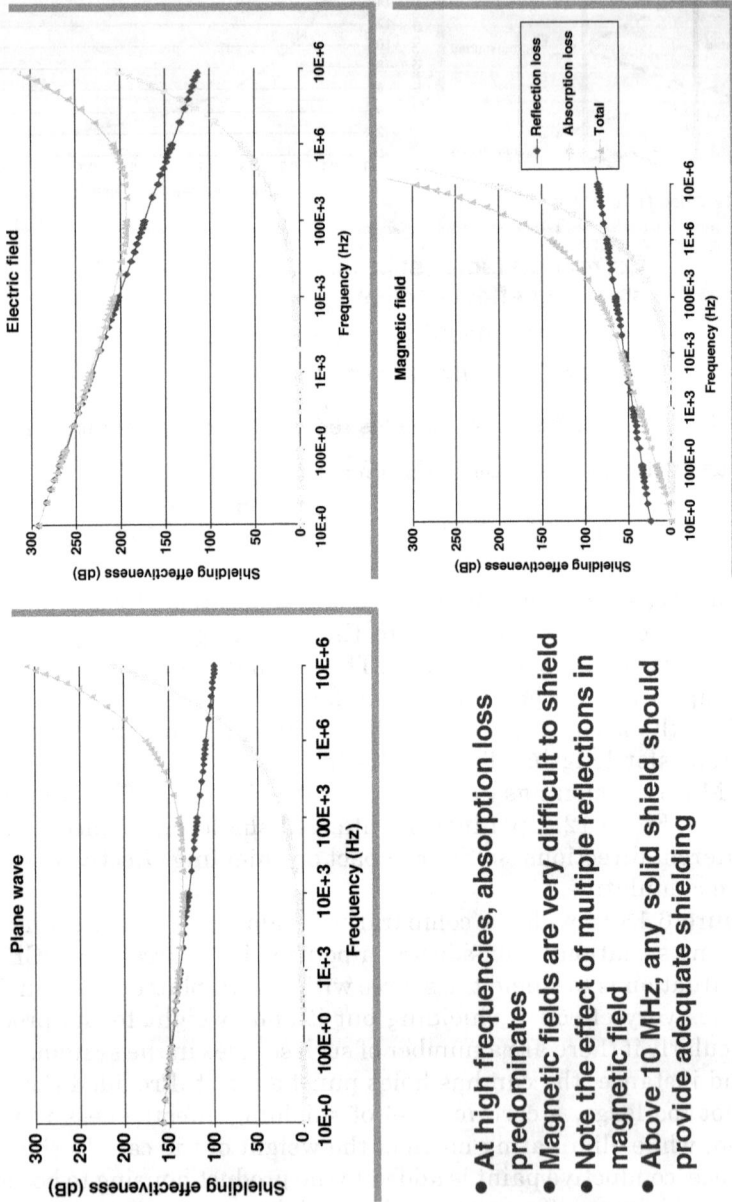

- At high frequencies, absorption loss predominates
- Magnetic fields are very difficult to shield
- Note the effect of multiple reflections in magnetic field
- Above 10 MHz, any solid shield should provide adequate shielding

Figure 6.7 Total shielding effectiveness.

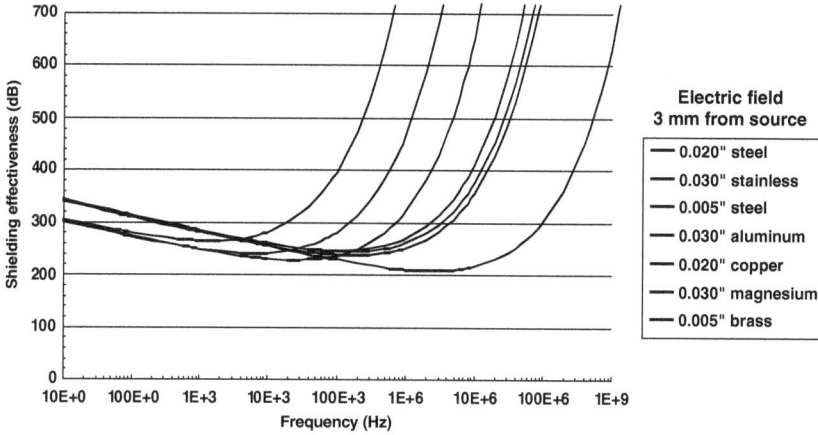

Figure 6.8 Shielding performance of various metals.

6.3 Thermal Management

Thermal management in a portable electronic device is concerned with the general issue of removing heat from the system. Heat is an unavoidable by-product when operating an electronic circuit. For many portable electronic products, the amount of heat generated is so small that it is simply absorbed into the surrounding components and passed into the

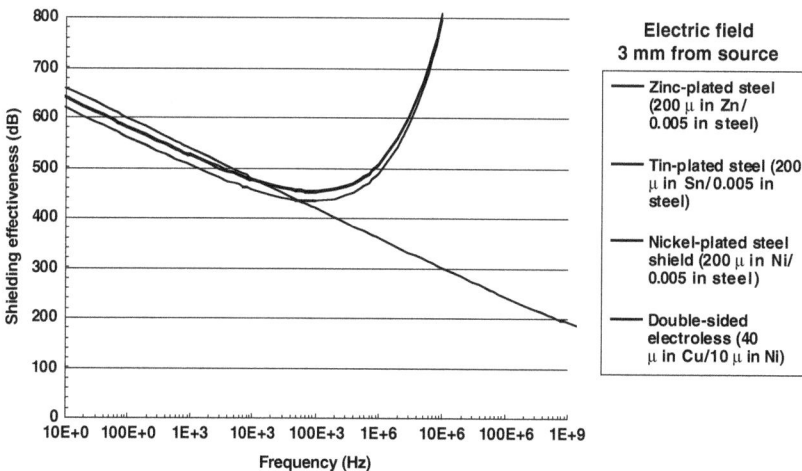

Figure 6.9 Shielding performance of plated metals.

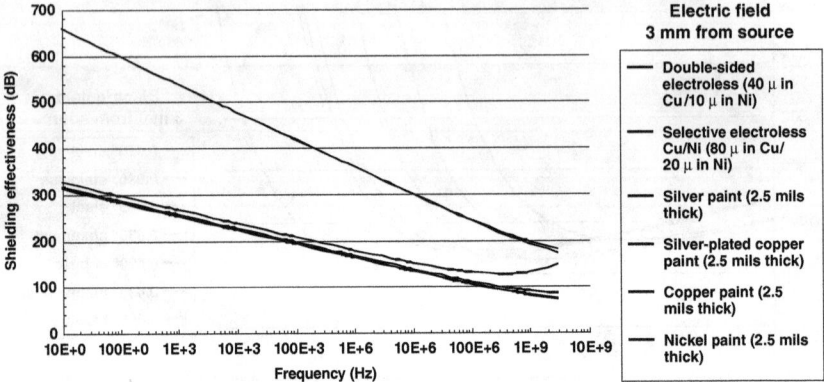

Figure 6.10 Shielding performance for conductive coated plastics.

ambient environment, un-noticed by the user. In other systems, the heat generated is substantial, and it is a challenge to keep the product from becoming uncomfortably hot to the touch. Thermal problems can become so severe that the system electronics stop functioning, or may even ignite.

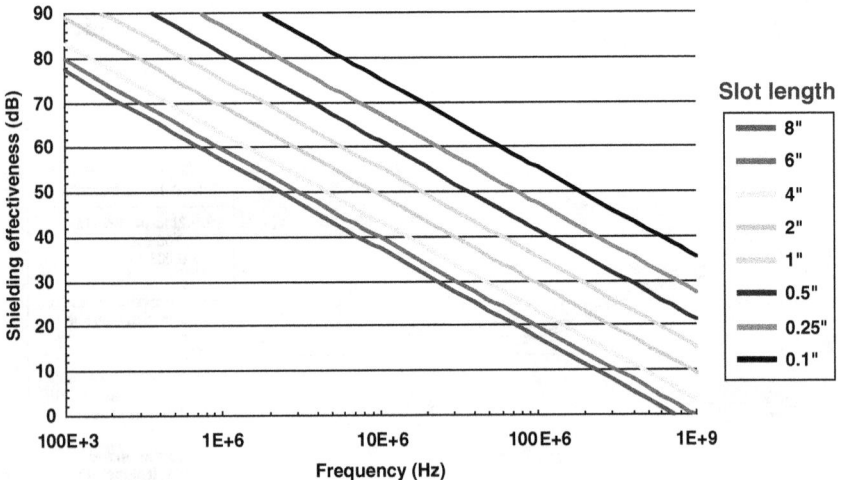

$$SE = 20 \log \left(\frac{\lambda}{2L} \right) dB$$

SE = Shielding Effectiveness (dB)
λ = wavelength
L = maximum dimension of slot

Figure 6.11 Shielding effectiveness of various slot dimensions.

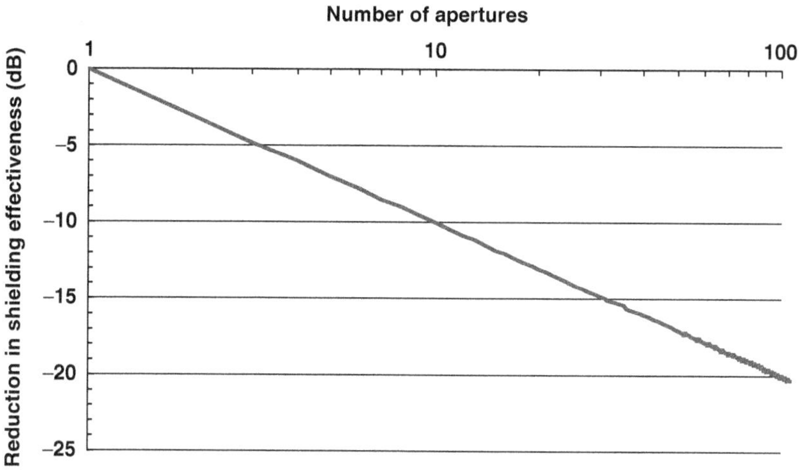

$$SE = -20 \log \sqrt{n}$$

SE = Shielding Effectiveness (dB), n = Number of apertures

Figure 6.12   Impact of multiple openings on shielding effectiveness.

6.3.1   High-level thermal analysis

For many classes of portable products the primary thermal concern may simply be understanding how hot the product will get while in use. A gross assessment of the product skin (outer enclosure) temperature can be a useful starting point in performing a first-order thermal analysis.

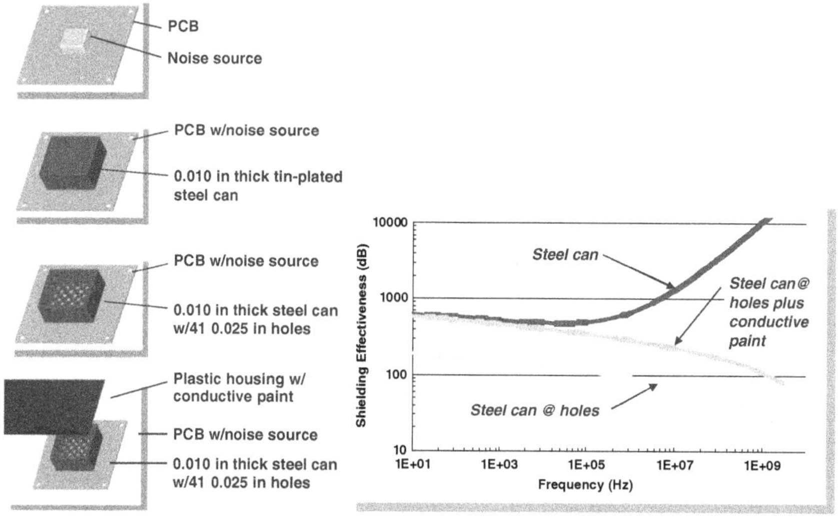

Figure 6.13   Comparison of shielding techniques.

Using the skin temperature as a reference point, we can analyze both internal temperature rises and the total external heat dissipation capacity.

Let us begin by considering how much heat can be expelled from a portable electronic product at a given skin temperature. The basic heat transfer equation is

$$Q = h A \, \Delta T$$

where Q = heat transfer (W)

h = coefficient of heat transfer [W/(cm^2 °C)]

A = surface area of the portable electronic device (cm^2)

ΔT = the temperature difference between the portable electronic device and ambient (°C)

Consider a portable electronic device such as a PDA, which has external dimensions of 8 cm × 12 cm × 1 cm. Suppose an analysis of all system components shows that the system's average power consumption while in use is 1.2 W. The important question to ask is: How hot will the skin temperature get when this product is in operation?

The answer to this question will allow the designer to determine whether the user will find the product to be uncomfortably hot, and to determine what the internal temperature rise of specific components will be if the internal mechanical structures are known.

To determine the skin temperature with accuracy, the value of h should be carefully determined. Mechanical engineering researchers have spent many years determining values for h under various conditions. The actual value for h can be significantly impacted by the following factors:

Ambient airflow patterns and velocity

Ambient humidity

Shape and orientation (vertical vs. horizontal) of product housing

Surface roughness of product housing

Temperature variations across the surface of the product

Physical contact with a table top or user's hand

Developing an accurate model for h is necessary only if the thermal situation for the product is marginal. A first order approximation may be developed by estimating a value for h. A reasonable value for this estimation might be somewhere between 3×10^{-4} and 7×10^{-4} W/(cm^2 °C) for typical indoor conditions in still air. So for this example,

$$Q = 1.2 \text{ W}$$

$$A = (8 \times 12) \times 2 \text{ sides} + (8 \times 1) \times 2 \text{ sides} + (12 \times 1) \times 2 \text{ sides} = 232 \text{ cm}^2$$

$$h = 5 \times 10^{-4} \text{ (estimated)}$$

$$\Delta T = Q/(h/A) = 10.3°C$$

So, in a room at a temperature of 25°C, the product would have a skin temperature of 35.3°C. This would be noticeably warm but certainly not too hot to handle.

Relative to temperature rises internal to the product, this skin temperature may now be used as the reference temperature for heat transfer. By estimating the thermal resistance between a particular electronic component and the skin, the designer may determine the temperature rise needed to dissipate the thermal energy generated by that particular component.

6.3.2 Thermal issues in notebook computers

The most advanced thermal management solutions are deployed in notebook computer systems. Many notebooks require active cooling of the microprocessor chip and very elaborate technologies have been developed to perform this function.

The primary objective in designing a notebook thermal solution is to keep the junction temperature of the microprocessor below the critical temperature that degrades performance or interferes with functionality. All CPU manufacturers specify a maximum junction temperature.

Within the notebook system there are also a number of other power hungry components such as the core logic and graphics chipsets, the hard drive, the power supply, and the display, all of which generate significant amounts of heat that need to be ejected from the system; this complicates the problem of cooling the CPU.

The thermal design engineer must design a system that enables the heat generated by the CPU to pass out of the system and into the ambient environment. The sequence of physical elements through which this thermal energy must pass is called the *thermal path*. The thermal resistance of this path must be reduced to the point where all of the power dissipated by the CPU can pass out of the system at a junction temperature that is at or below the maximum temperature specified by the manufacturer.

In creating this design, the thermal engineer must ensure that all other heat sources in the system are able to eject their thermal loads and that none of the surfaces of the product become uncomfortably hot to the touch. Internal fans are used to resolve both of these issues.

Thermal design power (TDP) is also specified by the CPU manufacturer. TDP is the maximum power level at which the CPU is designed to function and is therefore the upper limit for which the notebook thermal solution must be designed. Due to the conservation of energy, the power consumption and the TDP are the same. The power consumption of a CPU is proportional to the square of the voltage times the frequency, or

$$P = C\,v^2\,f$$

where P = power consumption
$\quad C$ = constant
$\quad v$ = CPU voltage
$\quad f$ = CPU clock frequency

For mobile CPUs this relationship poses an interesting dilemma. CPU performance is one of the major points of product differentiation for PC products. Increasing performance means increasing the clock frequency, which drives up power. Increased power has the negative effect of both reducing battery life and increasing the cost of the thermal solution.

Reducing the maximum operating voltage of a new generation of CPU products, while increasing the frequency, is the usual goal of CPU designers. This enables designers to reduce power while increasing frequency. The perceived market pressure to increase performance, however, is so great that the increases in frequency outweigh the reductions in voltage, and the CPU power consumption has gradually drifted upward. The result has been ever greater TDP values for notebook computer systems.

CPU manufacturers have attempted to deal with this problem in two ways. First, they have enabled notebook CPU chips with *voltage throttling* and *clock throttling*. These features enable the product manufacturer to reduce the actual power dissipation (and performance) if the system becomes too hot. Second, new chip architectures for mobile computing are taking advantage of features such as super scaler pipelining to increase performance without having to increase the clock frequency. Both of these developments can be exploited by the product designer to improve thermal management and battery life.

Regardless of these CPU developments, however, the notebook product designer is still faced with challenging TDP values in most cases. Thermal management in these products is further complicated by the desire to reduce the system form-factor. As the notebook product becomes thinner, there is less space available to create a low-resistance thermal path. As the overall surface area decreases, the average skin temperature must rise to dissipate a given amount of heat.

To understand the dynamics of heat removal in a notebook system consider the following heat transport formula.

$$P = (T_2 - T_1)/\theta_{2-1}$$

where P = power (W) in the form of thermal energy transferred from point p_2 to point p_1

T_2 = temperature (°C) at p_2
T_1 = temperature (°C) at p_1
θ_{2-1} = the thermal resistance between p_2 and p_1 (°C/W)

From these equations it can be seen that as long as a positive temperature difference exists between T_2 and T_1, any amount of heat can be transferred between points p_2 and p_1, given a small enough value of θ_{2-1}.

When conducting the thermal analysis of a notebook computer (or any portable product) we can write this equation as follows:

$$\text{TDP} = (T_{\text{die,max}} - T_{\text{external}})/\theta_{\text{die-external}}$$

$\theta_{\text{die-external}}$ is the thermal resistance of the thermal path from the surface of the CPU chip to the environment outside the product enclosure. $T_{\text{die,max}}$ is the maximum die temperature specified by the CPU manufacturer. T_{external} is the temperature of the external environment (room temperature). The designer's job, then, is to design a thermal path with a low enough $\theta_{\text{die-external}}$ to allow the full TDP to be dissipated when the chip temperature is no higher than $T_{\text{die,max}}$ and the external temperature does not exceed T_{external}.

There are a number of elements that may be a part of the thermal path with which the designer must contend, including a device package, a thermal transfer device such as a heat sink or heat pipe, and each of the interfaces between these elements, including the interface from the thermal transfer device to the ambient environment.

Most CPU die packages are constructed to facilitate low-thermal resistance, and metal or ceramic materials are often used. For aggressive thermal designs, a bare-die construction may be used to eliminate the thermal resistance of the package. A bare-die package leaves the back side of the CPU chip exposed so that it can be directly mounted to a thermal transfer device.

Attachment from the thermal transfer device to the CPU package, or bare die, is one of the most critical aspects of the thermal path. The goal is to maximize the contact surface area between these two elements. Thermal grease or compliant thermal films are employed to compensate for nonplanarity in the assembly. These are highly specialized materials that have very low thermal resistivity while maintaining the necessary mechanical compliancy.

The heat transfer element is usually a metal heat sink. The heat sink has fins, which lower the thermal resistance by maximizing the surface area available to transfer heat into the ambient environment. Since the heat sink is mounted inside the product enclosure, the heat transfer efficiency is reduced almost immediately as the air next to the heat sink rises in temperature. A small fan is usually employed to move cooler air from outside the enclosure past the heat sink to increase the rate of heat transfer.

In the most aggressive designs, more elaborate heat-transfer elements are used. A heat pipe is a device that lowers the effective resistance in the thermal path and enables the thermal path to be considerably longer. The heat pipe contains a working fluid (such as water or alcohol), which is sealed under low pressure. At low pressure, the fluid is able to transition to a vapor at relatively low temperatures. The heat from the CPU causes a phase change which absorbs the thermal energy with very good efficiency. As the vapor is transported away from the heat source by thermal microcurrents to other parts of the heat pipe, it condenses and releases its energy. An internal wick is employed to transport the fluid back to the CPU region.

Computation fluid dynamics (CFD) methods of analysis are integrated into thermal design tools and can be used to effectively model system temperature rises within 10 percent of actual values. The output from a CFD-design tool is shown in Fig. 6.14.

Figure 6.15 shows the construction of an integrated thermal module from a very compact notebook computer design. The fan unit, airflow containment, heat exchangers, and thermal interface material are all integrated into a single assembly that can be easily manufactured into the final product.

6.4 Mechanical Integration

The mechanical design of a portable electronic product includes not only the design of the custom mechanical elements within the system but also the physical integration of all the system components, including the electronic assemblies and various off-the-shelf modules such as displays and batteries. 3D CAD tools are used to manage this integration process. In some portable electronic products, such as digital pagers, the mechanical design is very straightforward, but in other products, such as electronically driven watches, the mechanical design can be incredibly complex.

6.4.1 Wristwatch example

Wristwatch products offer a prime example of the state of the art in mechanical design integration for portable electronic products. Electronics technology is used to provide extended functionality to

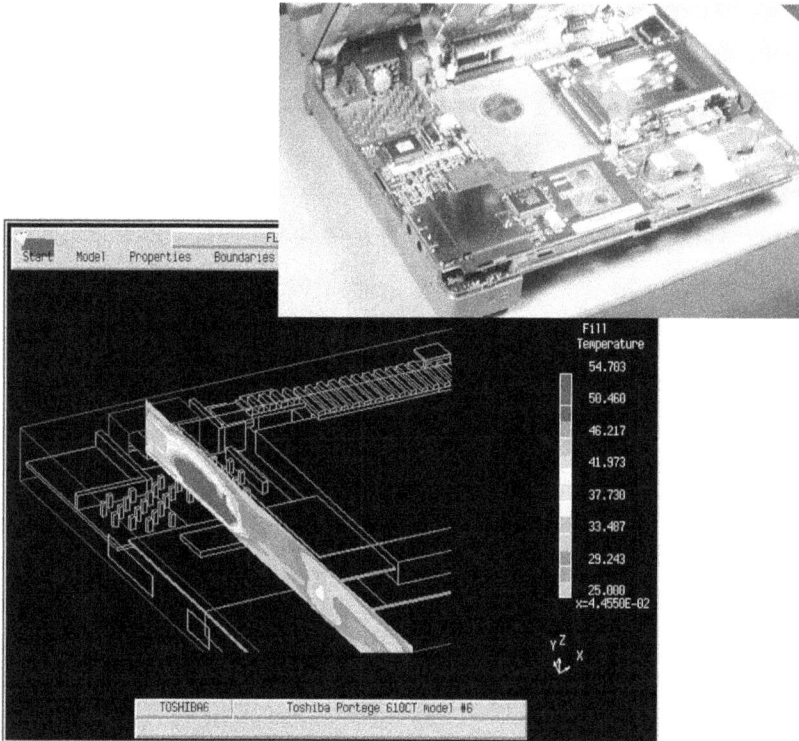

Figure 6.14 CFD thermal design tool.

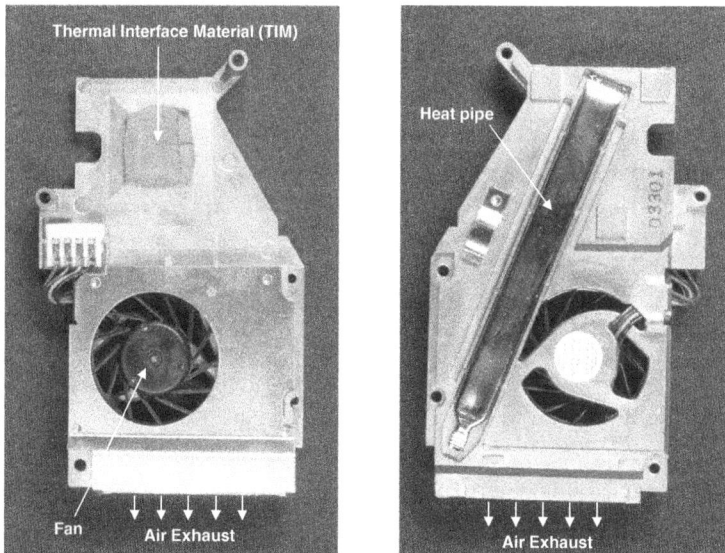

Figure 6.15 Integrated thermal solution.

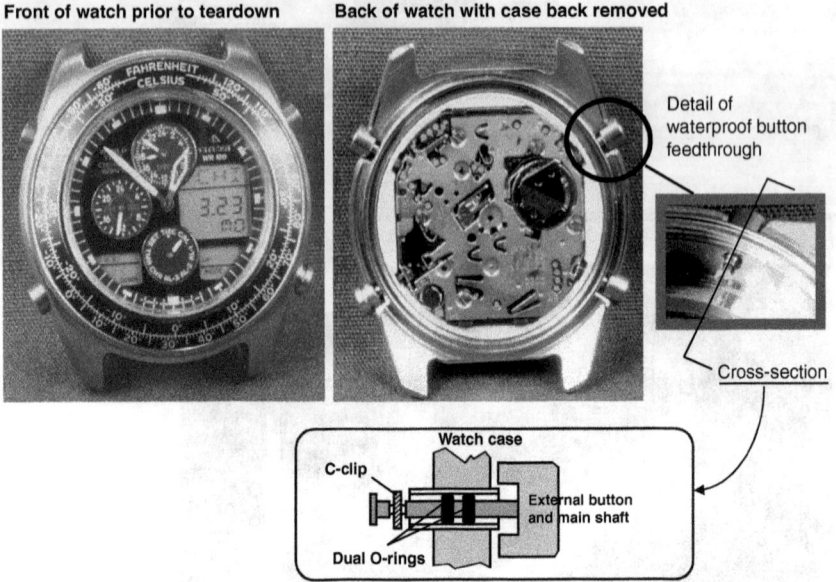

Front of watch prior to teardown   Back of watch with case back removed

Detail of waterproof button feedthrough

Cross-section

Watch case

C-clip

External button and main shaft

Dual O-rings

Figure 6.16   Citizen electromechanical watch.

both digital watches and watches that have mechanical movement. Figure 6.16 shows an example of a watch product that combines highly miniaturized mechanical and electronic packaging into the system design. This system uses a double gasket assembly to provide moisture resistance for the control button feedthroughs. This watch has the following features:

- Quartz digital-analog watch/chronograph with three alarms, timer, world time zone capability
- Analog time of day, UTC, mil time digital chrono, alarms, world time zone, calendar movements

The various parts of this watch were disassembled, as shown in Fig. 6.17. It is interesting to note the number and variety of parts in this system listed below:

- Single PCB, one IC, three 0805 discretes, three 0603 discretes, one SOT, one crystal, one coil, COB assembly
- LCD display with two zebra strip elastomer connectors
- Four metal microgears (three are compound assemblies with multiple gears and shafts)

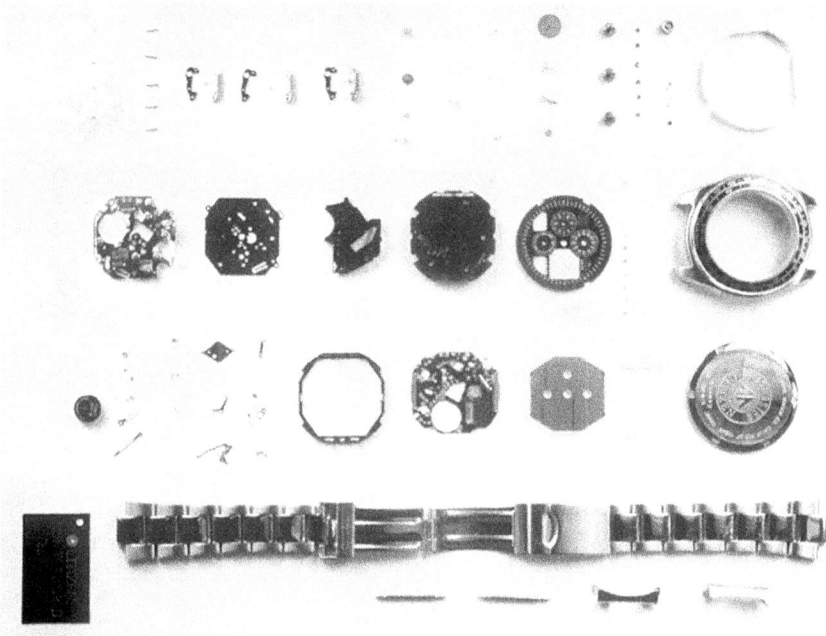

Figure 6.17 Parts of citizen watch.

- Eleven molded plastic microgears
- Nineteen molded plastic parts (one with metal inserts)
- Nine rotary-machined parts (three are compound rotor assemblies with magnets, gear, and shaft)
- Fifteen stamped metal parts
- Eleven etched metal parts (eight post-etch formed)
- Eight screws (two sizes)
- Eight seal gaskets

The custom LCD configuration is also worth noting and is shown in Fig. 6.18. Digital information is displayed on this LCD through a window in the analog faceplate. Mechanical feedthroughs for the analog movement are, in turn, provided for by holes in the glass of the custom LCD module.

Figure 6.19 shows how tightly the electrical and mechanical designs have been integrated into a single subassembly. Notice how both the electrical and mechanical systems have been carefully constructed around the circular battery compartment to achieve an efficient use of space.

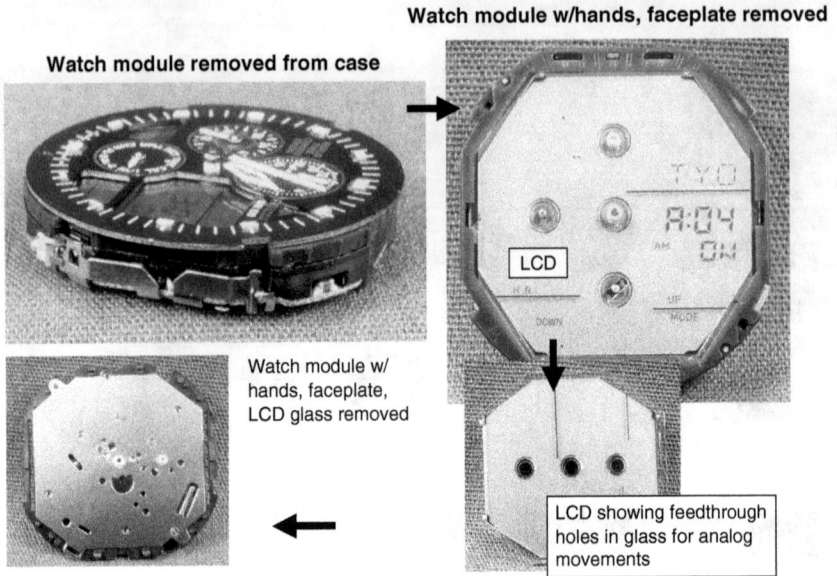

Watch module w/hands, faceplate removed

Watch module removed from case

Watch module w/ hands, faceplate, LCD glass removed

LCD

LCD showing feedthrough holes in glass for analog movements

Figure 6.18 Custom Watch LCD construction.

Figure 6.20 shows the electronic assembly detached from the system. Chip-on-board (COB) technology is used to minimize the circuit board area required to attach the integrated circuit. A plastic dam ring was placed around the IC chip prior to encapsulation on the circuit board. This dam ring serves as a retaining wall to limit the

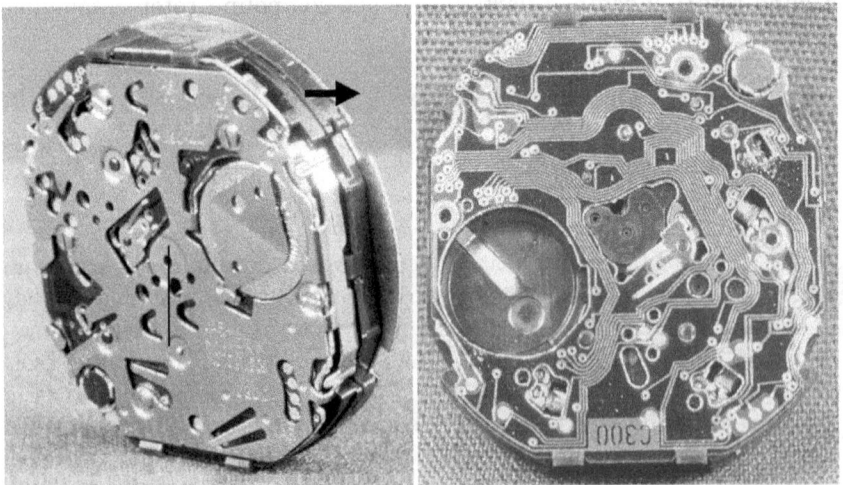

Figure 6.19 Integrated electromechanical watch module.

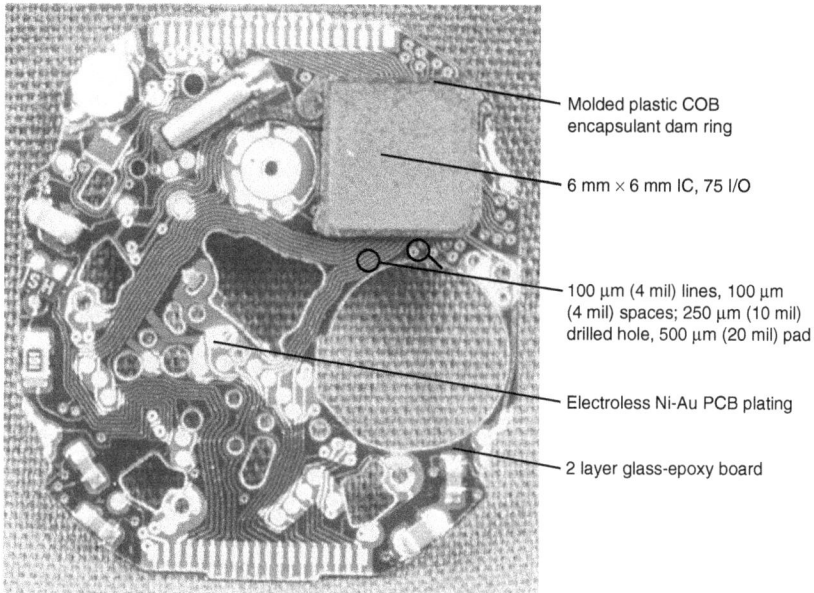

Molded plastic COB encapsulant dam ring

6 mm × 6 mm IC, 75 I/O

100 μm (4 mil) lines, 100 μm (4 mil) spaces; 250 μm (10 mil) drilled hole, 500 μm (20 mil) pad

Electroless Ni-Au PCB plating

2 layer glass-epoxy board

Figure 6.20 Electronic watch module.

lateral flow of the liquid encapsulant until it has undergone the curing process. A comparable packaged integrated circuit with 75 I/O would require at least four times the circuit board area to be attached to the system. By selecting 100-μm lines and spaces for the substrate design rules, the designer was able to route the design in the required area using a thin, two-layer, glass epoxy printed circuit board solution.

The mechanical portion of the watch module (Fig. 6.21) shows the miniature chassis onto which mechanical subassemblies and the circuit board are mounted. A zebra strip, z-axis connector is supported by this chassis to carry signals from the PCA through the mechanical assembly to the LCD mounted on the opposite side.

In Fig. 6.22 a mechanical subassembly has been detached from the integrated module. This subassembly performs the function of translating the motion from three micromotors to the analog dials on the watch face. When these gears are removed, we can see the mechanical substrate (Fig. 6.23) onto which the gears and motors are mounted. This ingenious design approach utilizes a single precision-injection-molded part to provide alignment and support of the mechanical components in the system. Some examples of these precision components are shown in Fig. 6.24.

Electronics side connection

Zebra-strip elastomer connectors for LCD interface to electronics
(Connectors pulled partially from feedthroughs for illustration)

LCD side connection

Lower half of analog movement molded plastic housing

Plastic retaining frame (provides force for PCB to elastomer connection)

Figure 6.21   Mechanical watch module.

This watch product, and other examples, infers that a number of capabilities are essential to producing high-function wristwatch products, including the following:

- Precision metal stamping (FP lead frames or better)
- Precision metal etching

Upper half of analog movement molded plastic housing

Analog movement motors

Analog movement gear cages

Front of watch module after faceplate, LCD, and coverplate removal

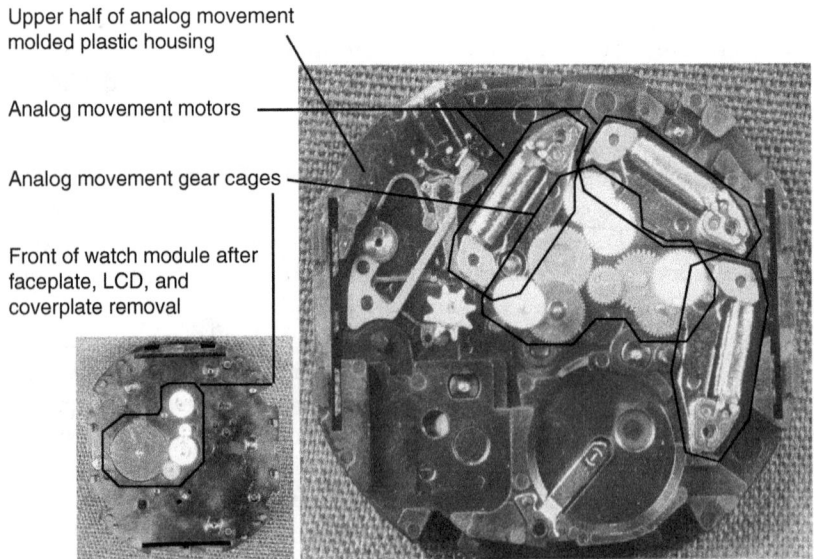

Figure 6.22   Analog movement gear train in mechanical watch module.

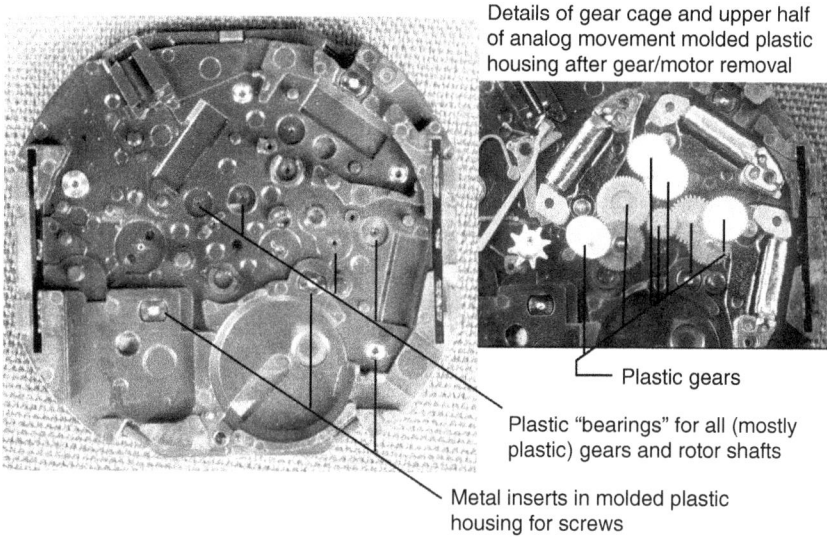

Details of gear cage and upper half of analog movement molded plastic housing after gear/motor removal

Plastic gears

Plastic "bearings" for all (mostly plastic) gears and rotor shafts

Metal inserts in molded plastic housing for screws

Figure 6.23 Mechanical substrate.

0.5 mm (20 mil) diameter screw

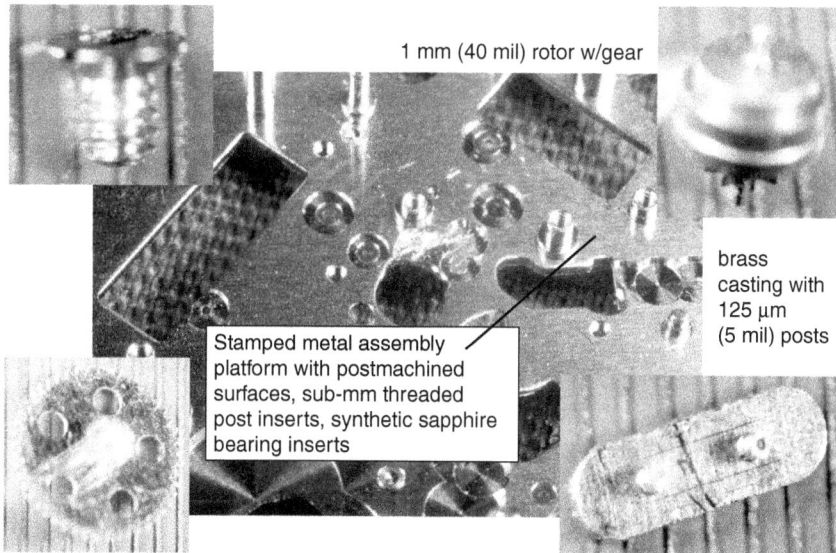

1 mm (40 mil) rotor w/gear

brass casting with 125 μm (5 mil) posts

Stamped metal assembly platform with postmachined surfaces, sub-mm threaded post inserts, synthetic sapphire bearing inserts

Compound gear with 50 μm (2 mil) teeth

Figure 6.24 Precision miniature mechanical components.

- Precision metal milling/machining (including bearings, threads, insets)
- Plastic molding (down to 125-μm features, low-friction surfaces)
- Metal forming/bending
- Rotary/screw machining (down to 125-μm features)
- Coil winding (~25-μm wire)
- Precision plating (including bondable gold)
- Wire bond/COB assembly (up to 104 I/O, three chips)
- Flip-chip assembly to laminate (including PCB, chip bumping, underfill)
- Fine-line PCB and complex depanelization/route
- Polymer thick film on Mylar
- Screen printing
- 3D CAD/mechanical design
- Custom IC design and manufacture
- Functional test and closed-loop trimming
- Custom LCD design and manufacturing (with internal feedthroughs)
- IR optics
- Software/firmware
- Sealing and waterproofing mechanisms
- Glass machining
- Flexible manufacturing/changeover
- High-speed microminiature part placement and assembly

Stacking electronic assemblies. Effectively stacking electronic assemblies is an important aspect of achieving the desired product form-factor. The *stack factor* (Fig. 6.25) is a gross metric for comparing the effectiveness of the mechanical integration of various electronic assemblies. Changing stack factor allows the designer to make trade-offs between electronic packaging technologies. Examining stack factor dynamics enables the designer to compare the benefit of high-density packaging technologies versus low-density packaging technologies. While the electronic packaging metrics discussed previously focus on planar density, the stack factor reveals the benefit of a particular technology selection's ability to reduce assembly thickness.

6.5 DFMA Analysis

Design for manufacturing analysis (DFMA) is a discipline developed to improve manufacturing efficiency and reduce the overall cost of a product. The basic tenets of DFMA include reducing parts count and

Design Characteristics

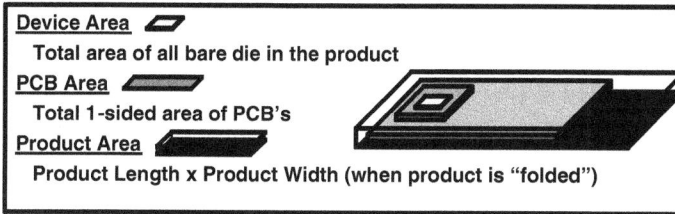

Device Area ▱
  Total area of all bare die in the product
PCB Area ▱
  Total 1-sided area of PCB's
Product Area ▱
  Product Length x Product Width (when product is "folded")

Electronic Packaging Metrics

$$\text{PCB Tiling Density} = \frac{\text{Device Area}}{\text{PCB Area}}$$

$$\text{Product Tiling Density} = \frac{\text{Device Area}}{\text{Product Area}}$$

$$\text{Stack Factor} = \frac{\text{PCB Area}}{\text{Product Area}}$$

Figure 6.25 Stack factor.

optimizing the part's design to be compatible with the manufacturing process.

Boothroyd Dewherst, Inc. (www.dfma.com) has developed a methodology for analyzing the manufacturability of a product. Their methodology lends itself particularly well to portable electronic products. Portelligent, Inc. conducts DFMA of portable electronic devices (using the Boothroyd Dewherst methodology) as a part of the information services that they provide to the electronic industry. Below is Portelligent's explanation of how they effectively conduct their DFMA activities. It is included here because it represents a definitive example of a disciplined approach to the use of DFMA methods.

DFMA Process—Overview and Discussion

Costs of injection-molded plastics and sheet metal parts are modeled using the Boothroyd Dewhurst (BD) design-for-manufacturing (DFM) software. Each mechanical element is entered into the BD DFM software package, providing measured/observed attributes of geometry, weight, size, thickness, material, and secondary operations from which the BD DFM software derives an estimated cost of manufacture.

The cost of final system assembly and integration is projected from Boothroyd Dewhurst design-for-assembly (DFA) tools. In effect, we rebuild the torn-down product, accounting for each of the placements, orientations, labelings, fastenings, etc. necessary to integrate all subassemblies into the complete product. Each step in final assembly corresponds to a specific estimate of time to perform the operation. Integrating each time-step, a total time for assembly is achieved. We then assume an hourly rate for the labor needed to perform final assembly operations and tabulate total final assembly cost.

As part of the DFA analysis, a DFA index is calculated. The DFA index is a benchmark that accounts for total part count and efficiency of part assembly, and creates a ratio between the optimal theoretical assembly time achievable with a given number of parts and the actual estimated assembly time.

The formula for DFA Index is as follows:

DFA Index = $N_{min} \times 2.93/t_{ma}$

N_{min} = theoretical minimum number of parts

t_{ma} = total assembly time as calculated from the reconstructive process and BD software estimates.

Assumptions:

- Single-station assembly
- All tools and parts are within easy reach
- Labor costs equal among directly-compared products

N_{min} describes (somewhat subjectively) the minimum number of parts that could be theoretically used to implement the product's mechanical aspects.

The 2.93 in the above formula reflects BD's empirically-derived "ideal" for per-part assembly times. For example, a cell phone could theoretically require only an upper case, lower case, display, battery, antenna, circuit board assembly, earphone, microphone, vibrator, and keypad for a total of 10 parts. At 2.93 seconds per part assembly, the "best" possible time for final assembly of our sample product would be (10 parts × 2.93 sec/part) or about 30 seconds. Of course phones frequently have additional parts to satisfy styling issues, shock/vibration resistance, interface characteristics, etc. Additionally our experience is that assembly times of 6-14 seconds per part typify many of the products we examine. So, if in our example the phone actually had 30 parts, with say 10 sec/part of estimated assembly time, a total assembly time of (30 parts × 10 sec/part) or 300 seconds would be needed. The resulting DFA index in this case would be 30/300, or 10%. In essence the higher the DFA index, the closer the product has come to achieving a "perfect" design for final assembly.

While the DFA is a subjective process where results might vary from user to user, when the analysis is performed in a consistent manner over multiple products, relative comparisons become valid.

In summary, this chapter has provided an overview of the mechanical issues that are relevant to portable electronic design. The product designer must take care to optimize the trade-offs between product form-factor, aesthetics, structural requirements, EMI requirements, thermal issues, and manufacturability. Each of these areas must be carefully considered to ensure that the product creates a good experience for the end user.

Software and Communications

We will discuss software and communications issues from an architectural point of view. While some simple portable electronic devices can be created without the need for software or external communications, higher complexity products will generally contain either embedded software or will be loaded with general purpose software applications and will have provisions for external communications protocols.

7.1 Software Hierarchy

Portable electronic devices that operate with a minimum of buttons to achieve a focused functionality are highly desirable, as discussed at several points in this text. In very simple products that perform a very limited functionality, such as a laser pointer, there is no need for a user interface beyond a single button, and there is no logic control in the system beyond. In such cases, the system does not have a software component.

In most products, however, some type of microcontroller is needed to provide a more complex set of logical operations. Thus, *embedded software* is needed to handle the complexity of user interaction. In these cases, the user will still interact largely with buttons or a simple character display and will not be aware of the role that software plays in the product functionality.

In the most sophisticated products, such as PDAs and notebook computers, the user will interact directly with application software and even the operating system. In these cases, the user considers the system software and the user interface to be identical.

There are a wide variety of conceptual representations to show how software components are organized within a portable electronic system. Figure 7.1 is a generalized hierarchy that we will use to illustrate the

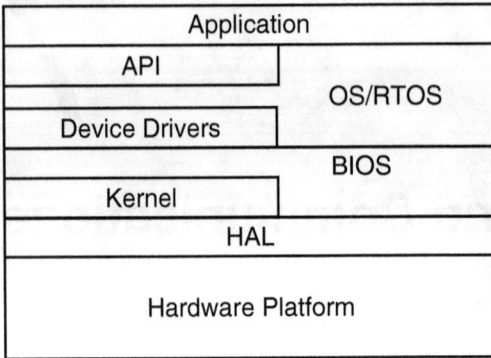

Figure 7.1 System software hierarchy.

basic concepts. This hierarchy is rather subjective, based on the particular processing architecture being utilized as well as the specific operating system and architectural philosophy being implemented.

7.1.1 Hardware platform

The hardware platform refers to the particular variety of microprocessor or microcontroller that is being used in the system. It may be further delineated by the selection of specialized logic that supports the CPU, or the variety of system-on-chip (SOC) solution.

7.1.2 Hardware abstraction layer (HAL)

The HAL is a layer of software that is used to create an abstract interface to the underlying hardware platform. A HAL may not be necessary for all portable electronic products, but it may be essential if the product designer wishes to leverage preexisting software elements at higher levels of the software hierarchy.

7.1.3 Kernel

The kernel coordinates the various processes within the system by providing scheduling and synchronization. This coordination consists of managing the execution of different threads by enabling their execution when the appropriate hardware resources are available.

7.1.4 BIOS

The basic input output system (BIOS) is used to enable basic communications between the CPU complex (hardware platform) and peripheral system elements that are required for higher level functionality.

The BIOS may also contain the CPU driver. For example, in a notebook computer, the BIOS must be loaded before the CPU can communicate with the hard drive to load the operating system.

7.1.5 Device drivers

Device drivers enable communication with additional system components not addressed by the BIOS. Device drivers may also enable a more complex feature set for specific devices than can otherwise be enabled through the BIOS.

7.1.6 OS/RTOS

The operating system (OS) or real time operating system (RTOS) provides the next level of abstraction above the device drivers. The OS/RTOS enables high-level system management and more abstract control of system resources. An OS is usually used in a general purpose computing platform such as a notebook PC, whereas a RTOS is associated with an embedded system implementation. Selection of OS/RTOS is a fundamental decision for the product designer, as it will tend to drive the overall system architecture options.

7.1.7 API

The application programming interface (API) is the set of hooks provided by the OS/RTOS that enable a programmer to control the system. While direct control of the system at deeper levels in the software hierarchy is possible, the API is the most convenient interface, provided it enables a sufficient level of control over the hardware resources.

7.1.8 Application

The application is the specific program that drives the user interface and controls the system at a high level to achieve the desired functionality. In fact, the application may be synonymous with the product functionality, as is the case in a notebook computing system. In an embedded system the application is more product specific.

7.2 OSI Network Communications Model

In developing the network communications design for a portable electronic product, the product designer will need to be familiar with the open systems interconnection (OSI) model (Fig. 7.2) for describing the hierarchy of networking functions within a system.

The lowest layer is Layer 1, also known as the physical layer (PHY). The physical layer is the direct interface to the communications hardware,

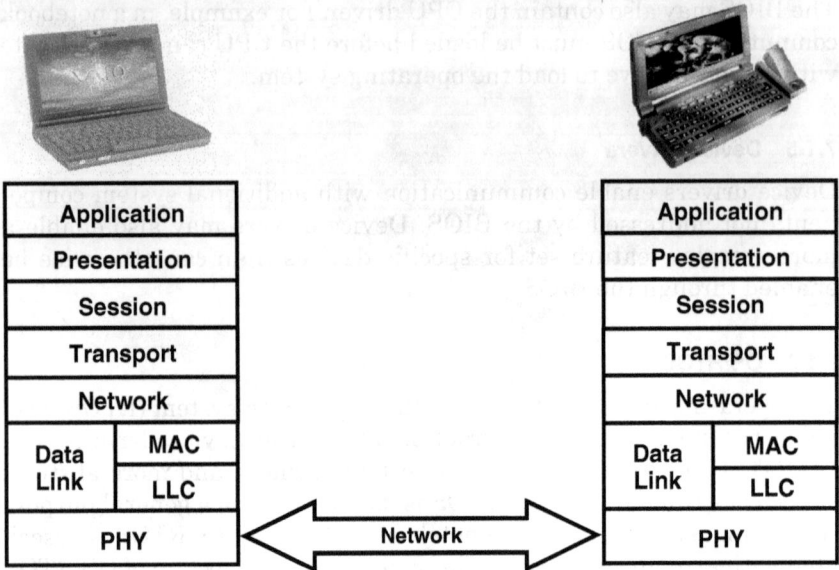

Figure 7.2   OSI network stack.

whether it is a wired electrical connection, a fiber-optical connection, a wireless RF connection, a free-space optical connection, or some other means of spatial information transfer. The physical layer enables all higher levels of the network stack to function without regard for which particular physical transfer medium is being used.

Layer 2 is the data link layer. In this layer, the data are encoded and decoded into data bits that can be effectively processed at higher levels of the stack. Transmission protocols, physical layer errors, frame synchronization, and flow control are handled at this layer. The data link layer is divided into two sublayers. The media access control (MAC) sublayer controls network access and transmit authorization. The logical link control (LLC) sublayer handles flow control, error checking, and frame synchronization.

Layer 3, the network layer, handles routing and switching to create virtual paths in the network (virtual circuits) for transmitting data from node to node. Forwarding, routing, addressing, internetworking, error handling, congestion control, and packet sequencing are all functions of this layer. Internet communications, for example, use internet protocol (IP) as the network layer interface.

Layer 4 is the transport layer, which provides for transparent data transfer between end systems. This layer is responsible for flow control and end-to-end error recovery, guaranteeing complete data transfer.

In Internet communications, transmission control protocol (TCP) resides at the transport layer and establishes the connection between two network hosts through "sockets," which are identified by an IP address and port number.

Layer 5 is the session layer, which establishes, manages, and terminates the connection between applications running on the connected hosts.

Layer 6 is called the presentation layer because it transforms the data to a format that the application can accept. This layer performs formatting and encryption of data so that it can be sent across the network.

Layer 7 is the application layer. The application layer provides services to support a specific application and creates an appropriate level of abstraction of the underlying stack to support efficient applications development.

7.3   Communications and System I/O

Standardized external connectors are important in enabling the market for portable electronic devices. These external connectors are usually incorporated into the product design for the purpose of either connecting to some type of network or a peripheral device or, in some cases, to facilitate the connection of the portable electronic product as a peripheral device to a PC or some other type of host system.

Serial and parallel port connections were the most common such external connectors for many years, enabling notebook and desktop PCs to connect to any printer or other external device. The development of these standard connections was critical to the early growth and development of the PC industry.

The notebook computer market was given a big boost in the early 1990s with the development of the PCMCIA standard. PCMCIA was developed to enable a low-profile peripheral device, such as a modem or solid-state disk drive, to be connected to a notebook computer as an internally mounted subsystem. Since its introduction, the PCMCIA standard has made it possible for a wide variety of peripheral and network products to come to market in a PC card format. Examples include add-on-dial modems, Ethernet cards, CDPD wireless modems, 802.11 wireless modems, and solid-state disk drives. New products that are introduced initially as PCMCIA cards, often become standard notebook PC features and are ultimately integrated onto the notebook PC motherboard, making the original PCMCIA product obsolete.

More recently, the USB connector has been important in creating market opportunities for new products. Most of the PDAs on the market today are able to connect, via a docking station, to a host PC through a USB connector. Many digital still cameras have a USB connection built in and can connect directly to a PC to download digital photographs. Compact solid-state memory (Flash) sticks would not be viable without a high-speed standardized USB connection suitable for communicating with peripheral mass storage devices.

The USB connector is actually an external implementation of a standardized bus, which can also be used to connect notebook PC functions that are mounted internal to the notebook housing. Another such standardized notebook system bus is the PCI bus. External PCI connections are uncommon on portable electronic devices. PCI connectors are frequently used to add modular functionality inside a desktop PC chassis. The miniconnector was developed so that the same type of modular functional additions could be made within a notebook product based on the same silicon implementations used in the standard PCI solution. Mini-PCI connections are becoming the method of choice for adding 802.11b wireless LAN capability into a notebook PC at the factory. This frees up the PCMCIA connectors in a notebook to be used for other functions.

Most notebook computers now come with an RJ-11 connector for wired Ethernet capability. The ability to make an Ethernet connection can be an important market enabler for a portable electronic device, since Ethernet is the most common digital network connection in use today. The very presence of an Ethernet connector on a product strongly implies (but does not guarantee) that the device supports TCP/IP protocols and can communicate with other devices.

Speaker and microphone connections are also standardized on many products, but custom connectors are also occasionally used. The use of custom headphone connections has become more common in recent generations of portable digital music products. Naturally, the prices (and margins) on these headsets are higher than would be the case if all vendors were using the same connector.

Power connectors are also notoriously customized in portable electronic products. Portable electronic devices do have a variety of power schemes, even within the same category of products. Plugging the wrong type of power supply into a product can result in catastrophic failure of the power distribution or recharge circuitry. Users are frustrated, nevertheless, when they have to purchase expensive custom power supplies.

Wireless standards that support external system I/O are becoming very prominent. Wireless technologies offer the opportunity to reduce materials and assembly costs related to physical connectors. They also, in principle, make for a more convenient usage model to the end user, since cables no longer need to be carried around and plugged or unplugged.

The IRDA standard was implemented in many notebook systems in the 1990s. This infrared wireless communications standard for notebook computers and other portable devices was intended to enable devices to connect to each other automatically when they were in physical prox-imity of each other. Thus, an IRDA-enabled printer sitting next to an IRDA-enabled notebook would be connected and able to communicate. Two IRDA-enabled notebooks sitting next to each other would be able to exchange files.

The cost of building IRDA into a product was only a few dollars, and millions of product units were shipped with this standard built in. Still, IRDA was unsuccessful as a standard. IRDA did require line-of-sight between the IR ports on each device, but this was not the reason for fail-ure of the standard. IRDA failed because the system software was not fully developed to support seamless interaction between two IRDA-enabled devices. Using IRDA usually required a rather technically involved configuration process to connect any two particular devices. With the exception of a few enterprise users who synchronized PDAs to PCs, IRDA went largely unused and now is rarely designed into a note-book system.

Bluetooth technology was developed to achieve a similar function to IRDA. Bluetooth is surely an improvement in that it relies on a wire-less RF link rather than an IR link, thus removing the line-of-sight requirement. Bluetooth is positioned as a true personal-area-network technology capable of replacing many of the cables that encumber cur-rent portable electronic systems.

Bluetooth standards are spreading slower than anticipated, however. The bandwidth of the initial Bluetooth standard is too slow to support many of the types of applications that portable electronic products will need, such as streaming video. Bluetooth was initially promoted as an extremely low-cost wireless I/O solution, but this has not materialized fast enough for many of the envisioned applications to be successful. Bluetooth is gaining some traction for applications such as a cellular phone to headset linkage and to connect a keyboard or mouse to a PC. Bluetooth is overkill for some such applications, however. Wireless key-boards can be more cost-effectively produced with a custom solution that does not have many of the unneeded features of Bluetooth. Just as manufacturers are motivated to make custom headset and power adapter plugs, many will be motivated to avoid Bluetooth.

So Bluetooth is the current solution for replacing wired external I/O connections in portable systems with a wireless connection. 802.11 wire-less standards similarly support the replacement of a wired Ethernet connection. 802.11 has been extremely successful as a standard for *wire-less Ethernet* connections, gaining rapid market acceptance in a PCMCIA card implementation that enables the user to upgrade a note-book or palm-top PC with wireless Ethernet connectivity. This is migrat-ing to an internally mounted mini-PCI implementation, as more notebooks start to ship with built-in 802.11 capability.

7.4 Wireless Standards

Wireless communications standards have been essential to portable electronic products since the introduction of the first transistor radios. National regulations relative to the licensing of radio spectrum have a tremendous impact on the viability of the business models that support wireless communications. Since standardization (agreed-upon language) and communication (information transfer) are nearly one and the same in this particular context, it is no wonder that nations and corporations expend a huge amount of energy to influence or control wireless communications standards. Table 7.1 gives a few of the wireless standards that impact portable electronic products.

At the time of this writing a true revolution in wireless standards is taking place, the results of which have yet to be fully realized. Wireless local area network (WLAN) systems based on the IEEE 802.11 standards are being adopted very rapidly. These represent the first widespread deployments of a truly high-speed wireless data network.

802.11 products began to appear in the marketplace in the mid-1990s. PCMCIA cards with 802.11 capability were developed, which enabled wireless functionality to be added to notebook computers cheaply and easily. Since 802.11 is an un-licensed band, the transmit power is limited to 1 W and the maximum practical range of these systems is less than 100 ft in an office or household setting. This means that the user must connect to a local base station that is, in turn, connected to the network. This makes the technology well suited to LAN applications.

The IEEE quickly realized that the bandwidth of these initial systems (<2 Mbps) was too limited for contemporary applications and developed the 802.11a, 802.11b, and 802.11g standards. Of these, the 802.11b standard was developed first and was quickly productized by numerous manufacturers. With the 11-Mbps bandwidth specified by this standard, wireless LANs began popping up in corporate locations and residences alike in large numbers. Often, the corporate installations were not performed by the IT staff but were installed by other office workers as a convenience to their own work environment. Low cost, ease of installation, and inherent compatibility with existing wired Ethernet networks made this grass roots type of deployment possible.

The effective bandwidth of an 802.11b connection is generally about half the rated bandwidth. (A pretty good rule of thumb for many of the current wireless data standards is *Actual ~50% of Rated.*) At about 5.5 Mbps effective bandwidth, 802.11b is well matched to most of the current requirements of enterprise and consumer networking but is showing signs of strain as more users attempt to stream high-resolution video across the wireless link. The 802.11a and 802.11g standards, both with rated bandwidths of 54 Mbps, have been developed to meet the needs of emerging user applications. Additional 802.11 related standards

TABLE 7.1 Wireless Communications Standards

Standard	System	Frequency band	Access method	Duplex method	Modulation format	Data carrier	Rate signal	Audio coding	Channel BW/SP	Max user	Power peak
AMPS	Cellular	800 MHz	FDMA	FDD	FM				30 kHz	600 mW	600 mW
GSM	Cellular	900 MHz	TDMA (8:1)	FDD	GMSK(.3)	270 kbit	13 kbit	RPE-LTP	200 kHz	250 mW	2 W
PCS-1900	PCS-TAG-5	1.9 GHz	TDMA (8:1)	FDD	GMSK(.3)	270 kbit	13 kbit	RPE-LTP	200 kHz	125 mW	1 W
IS-136	PCS-TAG-4	1.9 GHz	TDMA (3:1)	FDD	pi/4 DQPSK	48 kbit	8 kbit	VSELP	30 kHz	200 mW	600 mW
IS-95	PCS-TAG-2	1.9 GHz	CDMA (20:1)	FDD	OQPSK/QPSK	1.23 MChip	8/13 kbit	QCELP	1.25 MHz	200 mW	200 mW
W_CDMA	PCS-TAG-7	1.9 GHz	CDMA (64:1)	FDD	QPSK	4.096 MChip	32 kbit	ADPCM	5 MHz	200 mW	200 mW
IS-661	PCS-TAG-1	1.9 GHz	TDMA (32:1)	TDD	SE-QAM	781/5.0 MChip	8/32 kbit	CELP/ADPCM	5 MHz	9 mW	600 mW
DCTU	PCS-TAG-6	1.9 GHz	TDMA (12:1)	TDD	GMSK(.5)	1.15 Mbit	32 kbit	ADPCM	1.72 MHz	10 mW	250 mW
PACS	PCS-TAG-3	1.9 GHz	TDMA (8:1)	FDD	pi/4 DQPSK	384 kbit	32 kbit	ADPCM	300 kHz	25 mW	200 mW
PHS/PHP	PCS-Japan	1.9 GHz	TDMA (4:1)	TDD	pi/4 DQPSK	384 kbit	32 kbit	ADPCM	300 kHz	10 mW	80 mW
PDC	Cell-Japan	800/900-1,5 GHz	TDMA (3:1)	FDD	pi/4 DQPSK	42 kbit	8 kbit	VSELP	25 kHz	100 mW	300 mW
CT-2	CT-Europe	900 MHz	FDMA	TDD	GMSK(.5)	72 kbit	32 kbit	ADPCM	100 kHz	5 mW	10 mW
15.323	PCS-Voice	1.9 GHz	TDMA	TDD	Any Digital			Any	1.25 MHz	.1/BW	
15.321	PCS-Data	1.9 GHz	CSMA	TDD	Any				<10 MHz	.1/BW	
802.11-FH	WLAN	2.4 GHz	CSMA	TDD	GFSK(.5)	1-2 Mbit	1-2 Mbit		1 MHz		1 W
802.11-DS	WLAN	2.4 GHz	CSMA	TDD	DBPSK/QPSK	1-2 Mbit	1-2 Mbit		22 MHz		1 W
15.247-FH	CT-USA	900 MHz	FDMA	TDD	FSK	32 kbit	32 kbit	ADPCM	200 kHz	100 mW	200 mW
15.247-DS	CT-USA	900 MHz	FDMA/CSMA	TDD	BPSK	64 kbit	64 kbit	ADPCM	5 MHz	100 mW	200 mW
CDPD	WAN	800 MHz	FDMA	FDD	GMSK(.5)	19.2 kbit	19.2 kbit		30 kHz		600 mW

are also under development to improve network security and quality of service (QOS).

The cost to add 802.11 capability to a notebook computer or PDA has fallen below $20 and many such portable products are now being designed with 802.11 incorporated. The incremental cost to install a base station in the vicinity of an existing wired network connection is around $50, resulting in 802.11 wireless coverage zones, also known as *Hot Spots*, appearing in airports, hotels, restaurants, and other public places.

These Hot Spots are cost-effectively meeting the needs of many users with respect to mobile connectivity. One side effect of this trend is the rethinking of 3G (third generation) wireless network business plans. Wireless telecom service providers had planned to meet wireless mobile data needs using a capital intensive wide area network (WAN) solution, combining cellular service with high-speed data service over the same wireless network. 802.11 Hot Spots have forestalled the payback for a WAN approach, though Hot Spot service providers are also struggling to find a profitable revenue model. The most successful revenue model for Hot Spot operation so far has been to offer free wireless Internet access at a venue such as a hotel or coffee shop, as a means of attracting more customers.

In conclusion, this chapter has described some of the architectural issues surrounding software and communications in portable electronic products. Integration into existing software and communications infrastructures is essential if a portable electronic product is to be successful.

8

Cellular Phones

Cellular phones are produced in extremely high volume when compared to other types of sophistcated portable electronic devices, with over 400 million cellular phones being produced each year worldwide. Digital systems have replaced analog systems in most geographic regions, but global markets have yet to standardize on a single digital system. User expectations for basic handset functionality are fairly stable, but some manufacturers and service providers are working to increase the capabilities of cellular phones to include video telephony, GPS, and PDA-type functionality.

To facilitate our discussion of cellular phones as a class of portable electronic devices, we will examine the Audiovox CDM-8300 tri-mode phone as an example of system design shown in Fig. 8.1. The specifications for this product are listed in Table 8.1. This phone is considered tri-mode because it works on CDMA 1900 networks, CDMA 800 networks, and AMPS (analog) networks. It also has a GPS receiver that enables the phone to determine the GPS coordinates of its location. This product exploits data from GPS satellites and CDMA networks to locate a wireless handset to the accuracy of 5–50 m. The phone is equipped with a charging cradle pictured in Fig. 8.2, which has a two-color LED to indicate charging status. Synchronization of contact data with a PC is possible using a data cable available as an option.

The button layout provides for a typical cell phone interface, as pictured in Fig. 8.3. A numeric keypad is arranged on the lower portion of the interface. Green and red phone buttons provide call initiation and call termination, respectively. A joystick button enables the user to move up and down or side to side through menus displayed on the LCD. Many phones use directional arrows instead of a joystick. There are also a number of soft keys for which the function changes as the user moves

Figure 8.1 Audiovox CDM-8300 Tri-mode phone.

through different menus. This phone also features a speakerphone capability. The speakerphone audio is mercifully routed through the rear of the enclosure in the event that a user switches to speaker mode while the phone is still held near the ear. This system is quite compact as indicated by the dimensions shown in Fig. 8.4.

TABLE 8.1 Cell Phone Specifications

Produce type	Cellular phone
Produce name	Audiovox CDM-8300
Maufacturer & origin	Audiovox, Korea
Official release date	March, 2002
Operating frequency	CDMA 800/1900 MHz, AMPS 800 + GPS
Purchase date & price	01/02/03, $249
Technology	CDMA2000, AMPS, GPS
Talk & stand-by time	Talk: 3 h; Standby: 5 days
Power source	3.7 V Li-ion battery, 1000 mAh
Display	Monochrome 128×96 pixels
Web access	WAP UP 4.1 browser
Data rate	Up to 144 kbps
Voice activation	Yes
Memo recorder	Yes
SMS	Yes
Speaker phone	Yes
Text entry	T9 predictive
Phone book	300 names & 1500 numbers
Antenna	External
Weight (with battery)	90.7 Grams
Dimensions	$117.7 \times 45.3 \times 16.05$ mm

Figure 8.2 Charging cradle.

Figure 8.5 shows the snap-in battery pack removed from the product. This is a single-cell Li-ion battery with 1000 mAh of energy capacity and an operating voltage of 3.7. The form-factor is 68.5 mm × 44 mm × 7.76 mm and 28.1 g, yielding an effective energy density of 160 Wh/L or 132 Wh/kg. Opening the battery pack reveals the internal arrangement shown in Fig. 8.6.

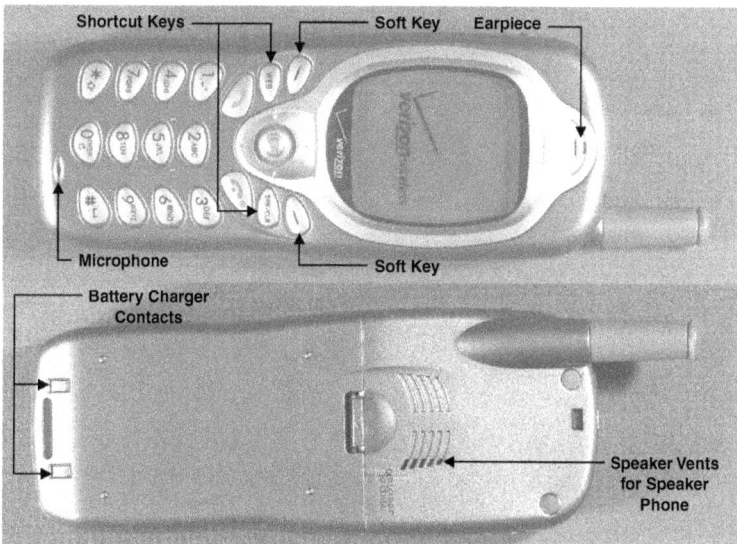

Figure 8.3 User interface.

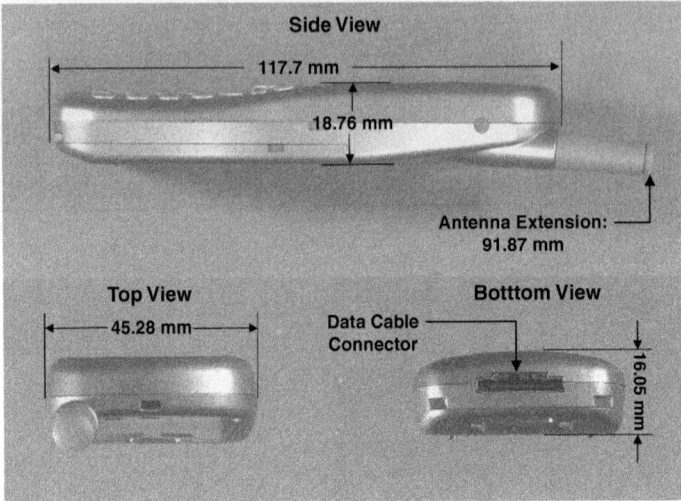

Figure 8.4 Product dimensions.

Notice that the battery cell does not consume 100 percent of the battery pack volume, impacting the energy density. It is almost always the case that the production dimensions of available battery cells will result in a slight reduction in effective energy density when the battery cell is implemented in a product design. This design utilizes a flex circuit to

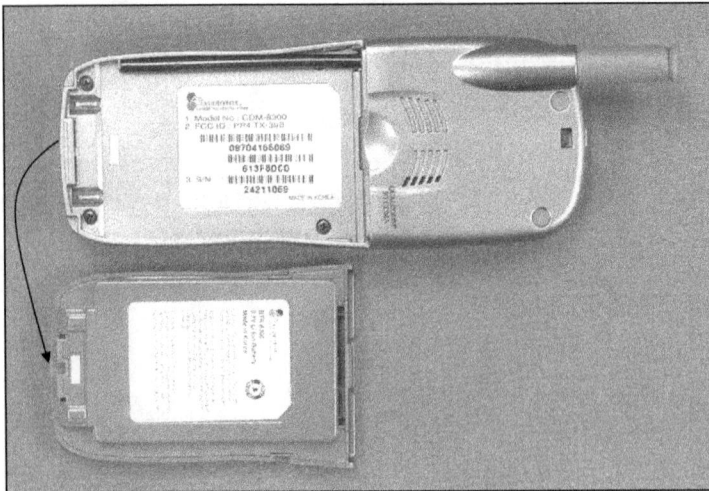

Figure 8.5 Battery removed.

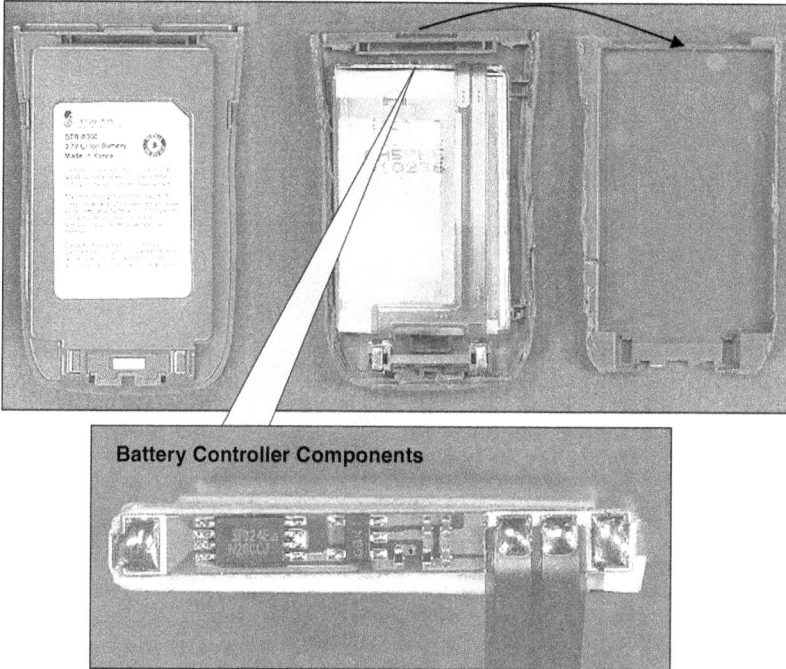

Figure 8.6 Battery disassembly.

make the connection from the battery cell electrodes to the battery casing electrodes. Polyimide flex is used because the polyimide material can withstand the substantial temperature rises due to high current drain without deforming. There is a small circuit board inside the battery pack, which controls the battery discharge rate and supports intelligent charging.

A power consumption profile for this product was generated and is shown in Fig. 8.7. The numbers on this chart indicate the following operational conditions of the phone:

1. Initiate power on, backlight on
2. Backlight off
3. Key-in phone number
4. Send call
5. Terminate call
6. Power off sequence

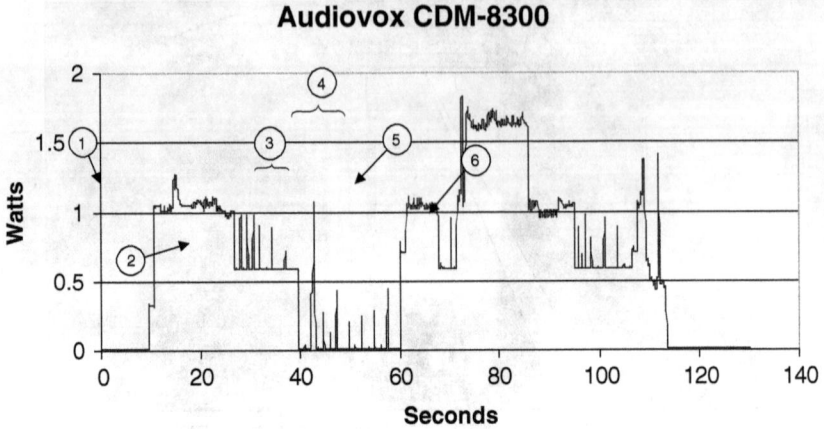

Figure 8.7   Power measurements.

This should give the reader some idea of the power levels typically encountered in cell phone operation. A simplified block diagram of the phone was estimated based on identification of some of the key system components and is shown in Fig. 8.8.

Figure 8.9 shows the major subassemblies that comprise the phone. Notice that the final assembly operation for a cell phone involves relatively few parts. The stack-up of these components is shown in Fig. 8.10. The entire assembly is held together by the six screws shown in Fig. 8.11. These screws are covered by rubber plugs to make the back surface of the phone smooth. The antenna screws into an internal connector that is soldered to the motherboard. Some contemporary cell phones have a snap-together design that uses no screws. This allows the user to switch faceplates or even swap out the phone electronics if there is a hardware failure.

Figure 8.12 shows the back enclosure removed, exposing side 2 of the main PCA. The speaker/vibrator assembly is fitted inside the back enclosure and is supported by a rubber gasket. Electrical connection to the main board is achieved using a metal-on-elastomer (MOE) connector. MOE connectors are often used for these types of press fit connections to connect components that reside in different subassemblies. The elastomer material maintains positive contact pressure even if the initial contact is overdriven.

The speaker/vibrator assembly is shown disassembled in Fig. 8.13. The assembly is held in place by a molded recess in the rear enclosure. Figure 8.14 shows a close-up of the internal construction of the speaker/vibrator device

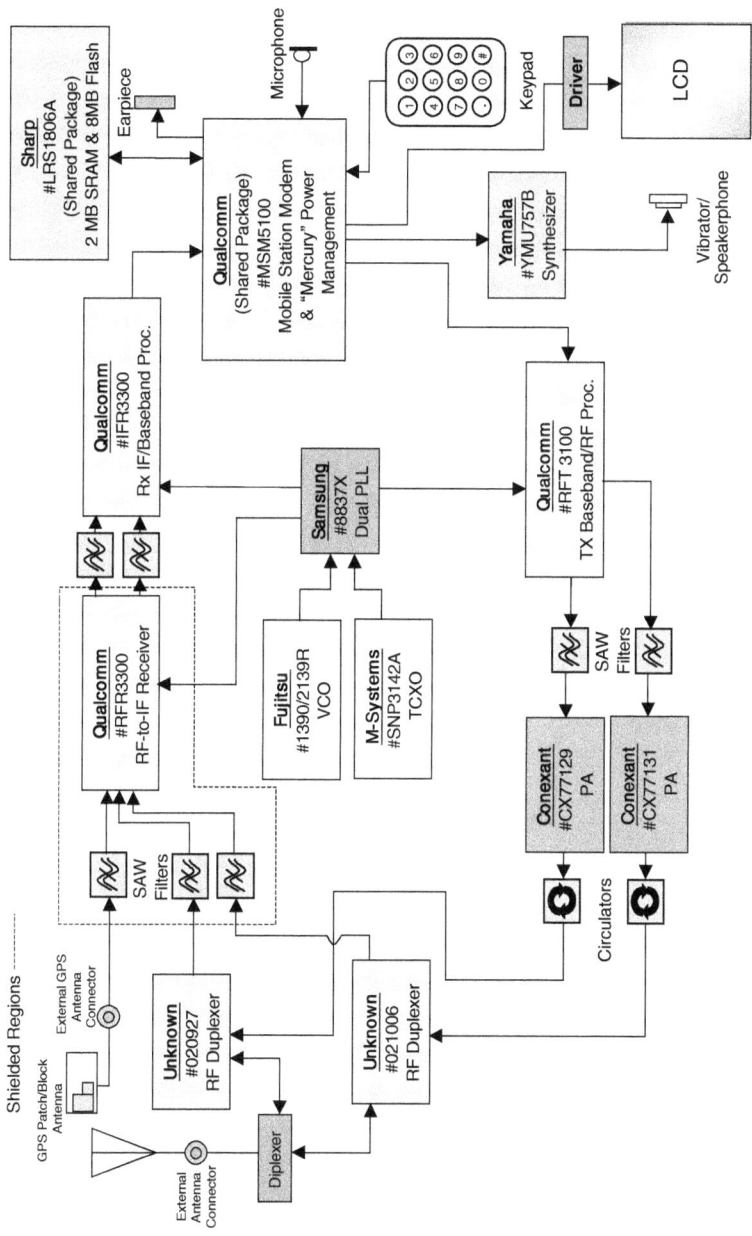

Figure 8.8 Simplified block diagram.

209

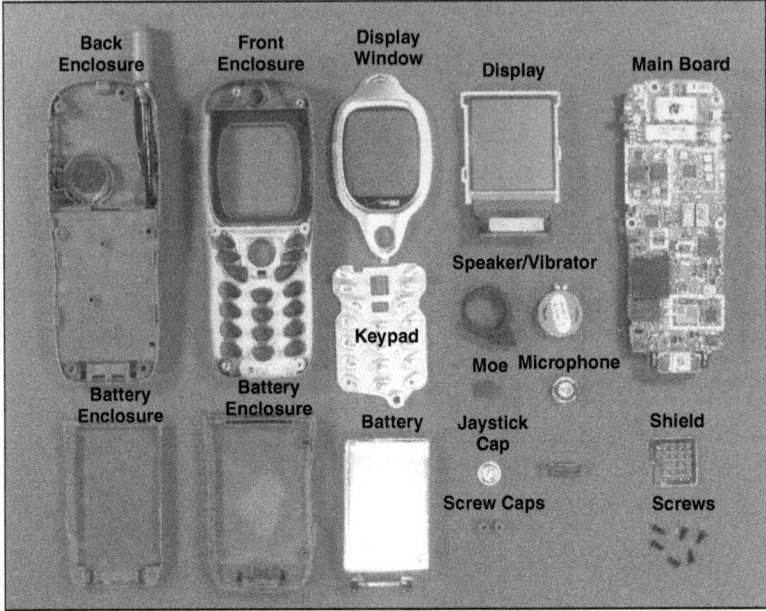

Figure 8.9 Disassembled cell phone.

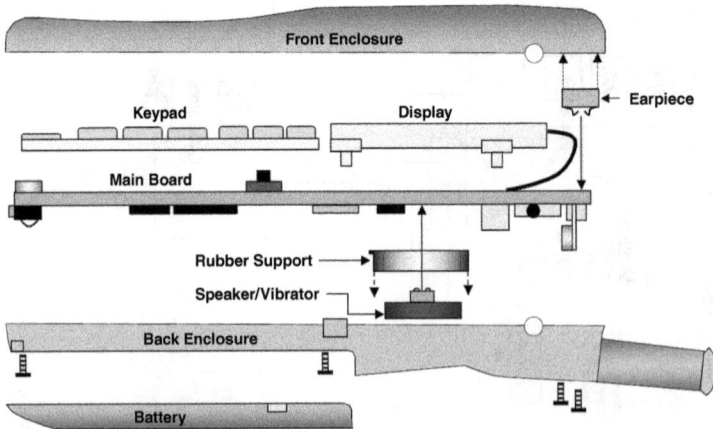

Figure 8.10 Component arrangement.

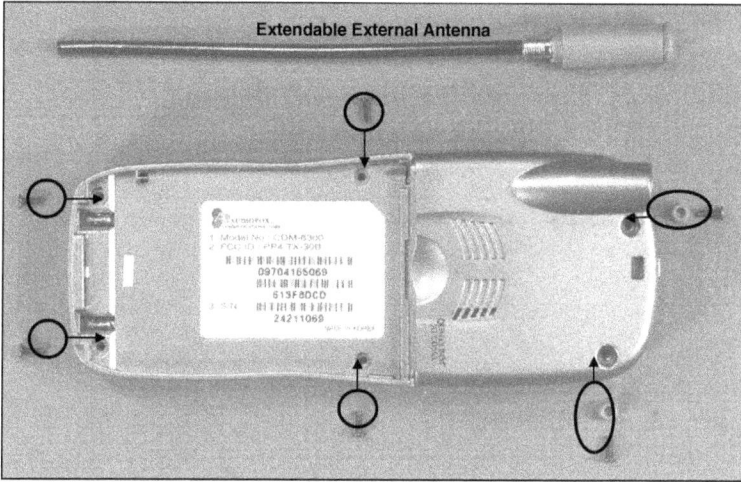

Figure 8.11   Back fasteners removed.

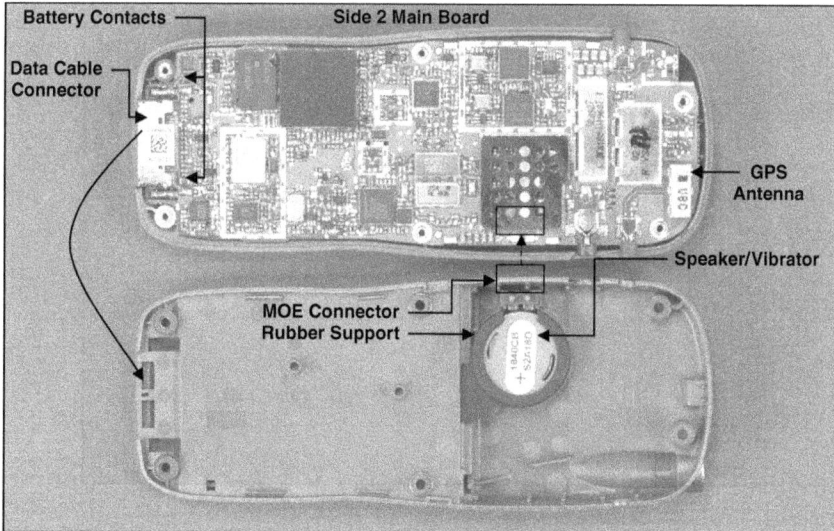

Figure 8.12   Back enclosure removed.

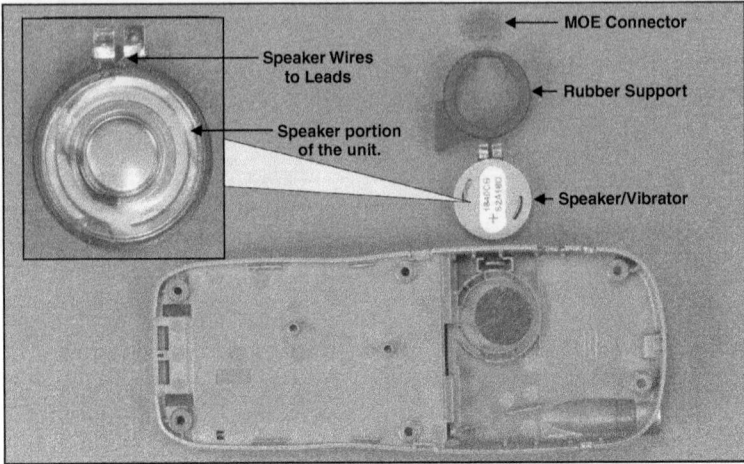

Figure 8.13 Speaker/vibrator assembly.

Figure 8.15 shows the removal of the front enclosure. The front enclosure is coated with a conductive paint to prevent RF emissions into the external environment. The front enclosure assembly contains a polycarbonate transparent window to cover the display and an earpiece speaker, both shown in Fig. 8.16. Removing the keypad cover reveals the

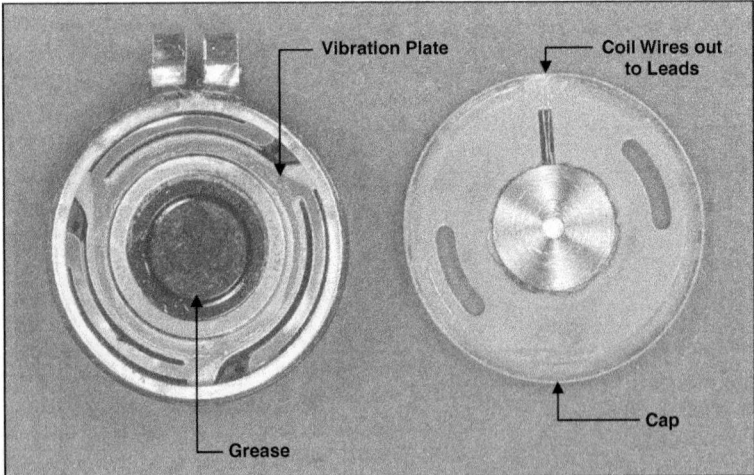

Figure 8.14 Speaker/vibrator assembly(close-up). The speaker/vibrator cap was removed exposing the vibrator plate and coil. Grease was placed in the middle of the vibration plate possibly to prevent the moving parts from sticking together. You can see in the above photo that there are small wires coming from the the the coil to the outer leads for power. The same is true for the speaker side of the module.

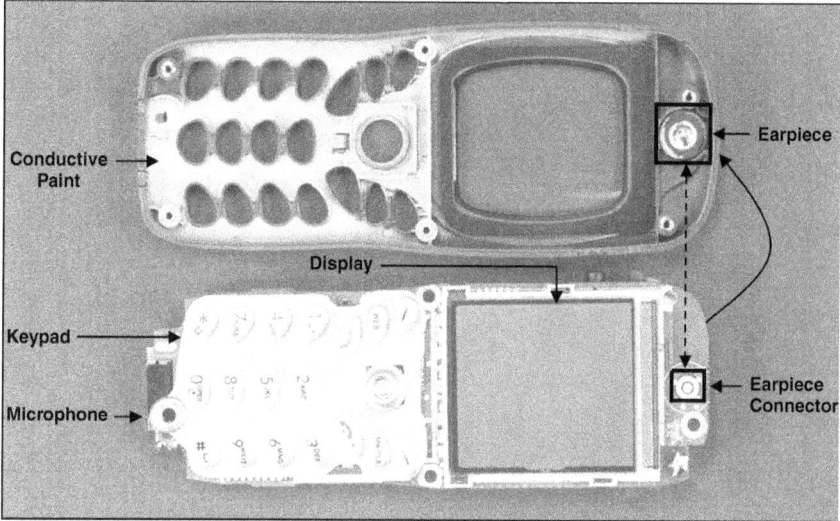

Figure 8.15   Front enclosure removed.

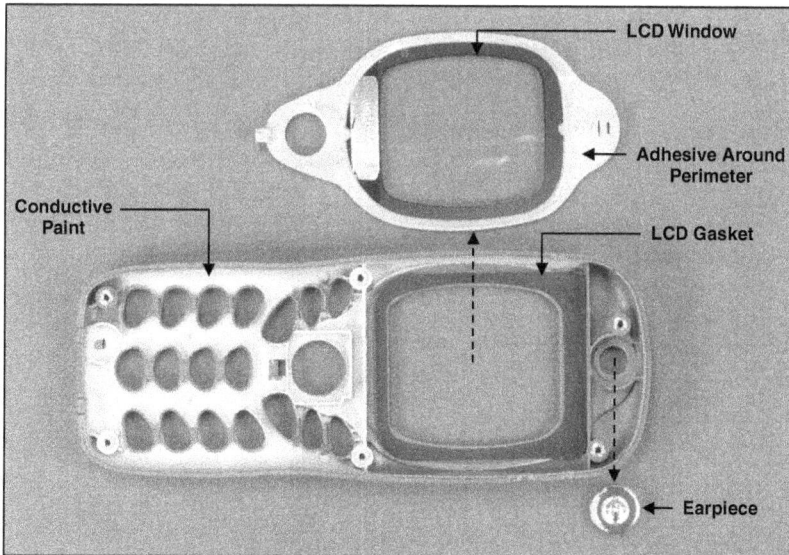

Figure 8.16   Display window removed. The polycarbonate LCD window is attached to the front enclosure with a pressure-sensitive adhesive and a locking tab. Also visible is the self-sticking foam LCD gasket. Metal inserts are used in all 6 of the screw locations.

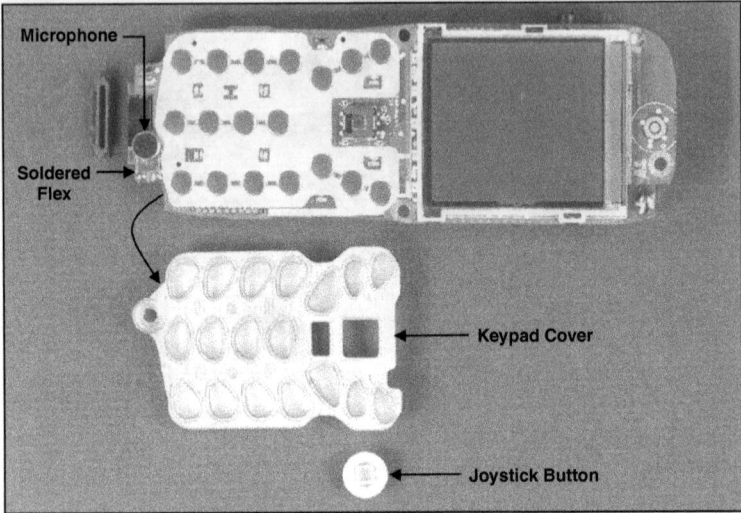

Figure 8.17   Keypad removed.

underlying keypad contact structure (Fig. 8.17). Unfolding the LCD reveals the display-drive chip, which is packaged in a tape-automated bonding (TAB) carrier. The input side of the driver TAB is solder bonded to the main PCA. The output side is bonded to the LCD, using aniso- tropic conductive adhesive (Fig. 8.18).

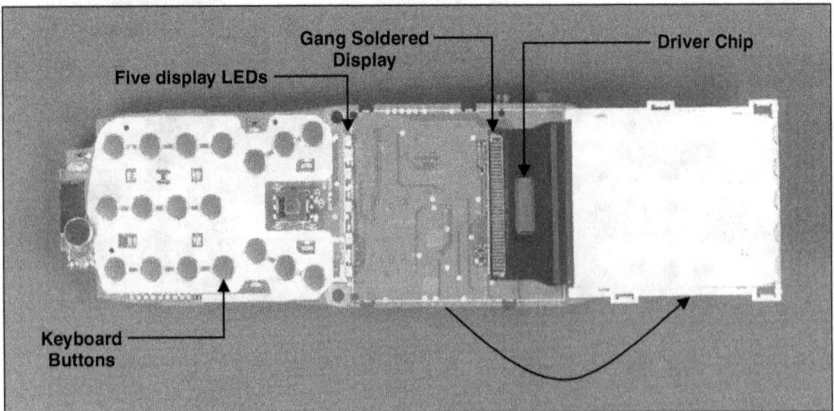

Figure 8.18   Display unfolded. Shown above is the display module unclipped from the main board and folded back exposing the gangsoldered flex connections. As shown, the display is backlit by five LEDs and the display driver chip is also visible.

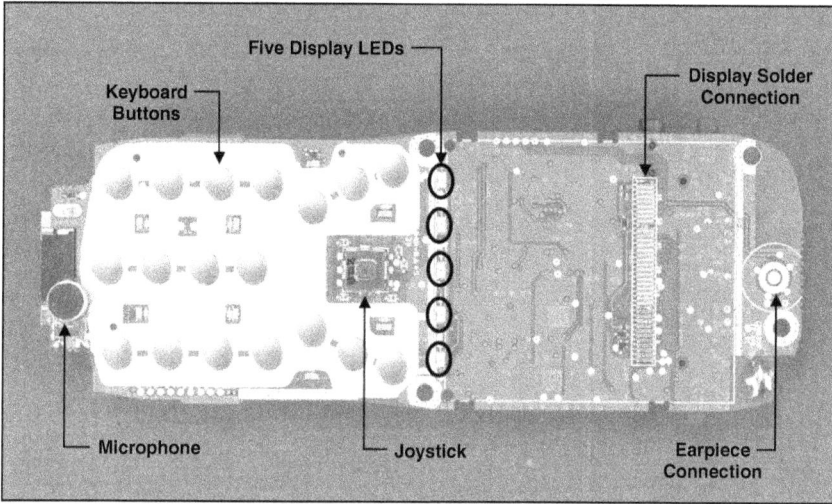

Figure 8.19 Main board side 1.

Looking at side 1 of the main board (Fig. 8.19), it can be seen that the keyboard buttons are actually preformed metal domes that are arranged in a sheet and laminated to the surface of the main PCA. These domes are deformed when pushed from above and create a short across adjacent electrodes formed by printed circuit traces on the PCA substrate. This configuration has the advantage of providing a tactile response to the user when the button has been fully depressed. Figure 8.20 shows

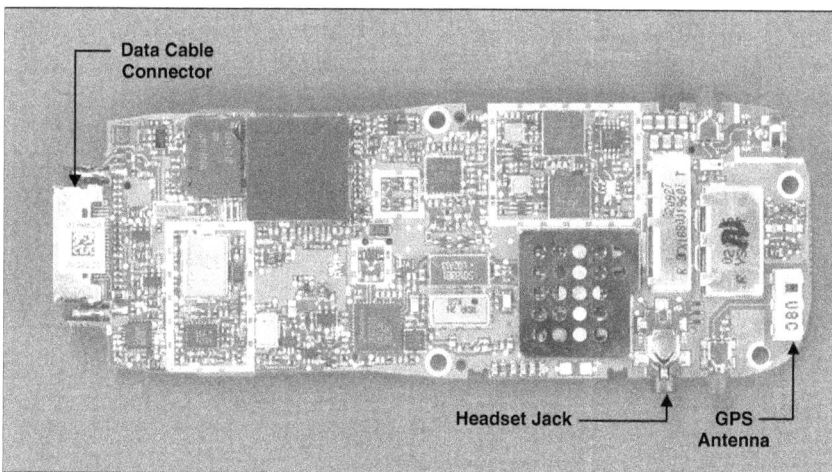

Figure 8.20 Main board side 2.

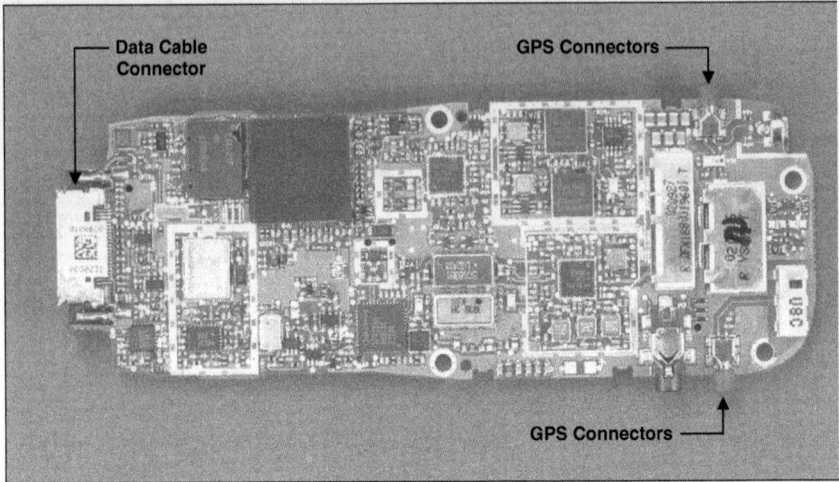

Figure 8.21   Main board side 2 (no shields).

the other side of this board. There is a section of RF circuitry that has been shielded with a rectangular can. Holes have been punched in the can to reduce the weight of the product. The size of the holes was determined based on the wavelength of the radiation being shielded. Figure 8.21 shows the board with the shielding can removed.

Figure 8.22 shows the dimensions of the active display area. The display module consists of the LCD, a backlight diffuser, and a plastic frame (Fig. 8.23). The five LEDs on side 1 of the main board (Fig. 8.19) provide the backlight illumination. The diffuser acts as a wave guide that captures the light and disperses it evenly across the rear of the LCD. The double-sided tape on the TAB driver assembly is used to keep the TAB frame attached to the main PCB during the solder-bonding operation.

Figure 8.24 shows the approximate side 1 board area required for various cell phone functions with this particular design. Functional area allocations for side 2 are shown in Fig. 8.25. Further details regarding the substrate technology are shown in Fig. 8.26.

The electronic assembly metrics for this system are as follows:

Main PCB

Substrate area (cm^2)   42

Metal layers   6

Circuit area (cm^2)   253

Routing density (cm/cm^2)   33

Number of components   501

Number of connections   1605

Component density (/cm^2)   11.9

Connection density (/cm^2)   38.1

Average pin count   3.2

Opportunity count   2106

Number of ICs   11

IC connections   444

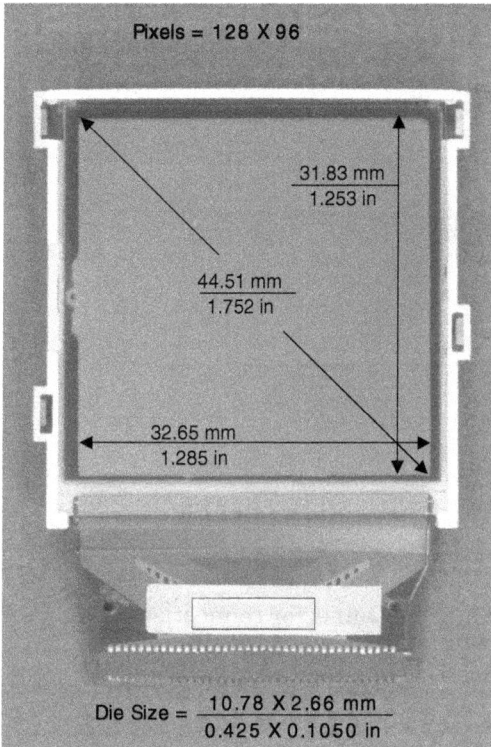

Figure 8.22   Estimated dimensions.

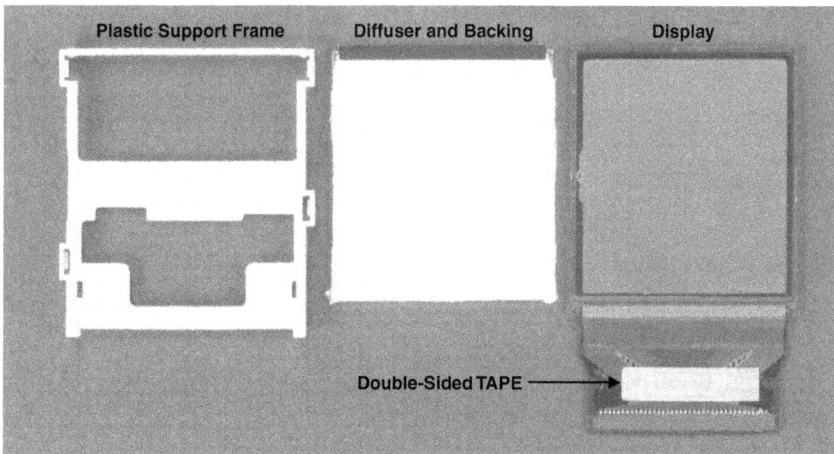

Figure 8.23   Display components.

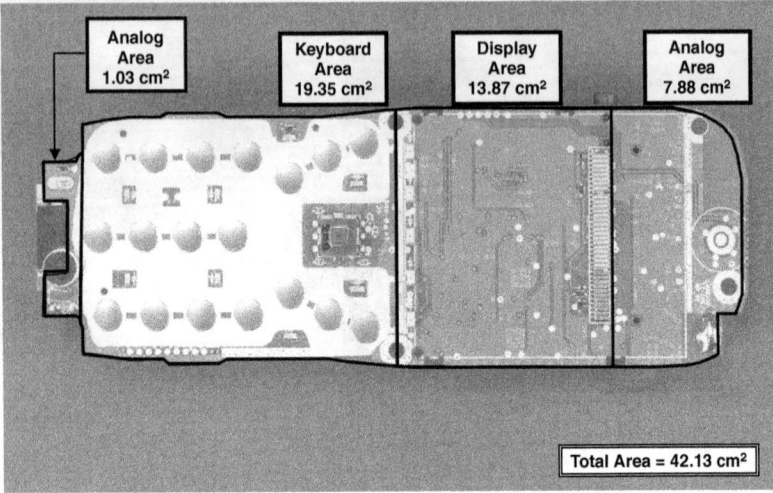

Figure 8.24 Main board—functional area photo side 1.

Modules (SAW filters, etc.)   23

Module connections   154

Discrete components   461

Discrete connections   962

Connectors   6

Connector connections   45

Die area (cm^2)   172

Substrate tiling density   4.1%

Footprint area (cm^2)   496

Die area/footprint area   0.35

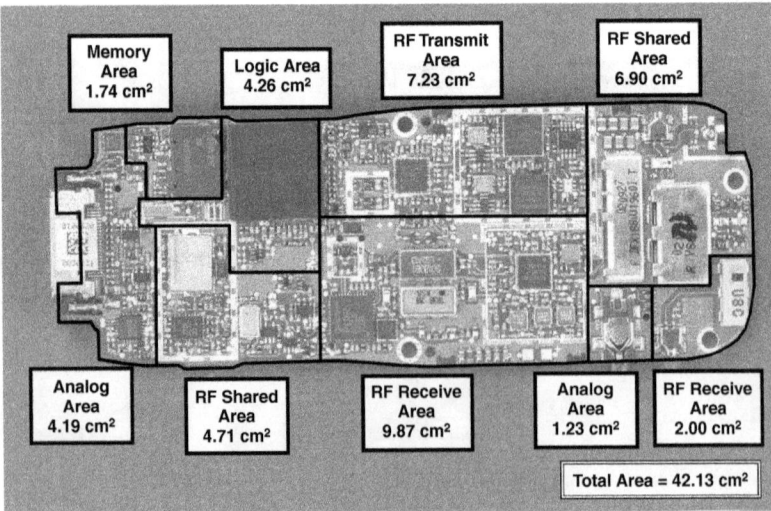

Figure 8.25 Main board—functional area photo side 2.

Main PCB
(*FR4 Substrate*)
(Build-up Technology)
Layers = 6
Finest Pitch = 0.203 mm/0.008 in
Narrowest Trace = 0.076 mm/0.003 in
Narrowest Space = 0.127 mm/0.005 in
Smallest Via I.D. = 0.152 mm/0.006 in
Smallest Via O.D. = 0.508 mm/0.020 in
Substrate Thickness = 0.930 mm/0.037 in
Assembly Weight = 24.1 g

Build-Up Technology

Main PCB – Side 1 Total Area =   cm²/in²

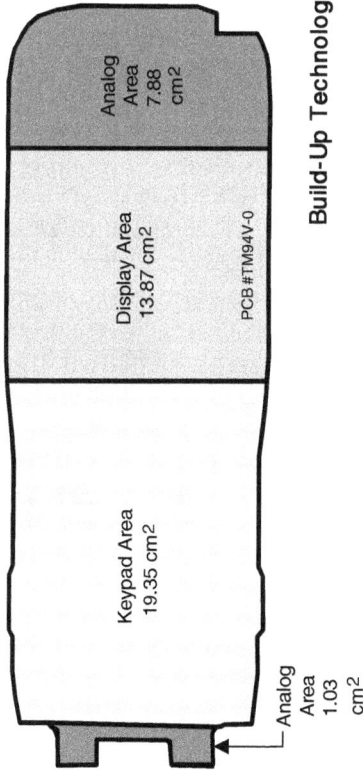

Keypad Area
19.35 cm²

Display Area
13.87 cm²

Analog
Area
7.88
cm²

PCB #TM94V-0

Analog
Area
1.03
cm²

Main PCB – Side 2 Total Area =   cm²/in²

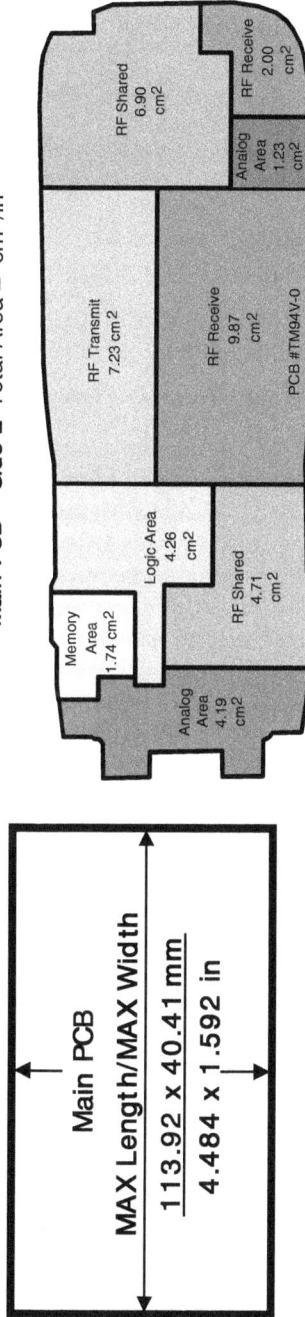

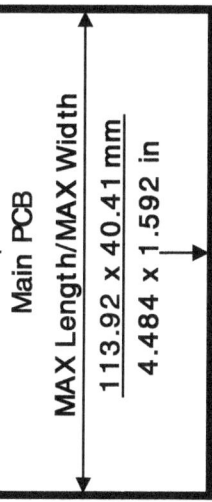

Memory
Area
1.74 cm²

Logic Area
4.26
cm²

RF Transmit
7.23 cm²

RF Shared
6.90
cm²

RF Receive
2.00
cm²

Analog
Area
1.23
cm²

RF Receive
9.87
cm²

RF Shared
4.71
cm²

Analog
Area
4.19
cm²

PCB #TM94V-0

Main PCB
MAX Length/MAX Width
113.92 x 40.41 mm
4.484 x 1.592 in

Figure 8.26 Substrate measurements.

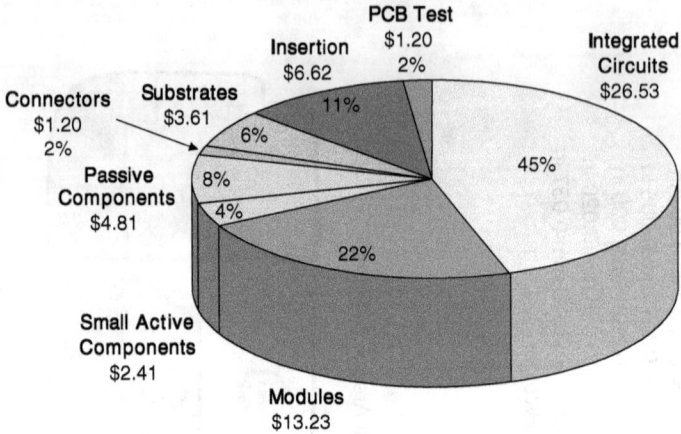

Figure 8.27 Electronic assembly cost estimates.

This particular cell phone design is typical of most in that it contains a single printed circuit assembly. Five hundred one electronic components are mounted to this PCA. The system is built around 11 IC components with a total of 444 connections. The ratio of discrete components (461) to IC connections is roughly 1 to 1. (This metric may be of interest in estimating the number of discrete components that will be needed for a new cell phone design, if only the key silicon devices and their IO count are known.) Only 45 connections actually leave the PCA.

Figure 8.27 shows the estimated cost of the electronic assembly based on the Portelligent cost model. ICs comprise 45 percent of the system cost. Modules that are mostly specialized RF components (Fig. 8.28) are rather expensive and comprise over 22 percent of the system cost. Substrate costs make up only 6 percent of the estimated system cost.

The results of a design for assembly analysis are shown in Fig. 8.29. This analysis shows that an estimated 26 steps are required to perform the final assembly of the product. Table 8.2 gives the estimated cost for

TABLE 8.2 Mechanical Parts Cost Estimates

Item	Description	Estimated cost
Front enclosure	Conductive painted ABS plastic, 8.2 g, 117.6 × 45.9 × 9.1 mm	$0.80
Bottom enclosure	ABS plastic 12.7 g, 135.2 × 43.1 × 18 mm	$0.46
Antenna	Wire (extendable), 3.5 g, 123.8 mm	$0.35
LCD window	Polycarbonate, 2.7 g, 39 × 38 × 1.25 mm	$0.48
Keypad	Elastomer, 3.2 g, 55.5 × 36.9 mm	$0.65
Miscellaneous	6 screws, labels, rubber buttons, etc.	$0.30
Total		**$3.04**

SAW Filter

GPS Antenna

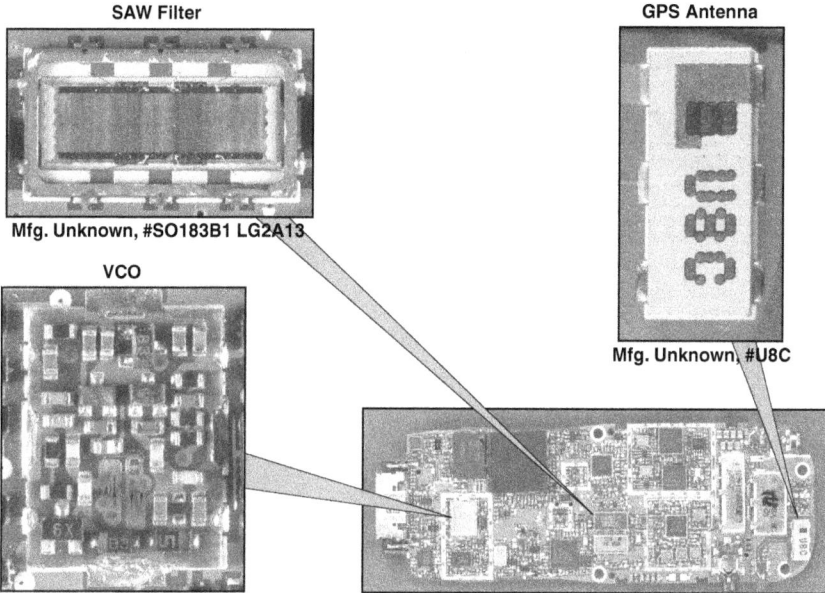

Mfg. Unknown, #SO183B1 LG2A13

Mfg. Unknown, #U8C

VCO

Fujitsu, #1390/2139R

Figure 8.28 Specialized RF modules.

DFA Index (%)	10.58%
Total Assembly Time (s)	249.28 s
Total Assembly Cost ($)	$0.69 (@$10/h, Taiwan)
Total Number of Steps	26

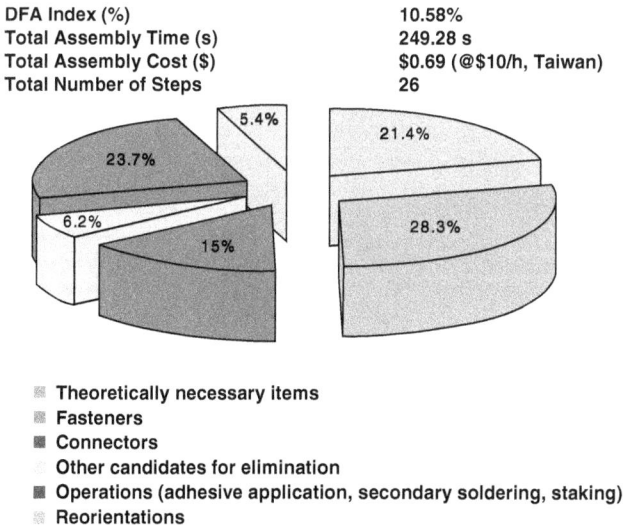

5.4%

21.4%

23.7%

28.3%

6.2%

15%

▨ Theoretically necessary items
▨ Fasteners
▣ Connectors
  Other candidates for elimination
▣ Operations (adhesive application, secondary soldering, staking)
▨ Reorientations

*From Boothroyd Dewhurst, Inc. DFMA Software

Figure 8.29 Design for assembly (DFA).

Total Cost	
Electronic Assemblies	$ 59.50
Display	$3.68
Housing/Hardware (DFM)	$3.04
Battery Pack	$4.86
Final Assembly (DFA)	$0.69
Total	**$71.77**

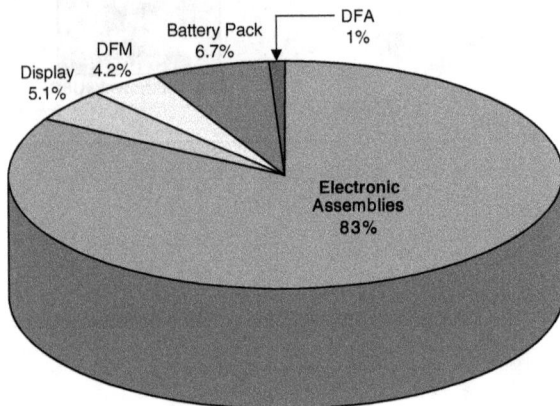

Figure 8.30 Cost estimate summary.

Item (CDM-8300)	Weight (g)
Enclosures	23.6
Front Enclosure	8.2
Back Enclosure (Incl. Ant.)	12.7
Display Window	2.7
Battery	28.1
Main Board	24.1
Display Assembly	6.4
Keypad	3.2
Miscellaneous	5.3
Earpiece	0.7
Speaker/Vibrator Assembly	3.8
Screws (6), rubber pieces, etc.	0.6
Total	**9 0.7**

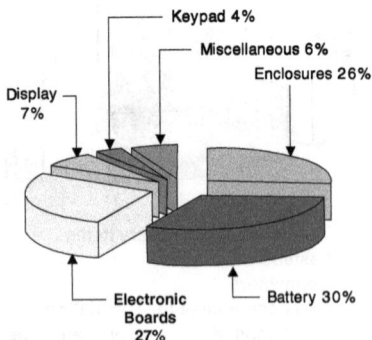

Figure 8.31 Component weights.

the mechanical parts in the system. An overall cost estimate is shown in Fig. 8.30. Most notable is the extent to which the electronics assembly cost (main PCA) totally dominates the product manufacturing cost.

Figure 8.31 shows the breakdown of weight distribution among the various parts of this cell phone product. It is interesting to note that two of the largest contributors to the system weight, the enclosure and the battery pack, comprise a relatively low percentage of system cost. This highlights the financial opportunity to reduce the weight of cellular phones with premium enclosure or battery solutions.

9

Portable PCs

Notebook computers are the most complex class of portable electronic devices being manufactured today. As general purpose platforms, notebook computers are designed to provide many different types of functionality at very high levels of performance. Given their current status in the marketplace as the ultimate portable electronic device, these products push the envelope on every critical design factor.

We will frame our discussion of notebook computers through examination of the Dell Latitude C800 as a case study, shown in Fig. 9.1. *All conclusions drawn and commentary provided about this design are opinions of the author and are subject to error for which the author bears sole responsibility. The manufacturer's claims and specifications regarding this product override any statements made by the author.* This particular system was purchased in January, 2001 at a list price of $2,539. The product specifications are as follows:

Processor	Intel Pentium III 850 MHz
Memory	64 MB (100 MHz SDRAM) expandable to 512 MB
Hard drive	10 GB, 9.5 mm (user-removable 20 GB & 32 GB options)
Modular bay	3.5" 1.44 MB diskette drive (standard floppy), 8X DVD drive (optional), 24X/10X variable CD-ROM drive (optional), CD read-write drive (optional), Zip 250 high capacity floppy (optional)
Fixed optical drive	24X/10X variable CD-ROM (standard), 8X DVD or CD read-write (optional)
Display	15.1" SXGA (1400 × 1050 resolution) TFT display, 16 MB video memory (upgradeable to 32 MB)
Modem	Internal 3COM mini-PCI modem (as shipped)

Figure 9.1   Dell latitude C800/*Notebook Computer*.

Graphics chip	ATI 3D video (rage mobility 128 AGP 4X)
Audio	ESS Maestro 3i controller, sound blaster emulation, Stereo 16-bit (A to D & D to A)
Keypad	QWERTY 87 keys, direct web access button, track stick and touch pad with entry buttons
Battery	14.4 V, 3800 mAh, lithium-ion
Weight	8.0 lb (with battery & floppy drive)
Dimensions	2.1" × 10.8" × 13.1"
I/O ports	IEEE1394 firewire port (for video camera), two USB ports. Serial (DTE): 9-pins, 16550 compatible, 16-byte buffer, parallel 29-pin, video 15-pin, S-video out. Infrared: IRDA compatible standard network & modem connectors, audio jacks, 200-pin docking connector, expansion dock
Software included	10 GB drive (8.8 GB free space)
	Millennium Edition (operating system)
	Internet Explorer
	Outlook Express
	Online Services
	AOL AT&T WorldNet

EarthLink

Prodigy

Office documents

My Documents folder

Music folder

Picture folder (8 photos)

My Network folder

My Briefcase folder

Windows Media player

DVD player

Games (12 simple games)

Model test

Accessories (tools, calculator, notepad, paint, etc.)

Dell accessories (Express Service Code)

Based on these specifications, this system was considered to be a full-featured three-spindle (three rotational storage media devices) box at the time of introduction.

Figure 9.2 shows the basic architecture of this system. This is a typical contemporary PC architecture consisting of an X86 architecture CPU (Pentium III), a North Bridge (GMCH), and a South Bridge (ICH2). These three key components determine the basic platform. The North Bridge and South Bridge together are commonly referred to as the chip set for the design.

In this type of architecture, the North Bridge provides the interface between the CPU and the rest of the system. The North Bridge contains the memory controller and interfaces directly to the system memory, the graphics controller, and the South Bridge. The South Bridge manages connections to all other parts of the system including peripheral devices and various types of busses.

All of the data that moves between system memory and the CPU is controlled by the North Bridge. The system bus, also known as the *front side bus*, and the memory bus are potential performance bottlenecks in this arrangement. This is why many processors include L1 and L2 cache memory to increase processing performance. The North Bridge also enables other system elements to access memory if needed.

This particular architecture is known as discrete *graphics* architecture because a separate graphics controller chip (ATI Rage) is used. If the graphics controller is built into the North Bridge, it is called *integrated graphics* architecture. A small amount of graphics memory is placed next to the graphics controller to reduce the latency of graphics

SDRAM
64-512 MB
M2V64S40DTP

100 MHz

PC Card
Slot

SynapticsT1004

Touch Pad
Controller

Track Port
Controller

Philips TPM T54

PCI
Controller
PCI 14451

Modular Drive
24x10x CD-ROM

Optional Drive
8X DVD 1.44 MB
3.5" Diskette

Docking
Connector

Bus Drivers
P15C32X245

Intel Pentium III
Processor - 850 MHz

System Bus

GMCH
Graphics & Memory
Controller Hub
FW82815

Hub Interface
(266 Mbps)

ICH2
I/O Controller
Hub
FW82801

Intel Chip Set
Accelerated Hub Architecture (AHA)

15" SXGA
Display

Driver

Graphics
Controller

ATI Rage Mobility 216M4

48LC2M32B2

SDRAM
8 MB

Hard Drive
10 GB 9.5 mm

Audio
Controller
ESS Maestro
ES1983S

Audio
Codec
ES1921S
AD1807JST

3COM Modem
Controller

EEPROM
Atmel 24C04N

TPA0202

Stereo
PA

Modem
connector

USB Ports, IEEE1394, Serial, Parallel Ports

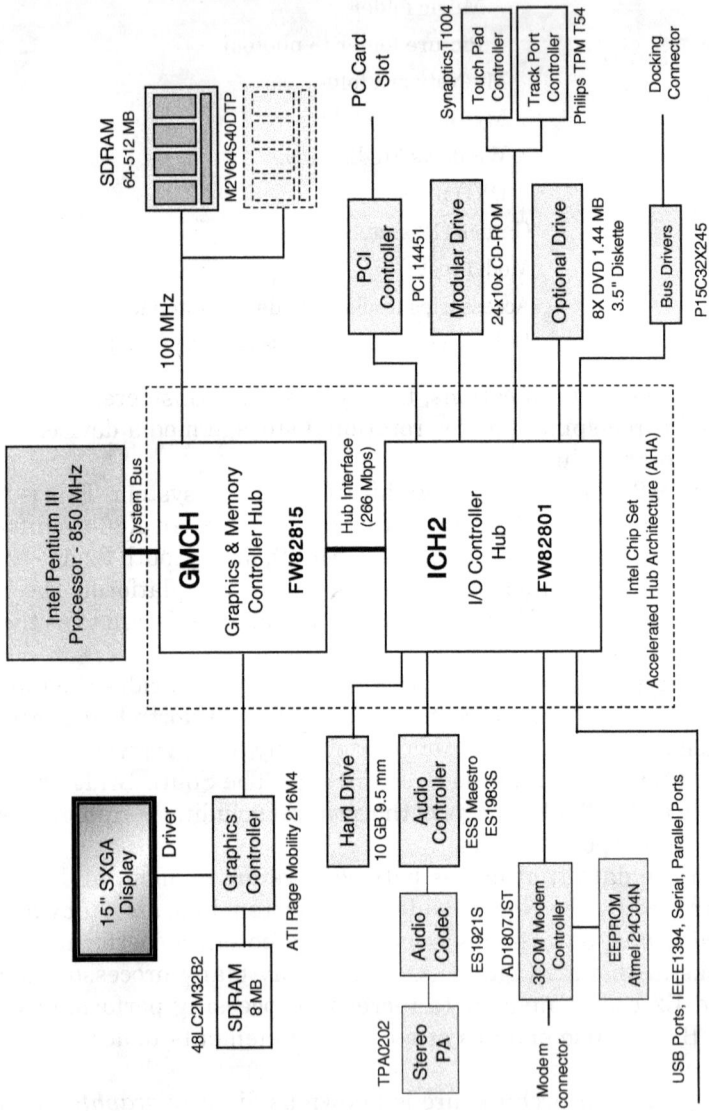

Figure 9.2 Basic block diagram.

processing and increase graphics performance. This dedicated graphics memory is also known as a *local frame buffer* (LFB). The graphics controller drives the display directly.

As system architecture evolves, each of the North Bridge interfaces (front side bus, memory bus, hub interface, and graphics bus) are systematically upgraded in various generations of products. So over time, the bandwidth specification for each of these key system interfaces will increase to the latest industry standard. Analyzing the load placed on each of these interfaces during different usage scenarios is the critical function performed by the system architect.

The selection of South Bridge determines many of the features and subsystems that the product will be able to support. First and foremost, it must have a hub interface that is compatible with the North Bridge. Second, it must support all the I/O interfaces needed to provide the peripherals and busses contained in the product requirements.

There have been various attempts by IC manufacturers to create single chip solutions that essentially eliminate the need for the South Bridge. While this seems like an attractive idea from the standpoint of system integration, the economics of this approach have not proven very effective. The I/O standards that the South Bridge must support tend to evolve very rapidly. Realizing this, most South Bridge manufacturers produce the South Bridge chip with a lagging generation of fabrication technology to reduce the cost if frequent redesigns are required. The North Bridge interfaces tend to evolve more slowly so it is possible to apply a more leading edge fab technology and take advantage of the smaller die size over the life of the part.

Another popular trend in PC architecture is to use a North Bridge with integrated graphics. This approach is starting to dominate for cost reasons although products that are positioned as high-performance graphics systems still use discrete graphics.

In search of further cost reductions, many manufacturers are now moving to a Unified Memory Architecture (UMA). UMA systems use the system memory for graphics and eliminate the cost of an LFB. Integrated graphics plus UMA solutions are currently the preferred architecture for notebook PCs.

Figures 9.3 and 9.4 show the top and bottom of the system motherboard for this notebook PC.

The user interface devices are shown in Fig. 9.5. Most notebook PC products have settled on this type of configuration with the touch pad mounted below the keypad. Mounting the air intake on this upper surface of the enclosure consumes prime real estate but ensures that the air intake is not inadvertently blocked by the user.

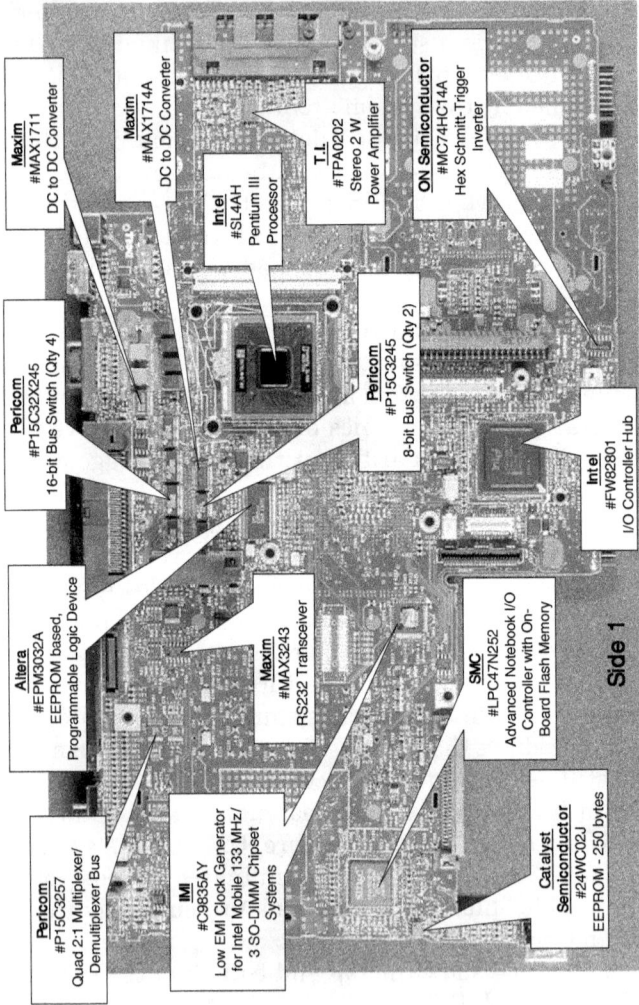

Maxim
#MAX1711
DC to DC Converter

Maxim
#MAX1714A
DC to DC Converter

T.I.
#TPA0202
Stereo 2 W
Power Amplifier

Intel
#SL4AH
Pentium III
Processor

ON Semiconductor
#MC74HC14A
Hex Schmitt-Trigger
Inverter

Pericom
#P15C32X245
16-bit Bus Switch (Qty 4)

Pericom
#P15C3245
8-bit Bus Switch (Qty 2)

Intel
#FW82801
I/O Controller Hub

Altera
#EPM3032A
EEPROM based,
Programmable Logic Device

Maxim
#MAX3243
RS232 Transceiver

SMC
#LPC47N252
Advanced Notebook I/O
Controller with On-
Board Flash Memory

Side 1

Pericom
#P15C3257
Quad 2:1 Multiplexer/
Demultiplexer Bus

IMI
#C9835AY
Low EMI Clock Generator
for Intel Mobile 133 MHz/
3 SO-DIMM Chipset
Systems

Catalyst Semiconductor
#24WC02J
EEPROM - 250 bytes

Figure 9.3 Main board (top).

230

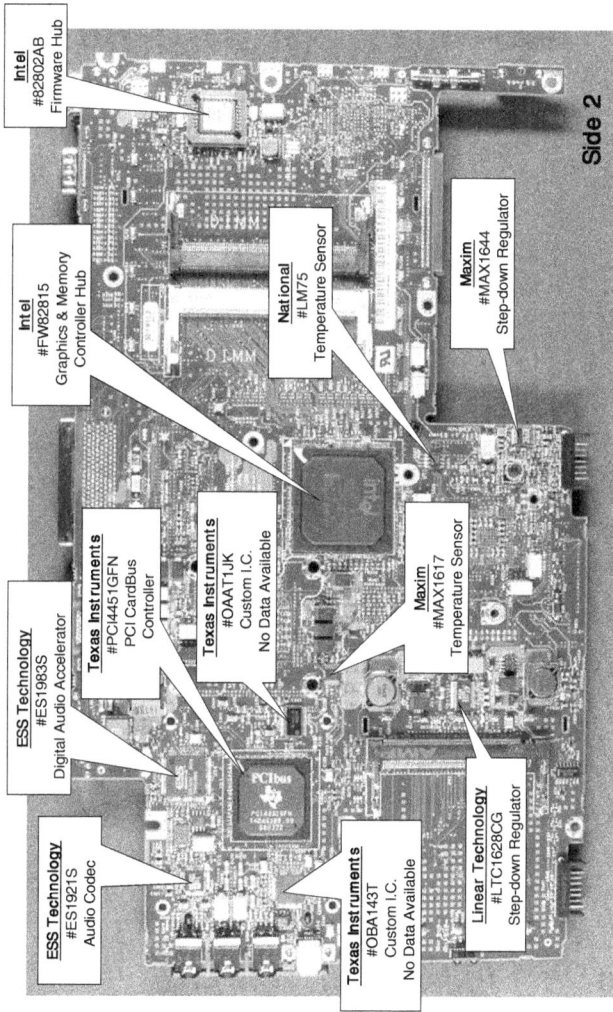

Figure 9.4 Main board (bottom).

231

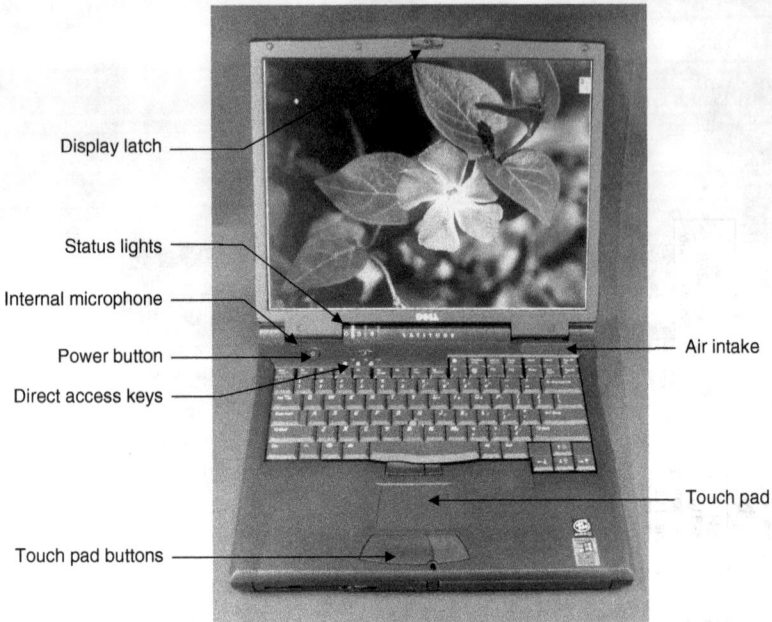

Figure 9.5 User interface devices.

Figures 9.6 and 9.7 show the external I/O ports and mass storage bays. It is noteworthy that notebook PCs must support a rather cumbersome list of legacy I/O capabilities. Most users will never use most of these I/O ports, but the general purpose nature of the PC platform dictates that all these standards be supported.

Figure 9.8 depicts the underside of the product, on which internal access ports for the system are mounted. The mechanical arrangement of key system components is shown in Fig. 9.9. While there is no industry-defined standard for the placement of components within a notebook, it is very common for the motherboard to be mounted in the upper half of the enclosure and for the battery bay and a mass storage bay to occupy the lower half.

Figures 9.10 and 9.11 show the battery pack and the battery control circuitry schematic. The battery has eight lithium-ion cells; each rated at 3.6 V and 1900 mAh, which are connected in a series-parallel combination. This provides an overall battery rating of 14.4 V and 3800 mAh. Lithium-ion battery packs contain a significant amount of electronics for reasons of safety and charging efficiency. There is also charging control circuitry on the motherboard that works in conjunction with the circuitry internal to the battery pack to minimize charging time, and to monitor battery life.

Left Side View

Right Side View

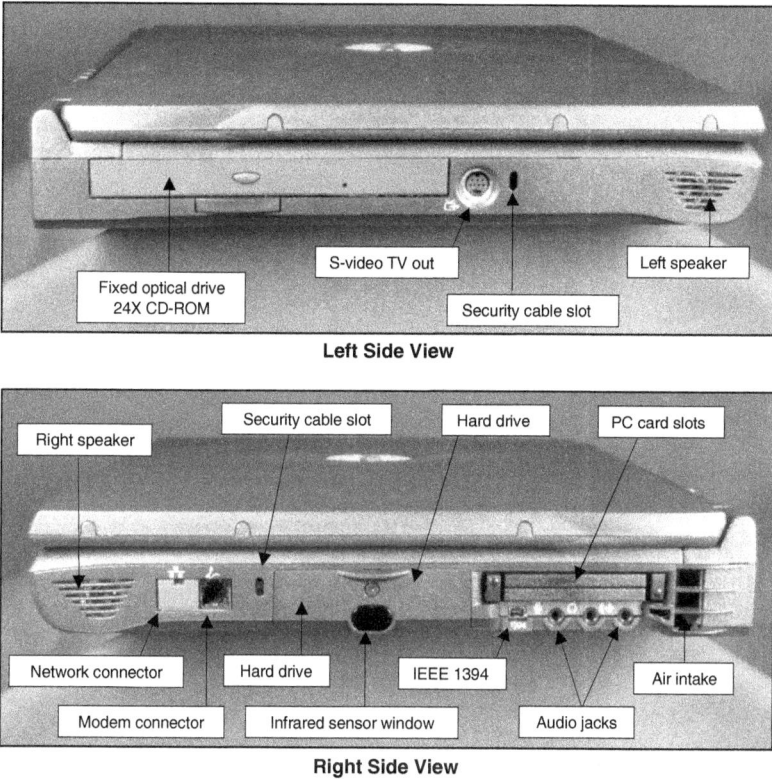

Figure 9.6   External interface (left & right sides).

Figure 9.12 shows a little more detail on how the modular bays and internal compartments are accessed. These ports are designed for rapid access to enable easy upgrading or servicing of the product. Usually a manual release latch or nested fit is combined with a single screw to enable rapid but intentional access. Likewise, keyboard removal is efficiently designed to enable rapid product assembly and easy servicing of the topside of the motherboard. Figure 9.13 shows how a single screw is typically used along with a nested fit to keep the keyboard in place.

Figure 9.14 shows an interesting aspect of the mechanical design. A copper bar is embedded in the body of the keyboard structural frame to facilitate heat transfer from the graphics chip mounted on the motherboard. The heat is spread into the aluminum keyboard frame where it is more easily dissipated. Figure 9.15 shows how a single screw enables release of the optical drive module. It also provides a close-up view of the dual inline memory module (DIMM) connector that is used to add

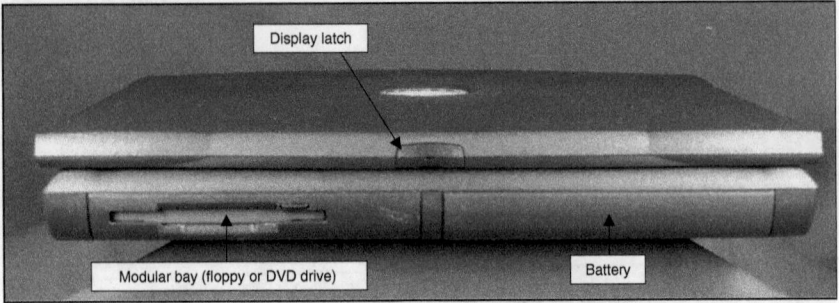

Front View

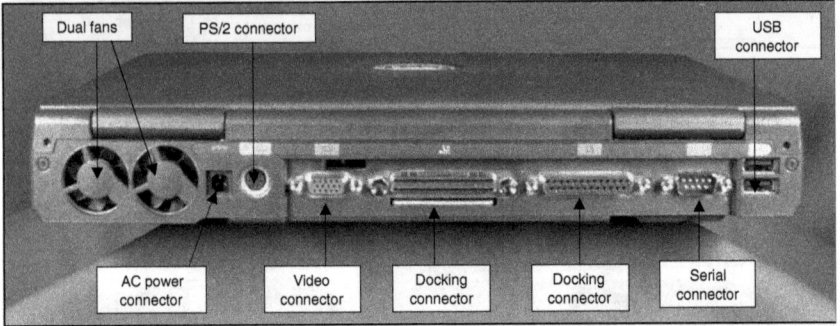

Back View

Figure 9.7 External interface (front & back views).

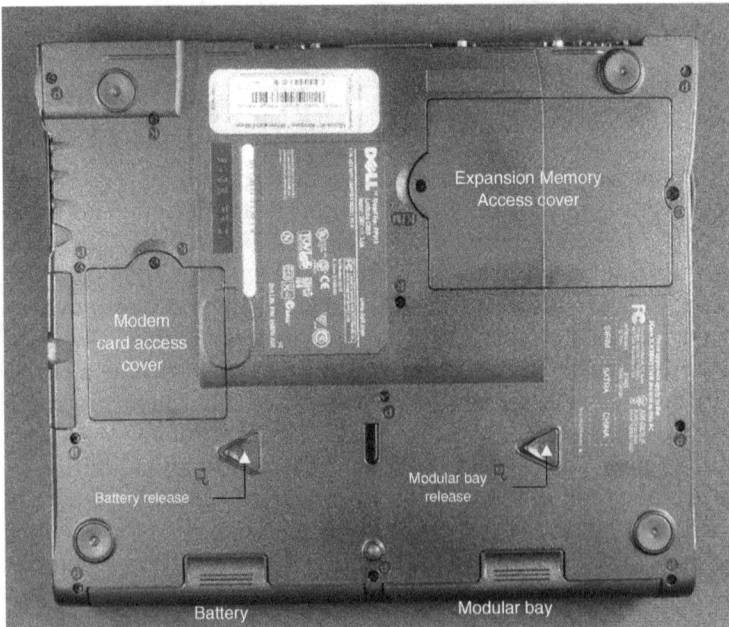

Figure 9.8 External interface (bottom view).

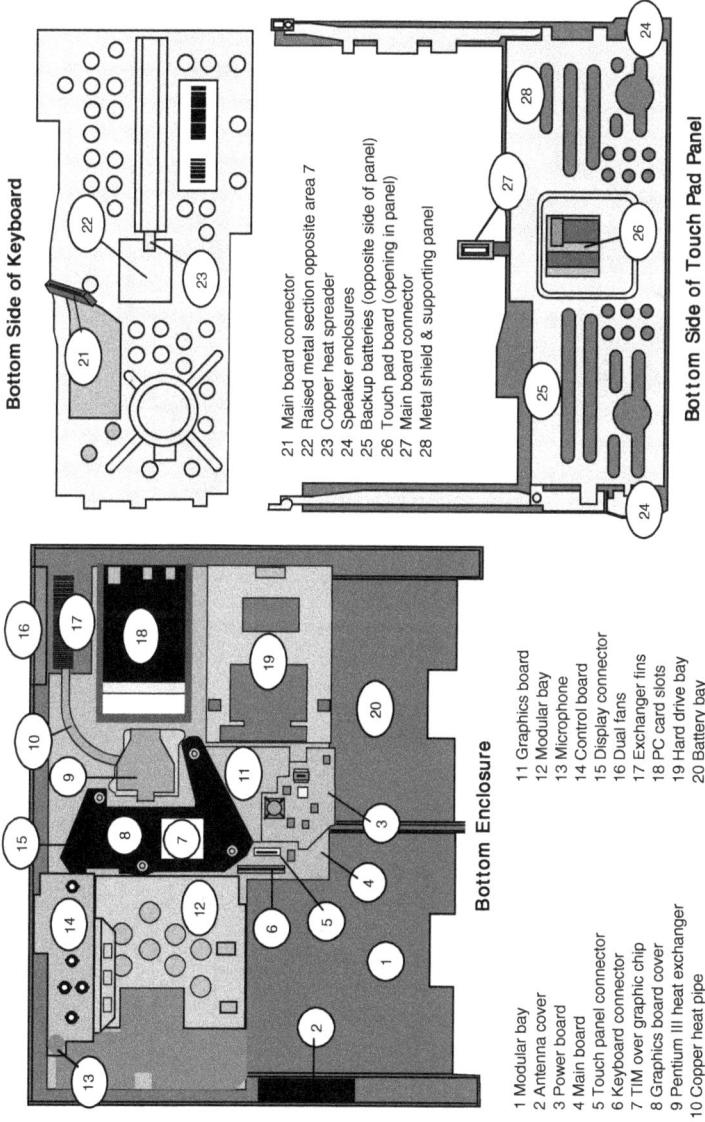

Bottom Side of Keyboard

21 Main board connector
22 Raised metal section opposite area 7
23 Copper heat spreader
24 Speaker enclosures
25 Backup batteries (opposite side of panel)
26 Touch pad board (opening in panel)
27 Main board connector
28 Metal shield & supporting panel

Bottom Side of Touch Pad Panel

Bottom Enclosure

1 Modular bay
2 Antenna cover
3 Power board
4 Main board
5 Touch panel connector
6 Keyboard connector
7 TIM over graphic chip
8 Graphics board cover
9 Pentium III heat exchanger
10 Copper heat pipe

11 Graphics board
12 Modular bay
13 Microphone
14 Control board
15 Display connector
16 Dual fans
17 Exchanger fins
18 PC card slots
19 Hard drive bay
20 Battery bay

Figure 9.9 Component arrangement.

235

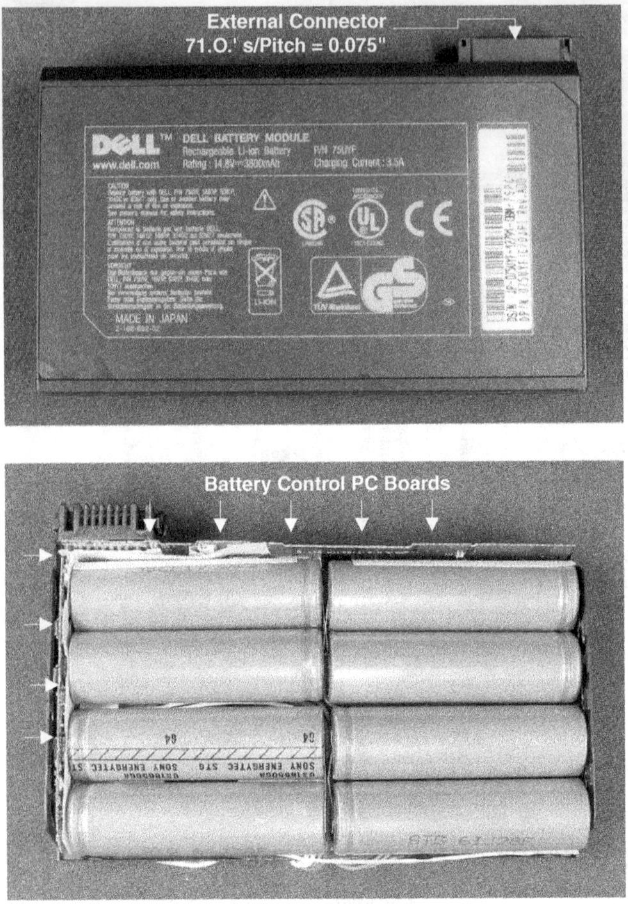

Figure 9.10 Battery pack.

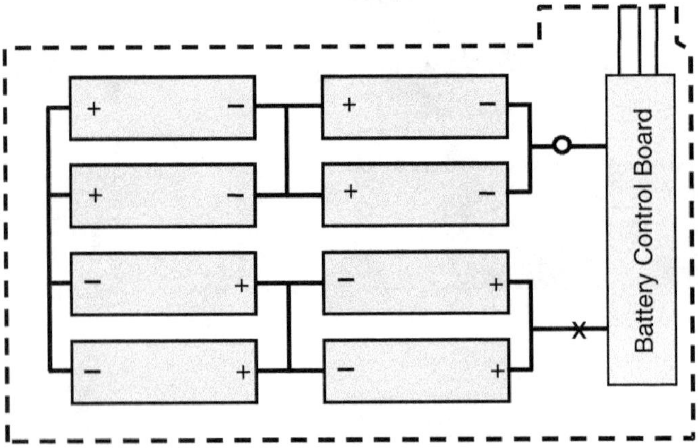

Figure 9.11 Battery control circuitry.

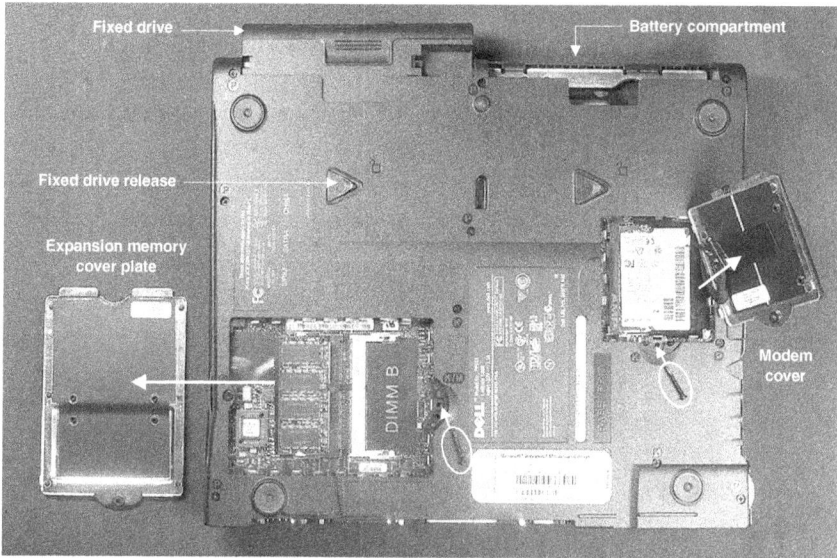

Figure 9.12 Fixed drive & bottom cover plates removed.

an optional increase in system memory. Figure 9.16 shows a similar scheme for the mounting of the hard drive. Each of the different drive modules are shown in Fig. 9.17 and the three drive bays are shown in Fig. 9.18.

Figure 9.19 shows the overall dimensions of the display module. This module is attached to the main enclosure by a couple of hinges. There is a single electrical flex connector that attaches the display to the motherboard. The display module is highly integrated and can be separately

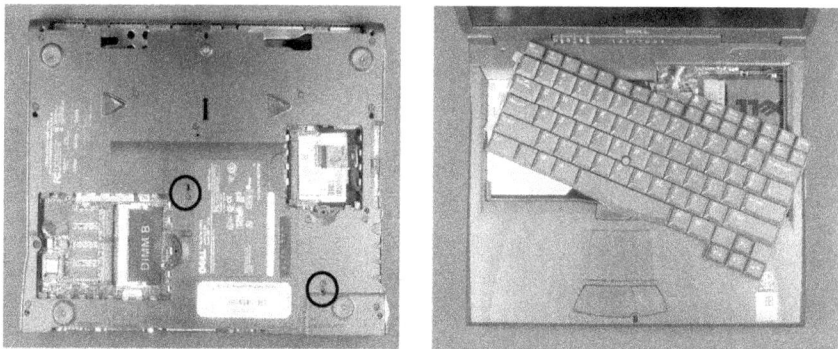

Figure 9.13 Keyboard removal.

Figure 9.14 Keyboard.

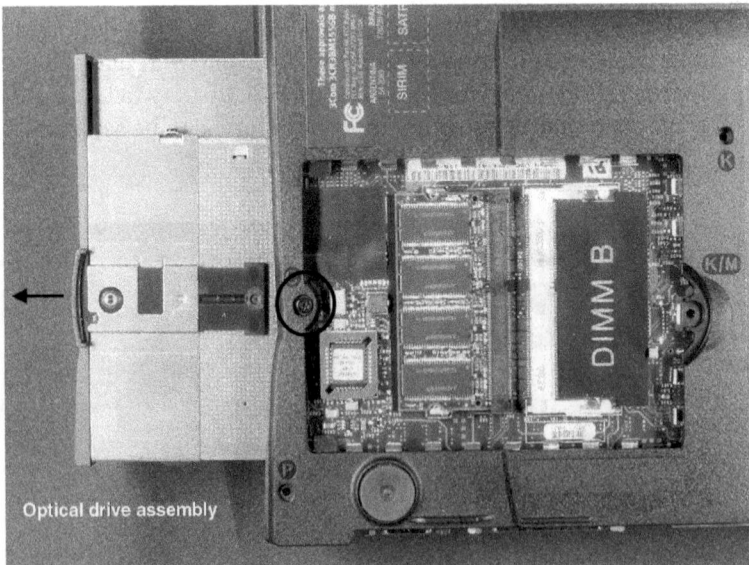

Figure 9.15 Fixed optical drive removed.

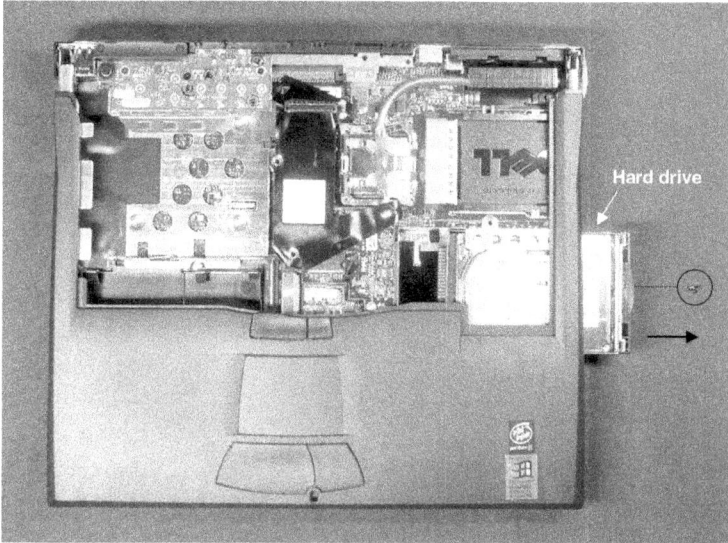

Figure 9.16 Hard drive removal.

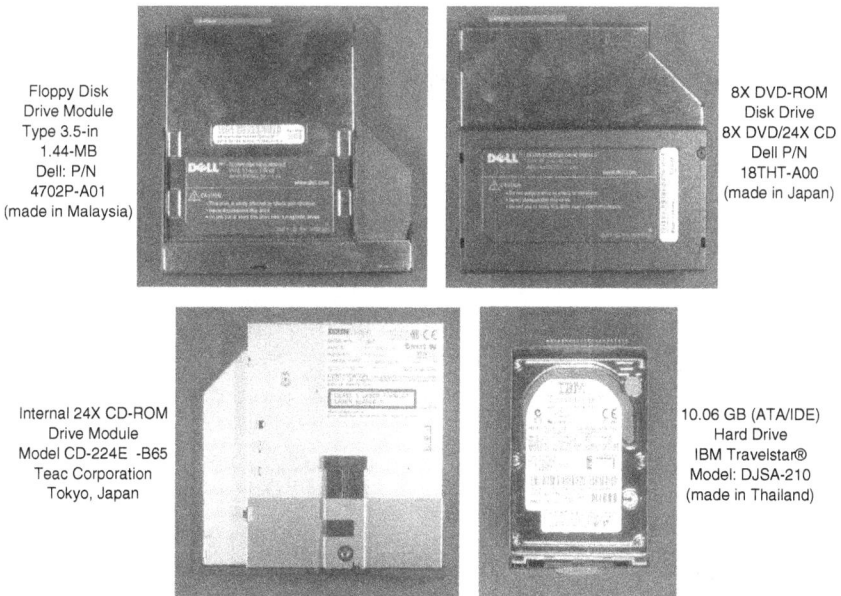

Floppy Disk
Drive Module
Type 3.5-in
1.44-MB
Dell P/N
4702P-A01
(made in Malaysia)

8X DVD-ROM
Disk Drive
8X DVD/24X CD
Dell P/N
18THT-A00
(made in Japan)

Internal 24X CD-ROM
Drive Module
Model CD-224E -B65
Teac Corporation
Tokyo, Japan

10.06 GB (ATA/IDE)
Hard Drive
IBM Travelstar®
Model: DJSA-210
(made in Thailand)

Figure 9.17 Drive modules.

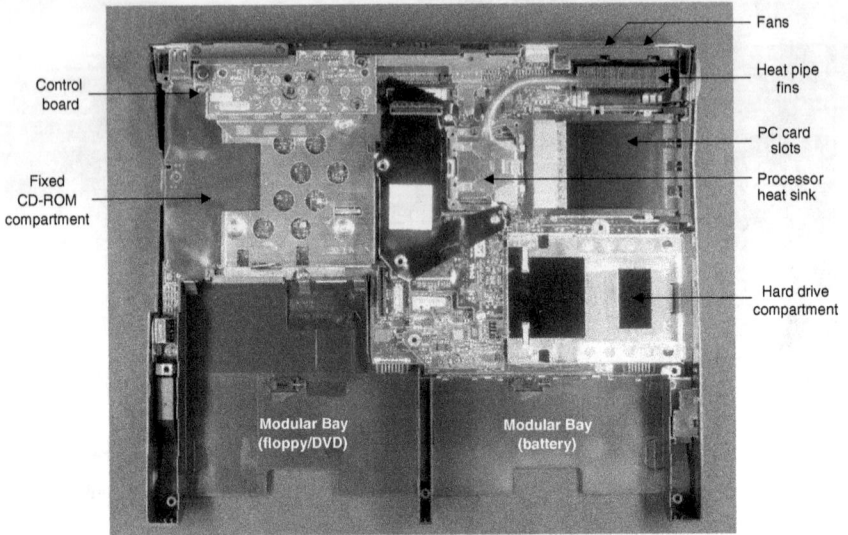

Top view of the lower enclosure with the touch pad enclosure removed.

Figure 9.18 Modular compartments.

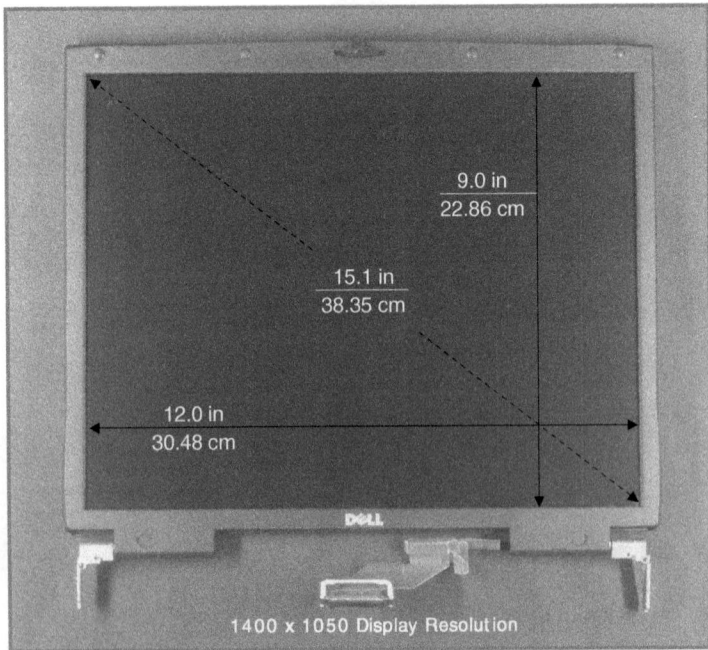

Figure 9.19 Display module.

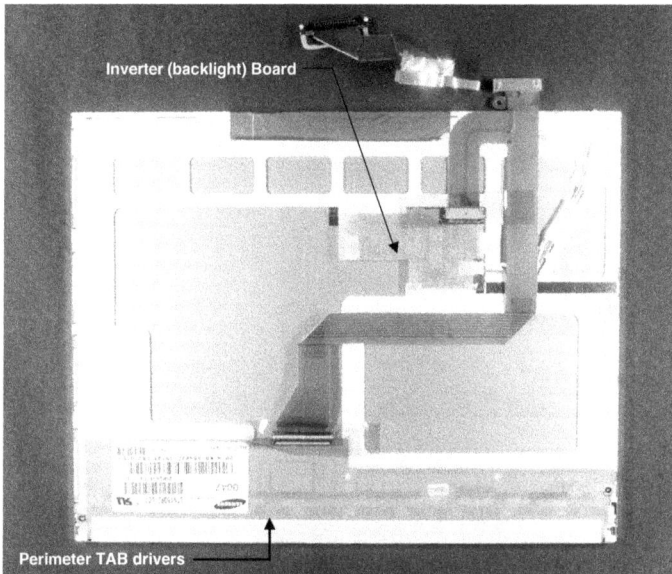

Figure 9.20   Display panel (back side).

tested before it is assembled into the final product. Figure 9.20 shows the display module with the external plastics removed. This represents the state of the display module as it is usually received from the display manufacturer with the display, backlight, and backlight electronics integrated into a complete solution. The design of the display flex interface shown in Fig. 9.21 is intended to allow the flex to accommodate the motion of the display hinge without undergoing a tight-bend radius that, over time, could cause fatigue in the electrical traces. The display module housing and frame (Fig. 9.22) are custom to the product design. The plastic housing has aluminum foil adhesively attached to its surface to prevent EMI emissions. The metal frame is stamped steel and is essential to provide rigid support for the LCD glass. Magnesium has been used in some notebook systems for this purpose, resulting in reduced weight, but significantly increasing cost.

The touch pad assembly is integrated into the upper panel of the main product enclosure (Fig. 9.23). The audio speakers are also part of this mechanical assembly. This is a further example of how groups of mechanical components are aggregated into subassemblies that can be quickly assembled during the final stages of product manufacture. The touch pad PCB is shown in Fig. 9.24. Touch pads are fairly sophisticated sensor devices and usually require unique drivers to support the silicon-enabled features provided by the device manufacturer.

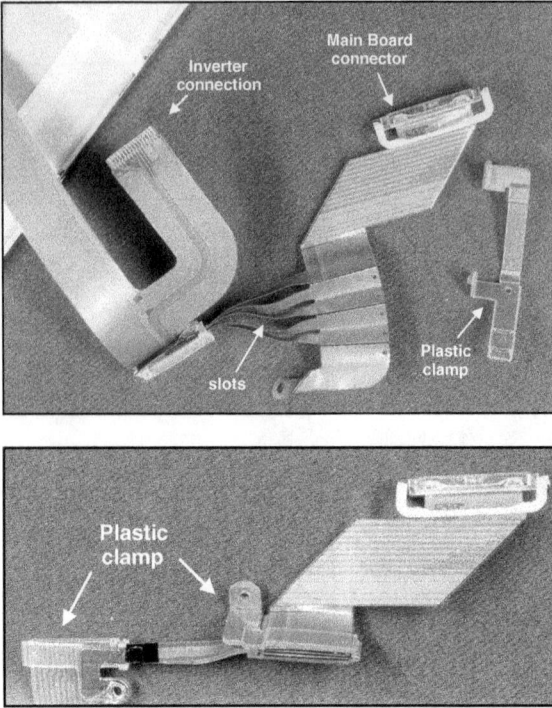

Figure 9.21   Display flex interface.

A view of the motherboard is shown in Fig. 9.25 with many of the obscuring structures removed. What is not easily discerned from the photograph is that there are a number of other PCA modules mounted on the motherboard, and parallel to the same horizontal plane as the motherboard. In desktop PCs, these modules are usually mounted

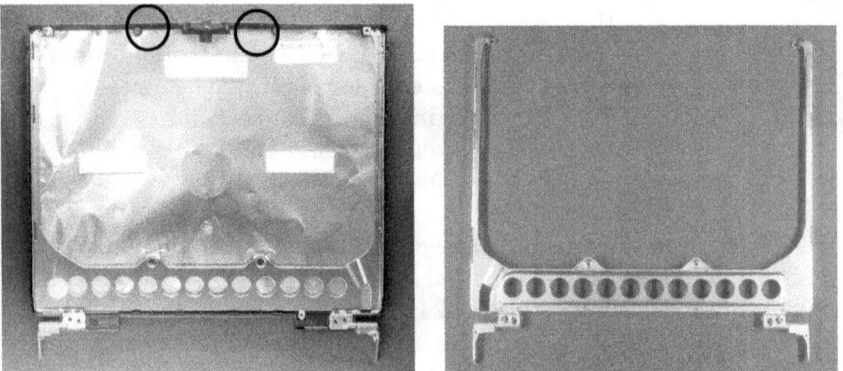

Figure 9.22   Display module housing & frame.

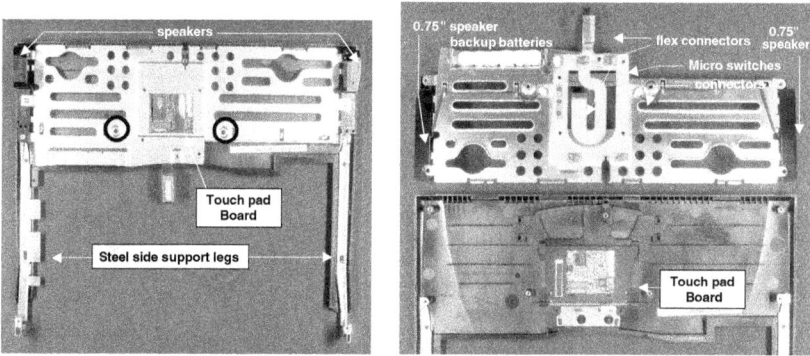

Figure 9.23 Touch pad enclosure.

perpendicular to the motherboard, but notebook systems are thin, so there is no room in the z axis to mount modules in a perpendicular fashion. For this reason, notebooks drive the development of miniaturized, parallel-mounted versions of PC module standards such as mini-PCI and small outline DIMM (SODIMM) connectors.

The graphics board is shown in Fig. 9.26 removed from the motherboard. Notice the surface mount connector that enables parallel attachment to the motherboard. This module contains the graphics chip and the LFB memory required for high-performance graphics. Since it is in a modular form, the notebook manufacturer has the option to upgrade

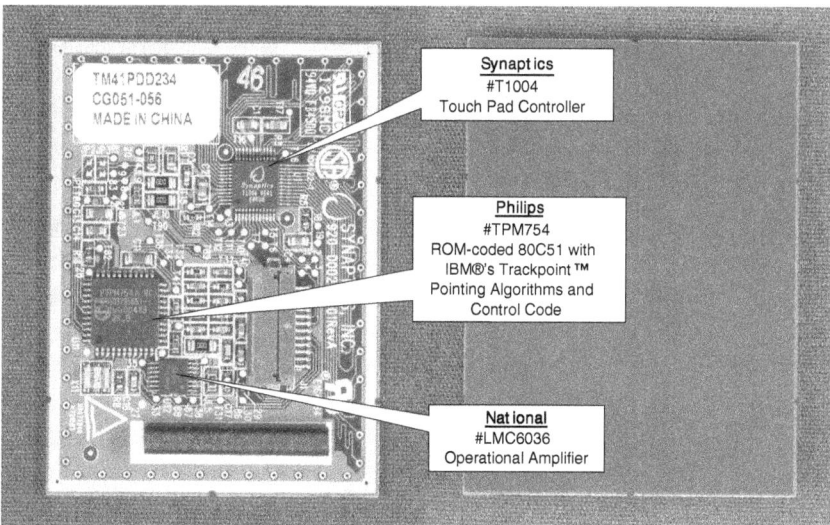

Figure 9.24 Touch pad PCB.

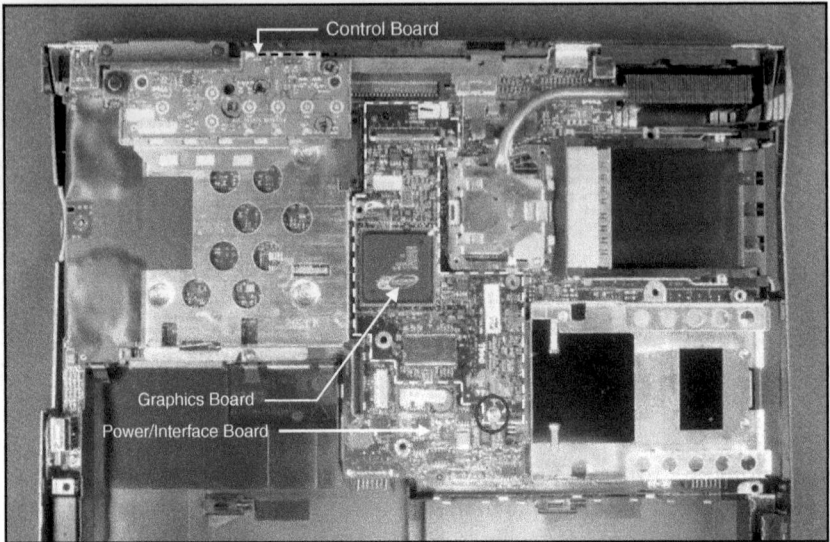

Figure 9.25 Motherboard (top).

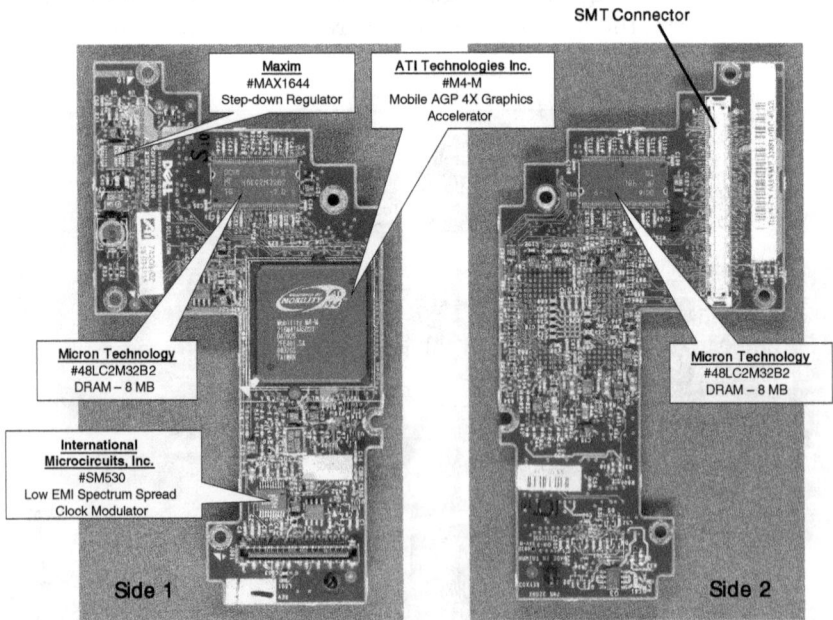

Figure 9.26 Graphics board (front/back).

the product with a different module from any vendor, provided the graphics to North Bridge interface is compatible.

The discrete graphics chip represents a significant portion of the system silicon. Discrete graphics chips will tend to have a quantity of transistors approaching that of the contemporary microprocessor. A CPU that has 70 million transistors, for example, may coexist in a system with a graphics chip that has around 40 million transistors. As the graphics processing becomes integrated into the North Bridge, the North Bridge will likewise comprise an even greater portion of the system transistor count. It is indeed possible that an integrated North Bridge of the future will have *more* transistors than the CPU and thus may dominate the system architectural design process.

As mentioned earlier, the graphics chip in this system requires an enhanced thermal solution that enables the heat from this chip to be transferred to the metal frame of the keyboard. Figure 9.27 shows a special film assembly that is placed on the motherboard to position a thermal interface film (TIM) between the graphics chip and the heat spreader. This TIM improves the contact between the two surfaces and decreases the effective thermal resistance.

The power interface board is shown in Fig. 9.28. This module also relies on a surface mount connector to make the parallel attachment to the motherboard. This module provides the intelligence for detecting the battery status and controlling the battery recharge.

The control board depicted in Fig. 9.29 combines a number of user interface devices. The built-in microphone (surface mounted) and the

Figure 9.27 Graphics chip TIM cover.

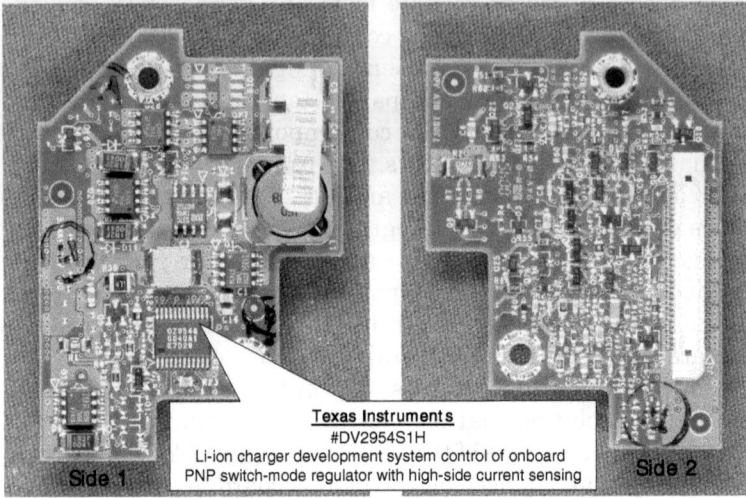

Texas Instruments
#DV2954S1H
Li-ion charger development system control of onboard
PNP switch-mode regulator with high-side current sensing

Side 1   Side 2

Figure 9.28   Power/interface board (front/back).

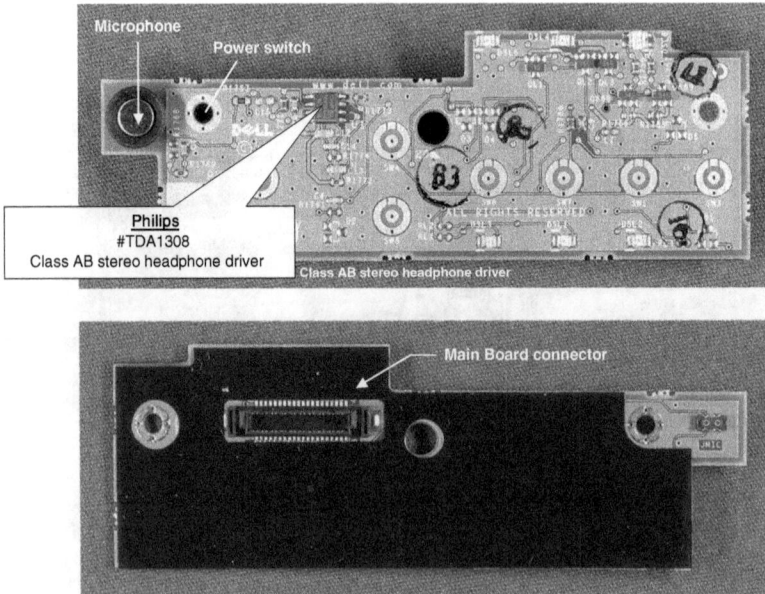

Microphone
Power switch

Philips
#TDA1308
Class AB stereo headphone driver

Class AB stereo headphone driver

Main Board connector

Figure 9.29   Control board (front/back).

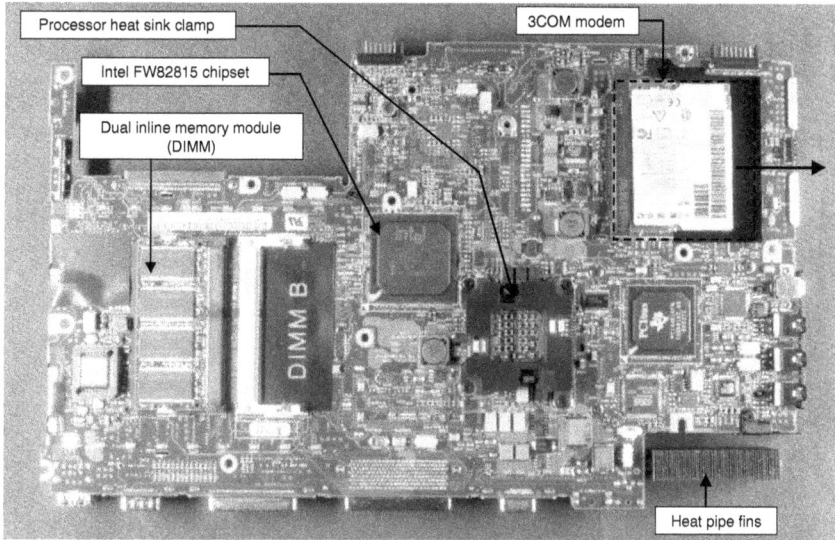

Figure 9.30   Motherboard (bottom).

stereo headphone driver are both mounted on this board. In addition, a number of switch contacts for system status lights and control buttons are incorporated into the board design.

Figure 9.30 shows the bottom side of the system motherboard. One notable feature is the processor heat sink clamp. This mechanical part is used to provide a backside mounting point for screws that are needed to clamp the heat sink to the top of the processor.

Again, there are a number of parallel-mounted modules that have been deployed. So, in a typical notebook design there are really three layers of PCA assemblies in the main enclosure; the motherboard, the topside modules, and the bottom side modules. The top and bottom side modules are positioned, of course, so that they do not directly contribute to the overall product thickness. Thickness-limiting elements, such as mass storage drives and battery packs, are mounted beside the PCA assemblies so that their thickness alone becomes the limiting factor to the overall thickness of the enclosure.

Figure 9.31 shows the DIMM card that contains the system memory. Both sides of this module have four 8 MB DRAM chips packaged in thin surface mount packages. Additional memory may be added to the system using the vacant SODIMM connector on the motherboard.

The modem module pictured in Fig. 9.32 is a mini-PCI standard form-factor. The mini-PCI connector allows the modem to be replaced with other optional functions such as a wireless LAN. The manufacturer has

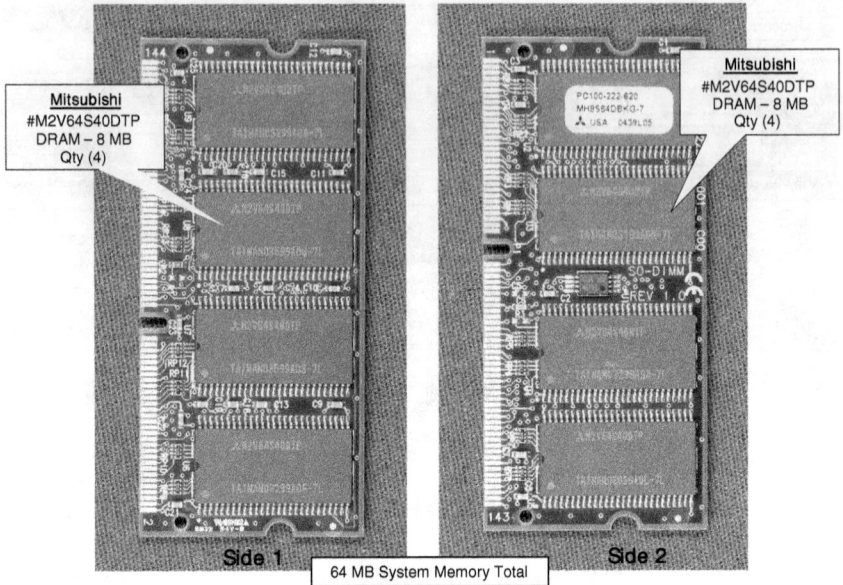

Figure 9.31 Dual inline memory module.

Mitsubishi
#M2V64S40DTP
DRAM – 8 MB
Qty (4)

Mitsubishi
#M2V64S40DTP
DRAM – 8 MB
Qty (4)

Side 1

Side 2

64 MB System Memory Total

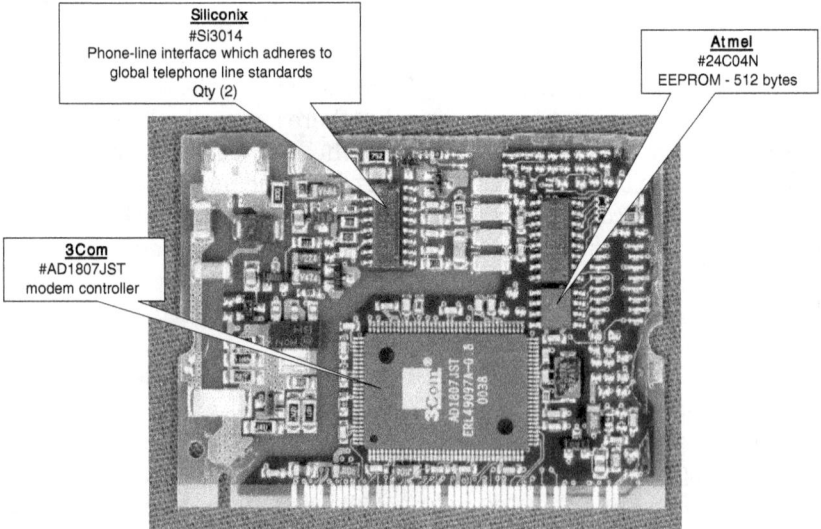

Figure 9.32 Modem PC board.

Siliconix
#SI3014
Phone-line interface which adheres to
global telephone line standards
Qty (2)

Atmel
#24C04N
EEPROM - 512 bytes

3Com
#AD1807JST
modem controller

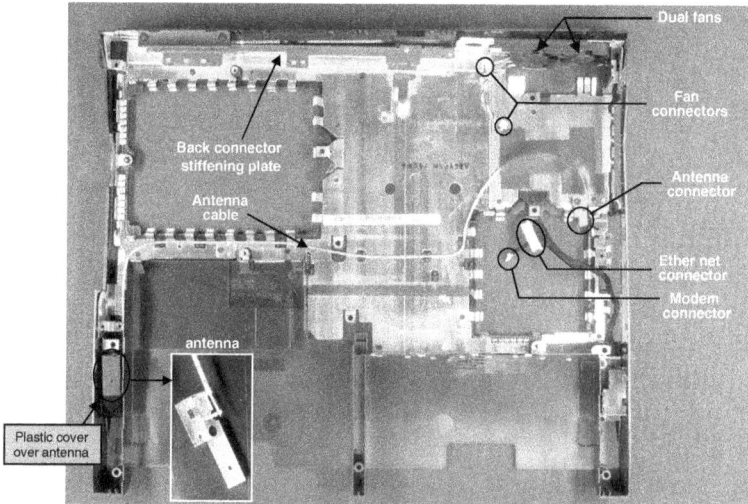

The metal shielding panel was heat staked inside the bottom (ABS) plastic enclosure. The entire assembly weighs 264 g, of which the back connector stiffening plate and dual fans weigh 34.5 g.

Figure 9.33   Bottom enclosure.

included an antenna in the design to support a wireless LAN mini-PCI card option.

The antenna is pictured in Fig. 9.33, which shows the bottom of the main product enclosure. This is a 2.4-GHz antenna that can be used for 802.11b or 802.11g wireless LAN cards. Since the antenna is an inexpensive component, the manufacturer has decided to include it in the standard design even though the wireless LAN was not the standard configuration at the time of introduction. The bottom enclosure also has two openings which enable the manufacturer or the user to add memory or change the mini-PCI card.

Most notebook computers require an extensive thermal design solution. Figure 9.34 shows the processor heat pipe, as it appears on the system motherboard. In this view, the heat pipe assembly is mounted to the top of the processor package. Figure 9.35 shows the heat pipe assembly after it has been removed from the motherboard. The processor interface plate, the heat pipe, and the heat exchangers are integrated into a single part to facilitate ease of assembly into the product. A TIM coupon is attached to the processor interface plate. The processor is in an array package with the die exposed, enabling a more direct thermal path between the die and the heat pipe assembly.

The processor package is in a pin-array configuration which is plugged into a matching 495 pin socket (Fig. 9.36). This socket is surface mounted to the motherboard prior to the installation of the processor. A socketed

Figure 9.34 Processor heat pipe.

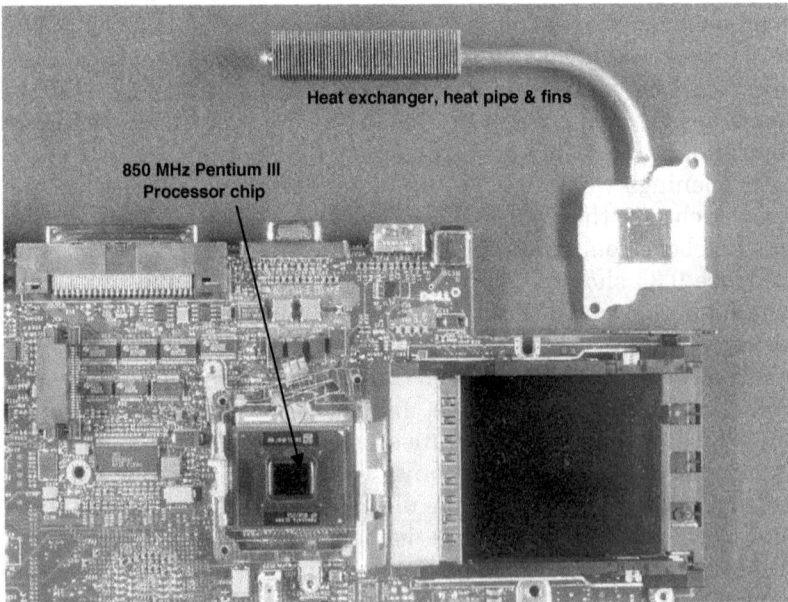

Figure 9.35 Processor heat pipe removed from board.

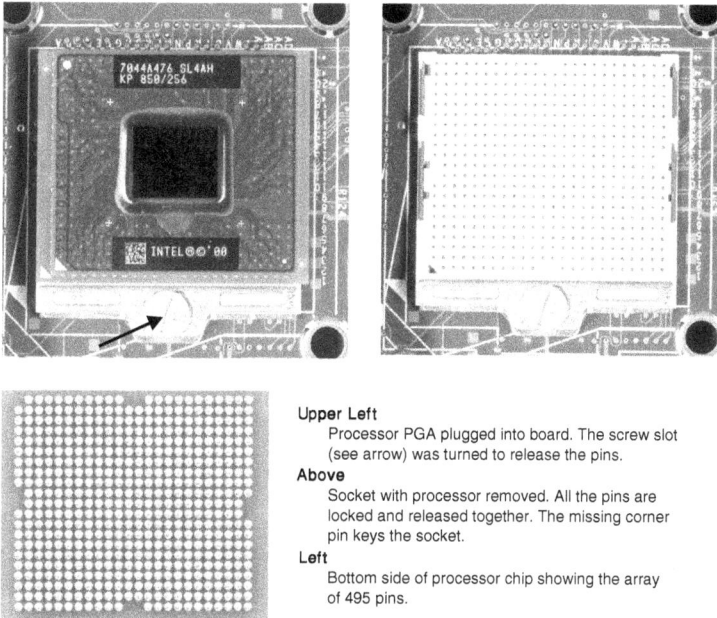

Upper Left
Processor PGA plugged into board. The screw slot (see arrow) was turned to release the pins.
Above
Socket with processor removed. All the pins are locked and released together. The missing corner pin keys the socket.
Left
Bottom side of processor chip showing the array of 495 pins.

Figure 9.36 Processor and socket.

processor strategy is used in almost all contemporary notebook products. There are two main reasons for this. First and foremost, the manufacturer will wish to *postpone* the insertion of the processor until just before it is shipped to the customer. This *postponement* is common practice in the industry and allows the manufacturer to avoid the inventory costs of owning the processors for a longer period of time. Second, over the course of the product life, there may be several *speed bumps* to the processor, due to process and yield improvements by the processor manufacturer. A socketed processor design allows the notebook manufacturer to populate the product with a faster processor at the time of sale, which may not have been available (or was more expensive) at the time of motherboard manufacture. This is an important feature when it is realized that processor prices for a particular speed grade can drop rapidly in the period of a few weeks. CPU socketing also enables "platforming" by allowing the manufacturer to offer the same basic product design with several different speed grades simultaneously in the market. Each of these speed grades can be positioned at different price points and accompanied with a different set of system configuration options.

Figure 9.37 is a thermal diagram summarizing the mechanical concept for cooling the processor. A lower clamp mounted on the bottom side of the motherboard is attached to the heat sink assembly and

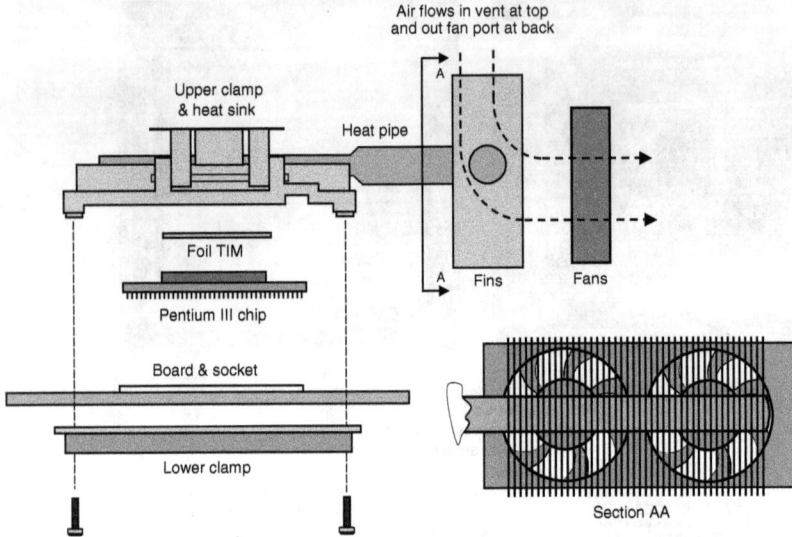

Figure 9.37   Thermal diagram.

tightened down. Sandwiched between the lower clamp and the heat sink assembly are the motherboard, the processor socket, the processor package, and a TIM coupon. Heat generated in the processor chip passes into the heat sink portion of the heat pipe assembly. The thermal resistance at this interface is reduced by the presence of the TIM. The heat moves from the heat sink into the heat pipe. The heat pipe conducts the heat to the heat fins. The fans force cool air from outside the product enclosure across the heat fins to increase the rate of heat transfer out of the system.

The concept of thermal design power (TDP) is important when matching a notebook computer form-factor to a particular CPU wattage. A CPU specification will include the maximum operating wattage. Higher wattage generally enables the CPU to operate at a higher frequency. Higher wattages also require that the thermal solution be able to move more power out of the system. The heat moving capacity of the thermal management solution is generally constrained by the maximum z-axis (thickness) dimension available for the motherboard/socket/CPU heat-sink stack-up.

This is one reason why "Thin & Light" notebook systems tend to have lower speed processors than "Full Size" notebook systems. A 50-W TDP box refers to a notebook system that has been engineered to handle a processor with a maximum power rating up to 50 W. Thus a 75-W box notebook design will usually be much thicker than a 25-W TDP box.

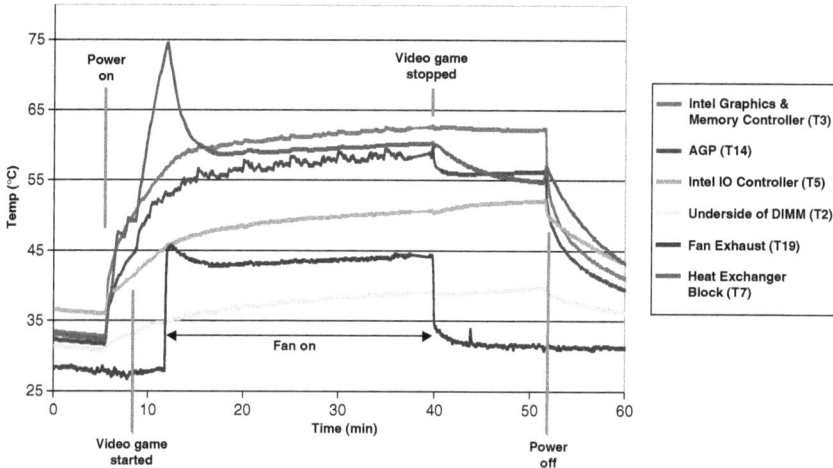

Figure 9.38 Thermal test results.

Mechanical engineers will analyze the thermal conditions within a product by attaching thermocouple sensors to various points in the system. Figure 9.38 shows the test results for a thermal analysis of this notebook system. Temperatures are recorded at key points in the system over a period of time during which the system is performing a heavy computational workload. This particular test was conducted while a high-performance video game was being played. Based on the temperature profiles provided, the thermal design engineer can infer the temperature at the semiconductor junction and determine if the critical die temperature (T_{DieMax}) is in danger of being exceeded. While built-in temperature sensors will keep T_{DieMax} from being exceeded during actual operation, it is important that the thermal solution makes sensor-activated system throttling (or in some cases, system shutdown) a rare event.

Another approach to addressing the thermal problem in a notebook computer is to simply reduce the power consumption. This has the added benefit of increasing the product battery life. Of course, the challenge is to reduce power consumption without creating a drop in performance that is problematic to the user.

The power consumption of the major system elements can be determined relatively easily by monitoring the total system power consumption and correlating that to various modes of system operation. Figure 9.39 shows the total system power consumption during system boot-up and stabilization. Peak boot-up power is about 40 W but the system stabilizes at around 20 W. The second chart in this figure shows

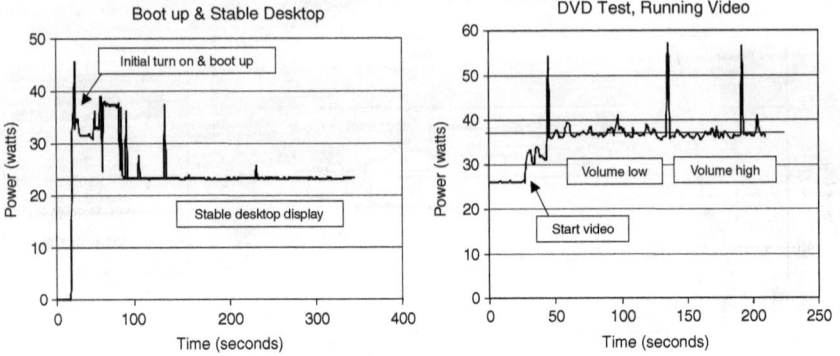

Figure 9.39 Boot up & DVD power.

that the system runs at about 40 W when playing a DVD, with occasional spikes above 50 W. Figure 9.40 shows the power consumption for the floppy drive and CD. In the chart on the left in Fig. 9.41 the total system power consumption is shown with the CPU working at 100 percent utilization. The display is on and is either presenting a static image or is being updated continuously. The right-hand chart shows the system power savings when the system is put into sleep and hibernate modes.

High-performance graphics can significantly increase total system power consumption, as shown in Fig. 9.42. The left-hand chart is the power consumption of a 3D video game in demonstration mode and the right-hand chart is during actual game play.

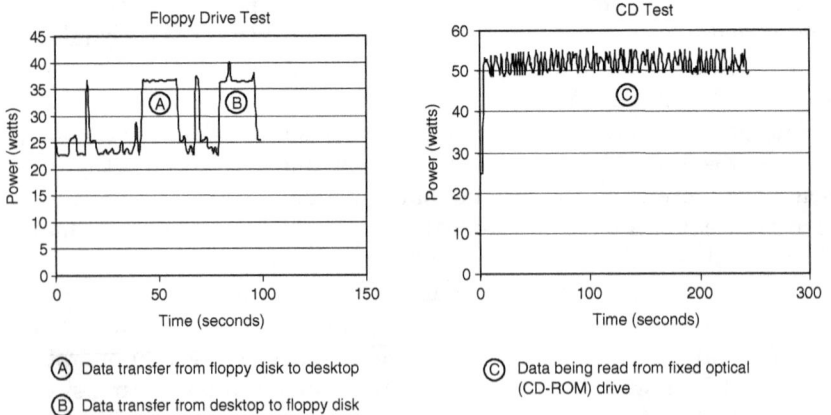

Ⓐ Data transfer from floppy disk to desktop

Ⓑ Data transfer from desktop to floppy disk

Ⓒ Data being read from fixed optical (CD-ROM) drive

Figure 9.40 Floppy and CD drive power.

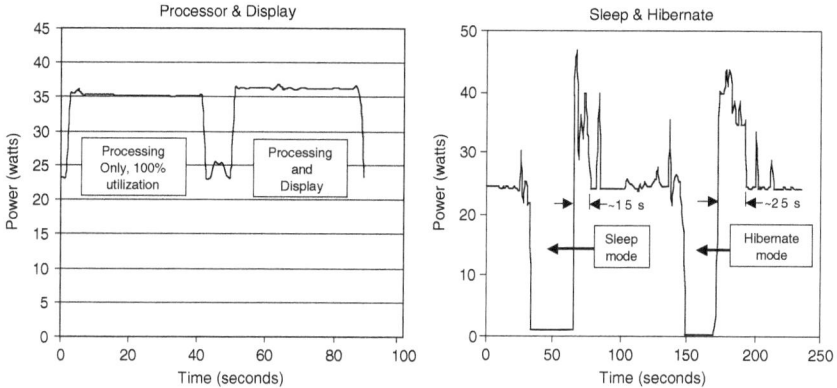

Figure 9.41   Processor & display/sleep & hibernate power.

Figure 9.43 shows a sequence of measurements conducted to isolate the power consumption of the display versus the display backlight. The results indicate that the display consumes 12 W of power, with 3 W going to the LCD and LCD driver chips, and the remaining 9 W going to the backlight and backlight inverter board.

This full-sized notebook system weighs 8.29 lb. Table 9.1 shows the breakdown of how each of the system components contributes to the overall product weight.

The single largest contributor to the system weight is the LCD module. This is no surprise since the display largely comprises two sheets of 15 in diagonal glass. The glass sheets must each be at least 1.5 mm thick or they will not have enough structural integrity. (One of the biggest benefits of a plastic display technology would be a significant reduction in product weight and an increase in product ruggedness.)

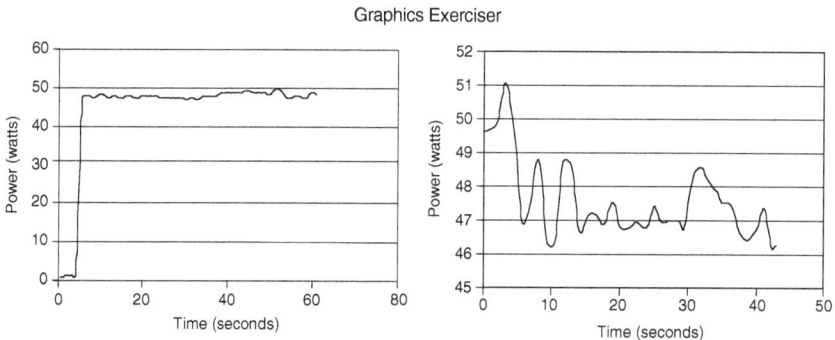

Figure 9.42   Graphics power.

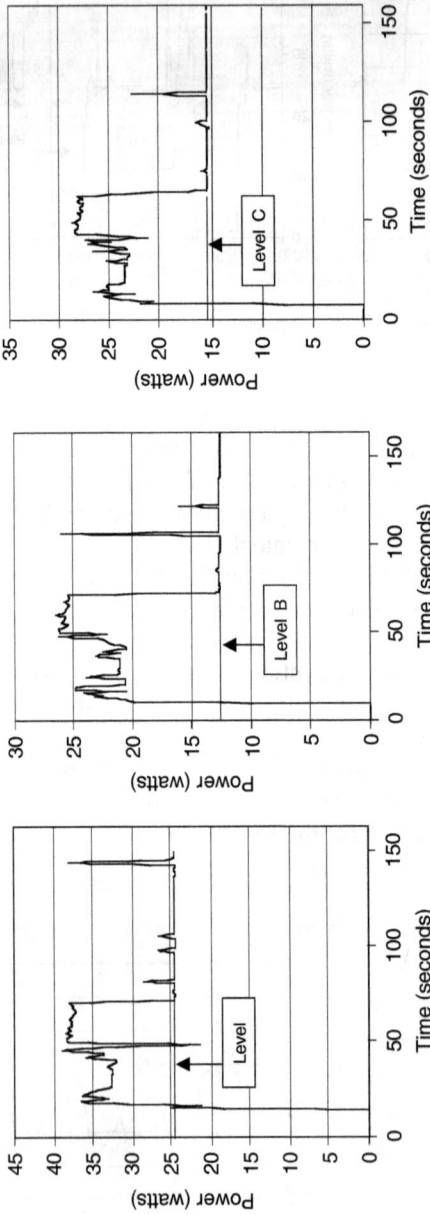

Total display assembly power consumption
(LCD + drivers + CCFL* + inverter board) = level A – level B = 24.5 – 12.5 = 12 W

Display inverter board + CCFL = level A – level C = 24.5 – 15.5 = 9 W

*Cold cathode florescent lamp

Figure 9.43 Display power.

TABLE 9.1   Component Weights

Component	Weight (g)
Eight cell Li-ion battery pack	420.8
3.5" 1.44 MB floppy drive	237.0
24X CD-ROM drive	231.1
8X DVD drive	344.9
Keyboard	173.9
Hinge cover plate (ABS plastic)	33.2
Display lid (ABS plastic)	237.2
Display frame & hinge	194.9
Display bezel (ABS plastic)	38.0
Display assembly (15.1" LCD & electronics)	698.0
Touch panel enclosure (ABS plastic, less frame, board, batteries)	127.3
Backup batteries	12.5
Touch pad board	8.8
Touch pad frame & speaker enclosures	74.5
Graphics board	35.4
Power/interface board	14.0
Control board	9.7
Graphic board cover (shield & TIM)	4.0
Control board shield	3.6
Heat exchanger (including copper heat pipe & fins)	45.9
Bottom enclosure (ABS plastic, less back connector frame & fans)	229.6
Back connector frame (aluminum, less fans)	22.0
Dual fans	12.0
Modem & memory expansion covers (ABS plastic)	64.0
3COM modem card	9.6
Memory board (DIMM)	9.8
Heat exchanger clamps (2 pieces)	30.5
Drive bay metal enclosure/shields (HD & CD-ROM)	66.8
PC card slots	25.3
Main board	334.1
Miscellaneous (80 screws, tape, springs, etc.)	21.0
Total	**3769.4 (8.29 lb)**

Batteries are the second largest contributor to the weight of the system, even with the recent advances in energy density that have resulted from lithium-ion technology. Again, reduction in total system power consumption is a highly leveraged improvement because it can also result in decreased system weight if fewer battery cells are needed to achieve the required product battery life.

The selection of PCA substrate technology becomes an issue for the designer at some point in almost all high-function portable electronic products. The substrate technologies for each of the modules in this notebook product are depicted in Fig. 9.44 (set 1), Fig. 9.45 (set 2), and Fig. 9.46 (set 3). In this system, the motherboard and the graphics board are 10-layer boards and the other modules are either two layers or four layers. It is now the trend in notebook design to attempt to reduce the motherboard to a four- or six-layer design to reduce overall system cost.

Main PCB
Total Area =
74.785 sq in/482.48 cm^2

Main PCB
(*FR4 substrate*)
Layers = 10
Finest pitch = 0.010
Narrowest trace = 0.005
Narrowest space = 0.005
Smallest Via I.D. = 0.012
Smallest Via O.D. = 0.024
Substrate thickness = 0.062
Assembly weight = 331.8 grams

PCB#41010

Power PCB
(*FR4 substrate*)
Layers = 4
Finest pitch = 0.012″
Narrowest trace = 0.006″
Narrowest space = 0.006″
Smallest Via I.D. = 0.012″
Smallest Via O.D. = 0.024″
Substrate thickness = 0.060″
Assembly weight = 14.0 g

Power PCB
Total area =
3.118 sq. in/20.12 cm^2

Memory PCB
Total area =
3.288 sq. in/21.21 cm^2

PCB#0439L

Memory PCB
(*FR4 substrate*)
Layers = 6
Finest pitch = 0.010″
Narrowest trace = 0.005″
Narrowest space = 0.005″
Smallest Via I.D. = 0.012″
Smallest Via O.D. = 0.024″
Substrate thickness = 0.040″
Assembly weight = 9.8 g

PCB#730RT

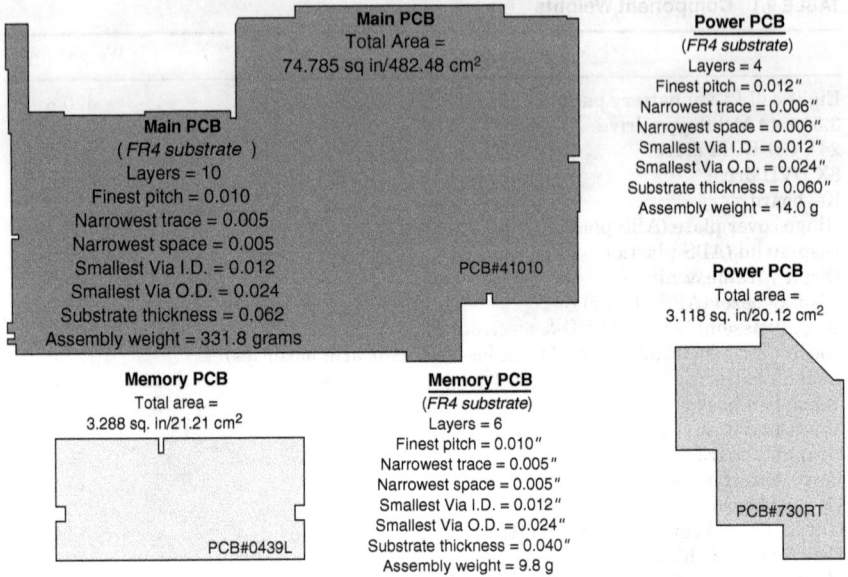

Figure 9.44 Substrate measurements.

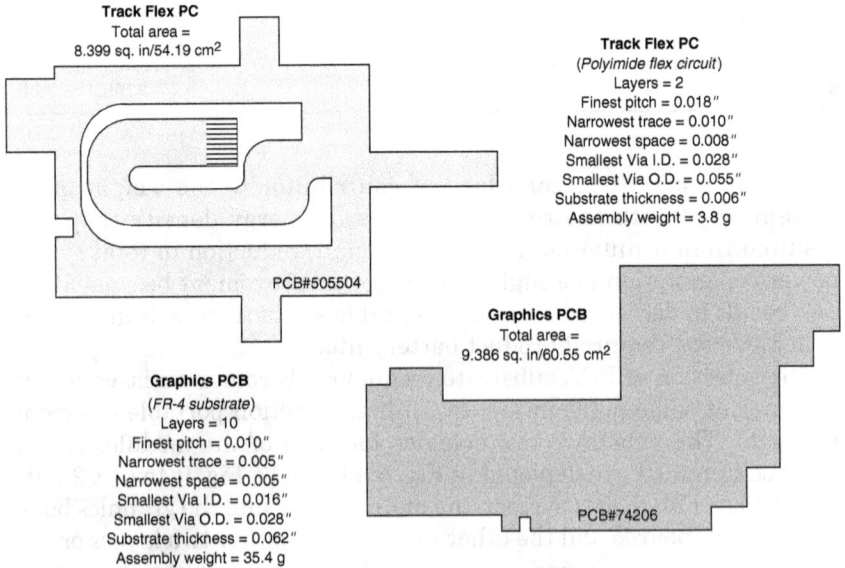

Track Flex PC
Total area =
8.399 sq. in/54.19 cm^2

PCB#505504

Track Flex PC
(*Polyimide flex circuit*)
Layers = 2
Finest pitch = 0.018″
Narrowest trace = 0.010″
Narrowest space = 0.008″
Smallest Via I.D. = 0.028″
Smallest Via O.D. = 0.055″
Substrate thickness = 0.006″
Assembly weight = 3.8 g

Graphics PCB
Total area =
9.386 sq. in/60.55 cm^2

Graphics PCB
(*FR-4 substrate*)
Layers = 10
Finest pitch = 0.010″
Narrowest trace = 0.005″
Narrowest space = 0.005″
Smallest Via I.D. = 0.016″
Smallest Via O.D. = 0.028″
Substrate thickness = 0.062″
Assembly weight = 35.4 g

PCB#74206

Figure 9.45 Substrate measurements.

Control PCB
Total area =
5.089 sq. in/21.21 cm2

PCB#79403

Control PCB
(*FR4 substrate*)
Layers = 2
Finest pitch = 0.013″
Narrowest trace = 0.005″
Narrowest space = 0.008″
Smallest Via I.D. = 0.024″
Smallest Via O.D. = 0.034″
Substrate thickness = 0.040″
Assembly weight = 9.7 g

Track Pad PCB
Total area =
4.909 sq. in/31.67 cm2

PCB#CG051

Track Pad PCB
(*FR4 substrate*)
Layers = 4
Finest pitch = 0.016″
Narrowest trace = 0.007″
Narrowest space = 0.009″
Smallest Via I.D. = 0.022″
Smallest Via O.D. = 0.040″
Substrate thickness = 0.034″
Assembly weight = 8.8 g

Modem PCB
Total area   =
4.016 sq. in/21.21 cm2

PCB#305C3

Modem PCB
(*FR4 substrate*)
Layers = 4
Finest pitch = 0.011″
Narrowest trace = 0.005″
Narrowest space = 0.006″
Smallest Via I.D. = 0.012″
Smallest Via O.D. = 0.020″
Substrate thickness = 0.040″
Assembly weight = 8.7 g

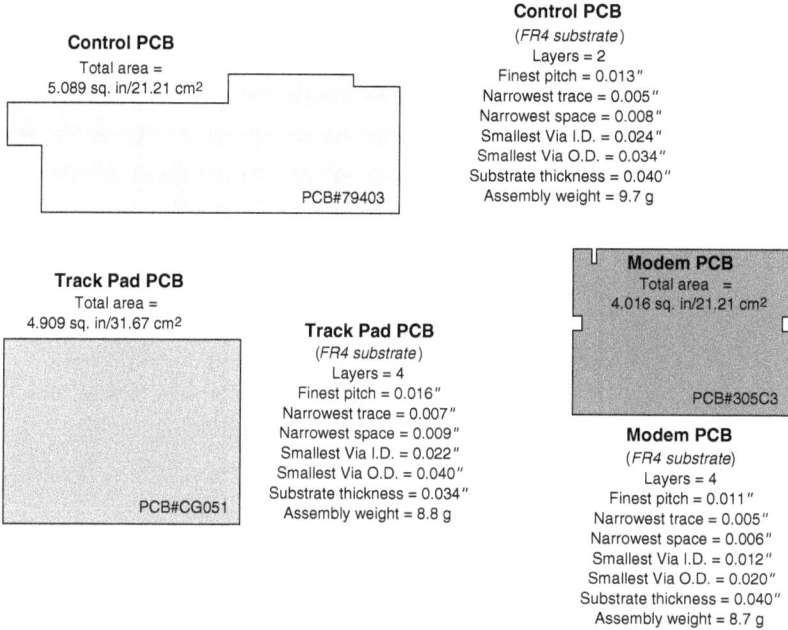

Figure 9.46 Substrate measurements.

The use of high-density modules for local escape routing and complex routing of closely coupled silicon can reduce the need for complexity at the motherboard level and so reduce overall system cost. These issues should be taken into consideration during the system-architectural partitioning and trade-off analysis phase of system design, as it becomes much more difficult to make changes later in the design process

Table 9.2 shows the electronic assembly metrics for the ICs in the notebook product. Notice that a large proportion of the system silicon area

TABLE 9.2   Notebook Computer IC Metrics

Board description	No. of IC's total	No. of analog ICs	No. of digital ICs	No. of IC IO's	Die area (in^2)	IC footprint area (in^2)	IC IO's/(IC footprint area)
Memory PCB	9	—	9	440	1.25	3.48	126
Power PCB	7	6	1	72	0.06	0.38	188
Track Flex PC	—	—	—	—	—	—	—
Graphics PCB	8	3	5	584	0.54	2.67	219
Control PCB	1	1	—	8	0.00	0.05	167
Track pad PCB	3	—	3	106	0.05	0.41	260
Modem PCB	4	—	4	168	0.08	0.80	210
Main PCB	50	16	34	2,827	0.81	10.00	283
IC total	**82**	**26**	**56**	**4,205**	**2.79**	**17.79**	**236**

TABLE 9.3 Notebook Computer Electronic Packaging Metrics

Board description	Board area	No. of PCB layers	No. of parts	No. of connections	PCB tiling density (die area/Bd) (%)	Connection density (conn./Bd area)	Routing density (trace length/Bd area)	Part density (parts/Bd area)	Average pin count (No. of conn./no. of parts)
Memory PCB	3.29	6	50	618	37.90	188	145	15	12.36
Power PCB	3.12	4	105	351	2.06	113	58	34	3.34
Track Flex PC	8.40	2	23	82		10	18	3	3.57
Graphics PCB	9.39	10	226	1,479	5.80	158	96	24	6.54
Control PCB	5.09	2	37	125	0.04	25	27	7	3.38
Track pad PCB	4.91	4	43	200	0.97	41	41	9	4.65
Modem PCB	4.02	4	94	364	1.89	91	56	23	3.87
Main PCB	74.79	10	1,427	7,956	1.08	106	73	19	5.58
System total	**113.01**	—	**2,005**	**11,175**	**2.47**	**99**	—	**18**	**5.57**

(the total area of all the system IC devices, excluding their packaging) is memory, with the graphics chip and the motherboard silicon (mostly CPU) accounting for a large part of the remaining portion. All of the silicon in this notebook product adds up to around 2.8 in^2 (or 1800 mm^2) of silicon. These numbers exclude the silicon in the LCD module that has a significant amount of silicon in the LCD driver chips.

The electronic packaging metrics for this product are given in Table 9.3. Due to the relatively low ratio of I/O to silicon area for memory device, the memory module has an extremely high average tiling density compared to the other modules in the notebook. The graphics module has the next highest silicon tiling density because it has two pieces of high-function silicon which have been packed tightly onto a 10-layer PCB along with the necessary discrete devices. The motherboard has a relatively low tiling density. Most motherboards do not require very tight packing of components because the substrate generally needs to stretch out to various edges of the enclosure to support the mounting of various external I/O connectors. Why, then, is this design a 10-layer board? Ten layers were probably required to enable the escape routing for the processor and chipset and to deal with local routing complexity between the processor and the chipset. A thorough analysis might reveal that the processor-chipset complex could have been routed in a separate module, eliminating the need for a large and expensive 10-layer motherboard.

Figure 9.47 shows the breakdown of the estimated system electronics costs, excluding the LCD module, based on the Portelligent cost model. Silicon is the largest single contributor to electronic systems cost. Bare substrate costs for notebook computers are also a significant

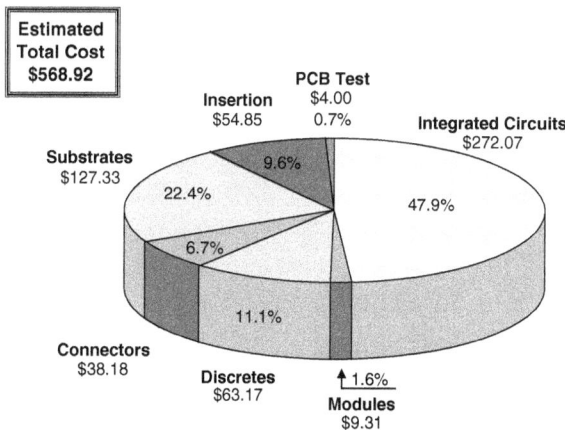

Figure 9.47 System electronics cost estimates.

cost contributor usually comprising 10 to 15 percent of the system cost. Most notebook systems will use four- or six-layer motherboards, which will significantly reduce substrate cost. The relative proportion of costs, in these analyses, are more useful to the system designer than the absolute costs, since these are estimates and are subject to rapid changes over time.

The notebook product designer's primary avenue for controlling the electronics cost is through the substrate partitioning and design. The choice of silicon is often determined by the required product feature set and is thus beyond the direct control of the designer. The substrate partitioning and design, however, is fully under the product designer's control and affects not only the substrate cost but also the cost of connectors, discretes, insertion (surface mount processing), and testing.

Estimates for the cost of other parts of the system are given in Table 9.4. Most of this cost is for mass storage drives. The remaining mechanical parts comprise a relatively minor percentage of the total system cost.

TABLE 9.4 Cost Estimates for Mechanical Components and Mass Storage Modules

Item	Description	Estimated cost
ABS plastic enclosures		
Bottom enclosure	(includes metal shield, antenna, connector frame)	$3.28
Touch pad enclosures	(includes metal support frame)	$2.85
Display lid	(includes shielding, support bracket, hinge assembly)	$5.13
Hinge cover plate	(Includes light pipes)	$0.80
Display Bezel	(Includes adhesive tabs)	$0.60
Heat exchanger	(includes heat pipe, fins, 2-piece clamp assembly)	$15.00
Dual fan assembly	(includes cables & connectors)	$7.50
Drive modules	3.5" 1.44 MB diskette drive[a]	$41.40
	8X DVD drive[a]	$101.40
	24X/10X CD-ROM drive[a]	$59.40
	10 GB 9.5 mm hard drive[b]	$94.20
Cable assemblies	Antenna coaxial assembly w/connectors	$1.35
	Modem cable assembly w/connectors	$1.25
PC card slot (2 card slots)	Complete assembly	$3.50
Keyboard	Complete assembly	$12.50
Graphics board cover	(3-layer cover, TIM)	$1.75
Metal enclosures/ shielding	(CD-ROM & HD bays, control board shield)	$3.00
Miscellaneous	(ABS plastic doors (2), screws (80), labels (7), springs (2), adhesive tape (7), & foil lining (1)	$2.50
Total		**$357.41**

[a] 60% of list price for item per Dell website on 2/14/01.
[b] 75% of price as listed at Access Micro (www. accessmicro.com) 2/20/01.

Notebook product designers spend an extensive amount of effort managing the mechanical design of the system, and much of this effort is focused on thermal management. Even though mechanical parts do not comprise a large part of the system cost, they are the part of the system that is most custom to a particular product, and the thin margins of competitive manufacturers are heavily impacted by the mechanical implementation. Mechanical tooling costs are also quite large for a notebook product. Custom injection molds and special tools for mechanical parts may approach $1 million for a single product design. These costs can be quickly amortized if the product is successful, but can make the product look very expensive if the product sales volume is not as high as expected in the business plan. So, while mechanical parts are theoretically a low percentage of the product design, they represent a high percentage of the product RISK in terms of sunk costs.

The final product assembly costs for this notebook were estimated using the Boothroyd Dewhurst methodology and the results are shown in Fig. 9.48. The relatively low score for this product (2.7 percent) indicates that there is significant opportunity to reduce the assembly time. However, since the total cost of assembly is such a low percentage of the total product cost, enhancements that would improve the DFA score are probably overridden by other system considerations that have a higher impact on overall product cost.

DFA index (%)	**2.7%**
Total assembly time (s)	**1443**
Total assembly cost ($)	**$12.02 (@$30/h)**

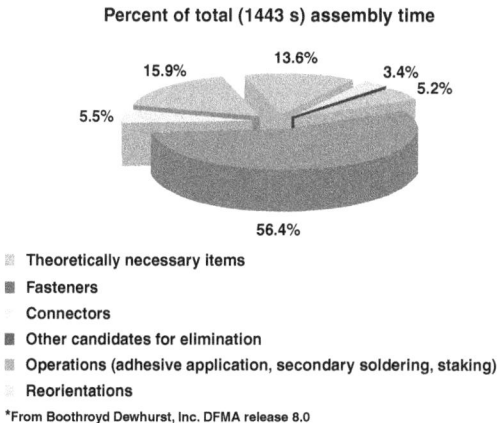

Percent of total (1443 s) assembly time

13.6%
15.9%   3.4%
            5.2%
5.5%

56.4%

▨ Theoretically necessary items
■ Fasteners
  Connectors
■ Other candidates for elimination
▨ Operations (adhesive application, secondary soldering, staking)
  Reorientations

*From Boothroyd Dewhurst, Inc. DFMA release 8.0

Figure 9.48   Design for assembly.

Dell Latitude C800 Laptop	Cost
Electronic assemblies	$568.92
Display	$398.85
Housing/hardware (DFM)	$357.41
Battery pack	$69.40
Final assembly (DFA)	$12.02
Total	**$1,406.60**

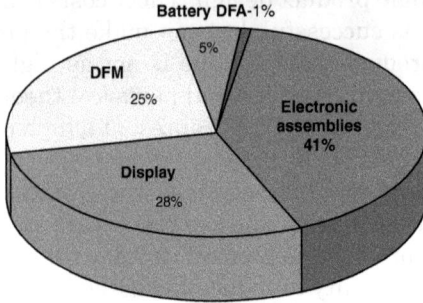

Figure 9.49 Notebook cost summary.

The product level cost summary is shown in Fig. 9.49, and rolls up all of the costs discussed above. The system electronics are the largest portion of system cost. The display module has the highest cost for a single component in the system.

10

Personal Digital Assistants

Personal digital assistants (PDAs) are perhaps the most interesting class of products currently being developed because of the dynamic nature of the market and the high degree of experimentation currently underway in PDA design. The PDA concept combines the potential for high functionality and an information-rich display size, with a much smaller form-factor than a notebook computer. As more functions are integrated into the PDA, it may displace the notebook computer as the platform of choice for consumer access to general purpose portable computing.

To understand the evolution of the PDA and portable electronics in general, it is instructive to look at the history of this product concept and the design of some early PDA products.

The concept for PDA products evolved from electronic organizer products developed by Japanese manufacturers in the 1980s. Electronic organizer products of this era, such as the Casio Boss and Sharp Wizard, were essentially beefed-up calculator products that enabled the user to store contact information, calendar appointments, and memos. Around 1987, Casio came up with the Casio IF-8000 (Fig. 10.1 Casio IF-8000). This product incorporated a touch sensitive display that allowed the user to input and store notes in the form of digitized pen input.

In 1991, Kyocera developed a very novel product called the Kyocera Refalo (Fig. 10.2). The Refalo product was constructed to resemble a ring binder style paper-based organizer. In fact, the product was designed to combine both traditional paper-based personal organization with digital technology. The left cover of the binding housed a larger display than a typical organizer product, with 240×320 pixels. This display was capable of pen input. Special electronic "pages" could

Figure 10.1   Casio IF-8000.

also be purchased and placed into the binder. These pages were actually keypads with electronics that supported specialized organizer functions. These pages communicated with the rest of the system using magnetic induction, coupled through miniature coils embedded in the binder rings and the pages themselves. The system also featured two

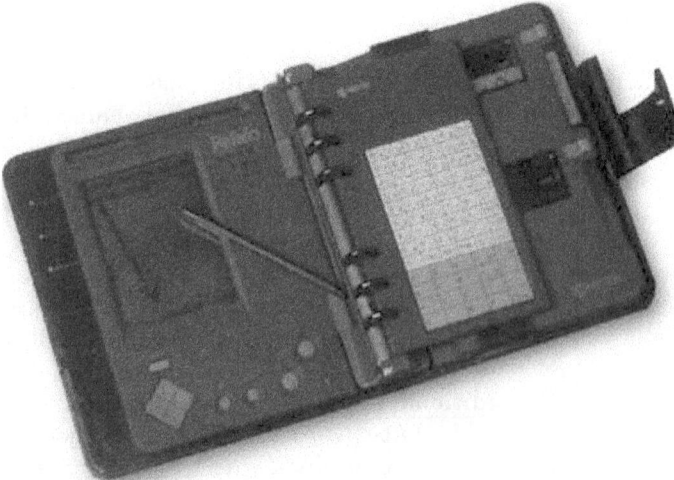

Figure 10.2   Kyocera Refalo.

Figure 10.3 Sony Magic Link.

memory card slots and an RS232 interface for synchronization with a PC.

Also in the early 1990s, the Sony Magic Link (Fig. 10.3) was introduced, using the Magic-Cap OS developed by General Magic. This product introduced the concept of a product that is very display centric, relying mostly on pen input rather than on a keypad.

In 1994, the EO Personal Communicator (Fig. 10.4) came to market, using the Penpoint OS from GO Corp. This product was available with

Figure 10.4 EO Personal Communicator.

Figure 10.5 IBM Simon.

a cellular phone attachment for voice calls and fax. The large form-factor and limited functionality resulted in a very short product life for this concept.

A very noble attempt to create a connected PDA resulted in the introduction of the IBM Simon (Fig. 10.5), also in 1994. This product, concept was more viable than the EO, but the brick-like form-factor ultimately resulted in the demise of this product concept as well.

After a number of such false starts, including a near miss by Apple with the Apple Newton product, the PDA market finally materialized in 1995 with the introduction of the Pilot by U.S. Robotics (Fig. 10.6). This product had the following attributes:

Low cost PDA
- Pilot 1000: $299
- Pilot 5000: $369 (5X memory)

Extremely small size and weight
- 4.7" × 3.2" × 0.7"
- 5.7 oz (including batteries)

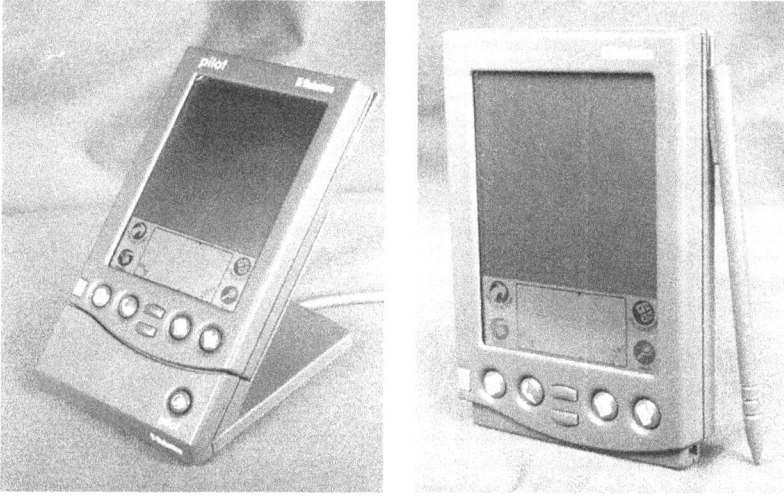

Figure 10.6   Pilot PDA.

Palm top organizer
- Date book
- Address book
- To do list
- Memo pad
- Calculator

Hot SyncTM
- Import to and export from PC
- Backup on PC
- Duplicate desktop software

Large display
- 2.5 in^2 touch-screen LCD
- Separate Graffiti touch screen

The key to the Pilot's success was the ability to synch with Microsoft Outlook, which had become, and still is, the de facto standard for enterprise-based personal information management. The Palm also achieved a very compact form-factor and extended battery life. The product also featured an easy-to-use synchronization and recharge cradle. Another key aspect of innovation for this product was the realization that a low-cost monochrome display was sufficient for delivering the essential functionality and that a modified, intuitive handwriting input scheme could be used to simplify the handwriting input processing requirements (Fig. 10.7).

Touch Screen Display Assembly

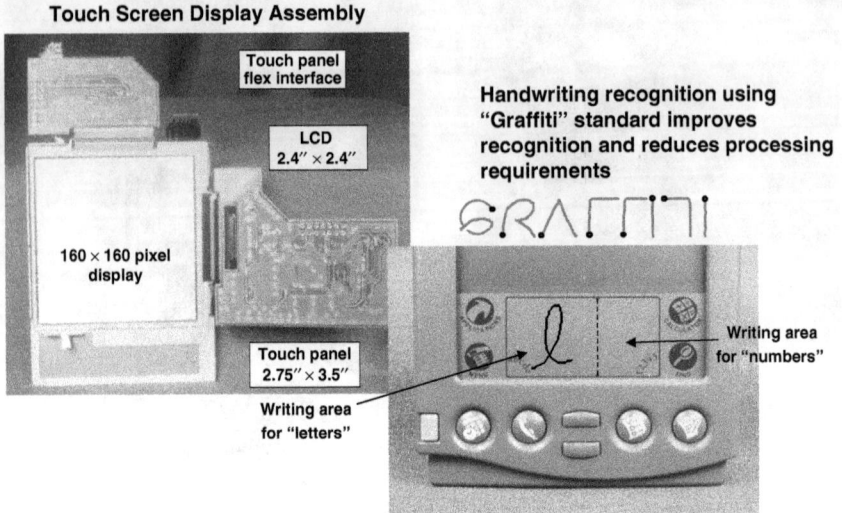

Figure 10.7　Pilot key technologies.

The first product to successfully combine PDA functionality and connectivity was the Nokia 9000 Communicator (Fig. 10.8). This product was introduced into the European GSM cellular phone market around 1997, and continues to be a successful product concept. The Communicator functions as a traditional cell phone, but can also be opened up to expose a small keyboard that facilitates personal organizer functions, email, and even web browsing. Key to the success of the initial product was an innovative mechanical packaging design that stacked a number of subassemblies while still maintaining a relatively thin overall profile. These subassemblies are pictured in Fig. 10.9. Figure 10.10 shows how these subassemblies are stacked within the product. Nokia was able to stack all these subassemblies into a thin form-factor by using aggressive electronic packaging technologies, as shown in Fig. 10.11.

Another early direction for PDA products was to incorporate electronic imaging capability. One of the pioneers of this concept was Sharp. Sharp introduced the Sharp Zaurus in the mid-1990s and continues to manufacture this product line. The original Sharp Zaurus was a high-function PDA that included a PCMCIA slot. Sharp developed an electronic still camera module that plugged into the PCMCIA slot, resulting in an imaging-enabled PDA, as pictured in Fig. 10.12. This product had the following attributes:

Figure 10.8 Nokia 9000 Communicator.

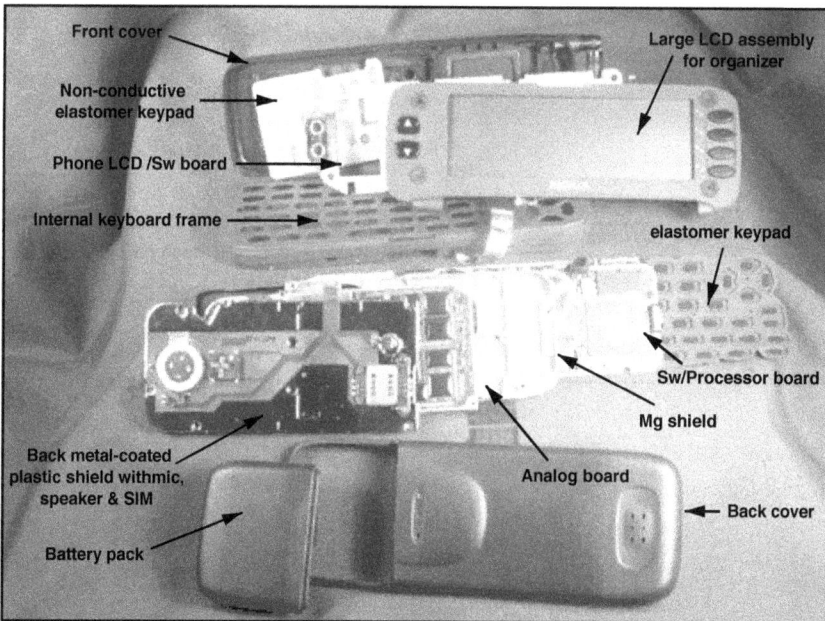

Figure 10.9 Nokia 9000 subassemblies.

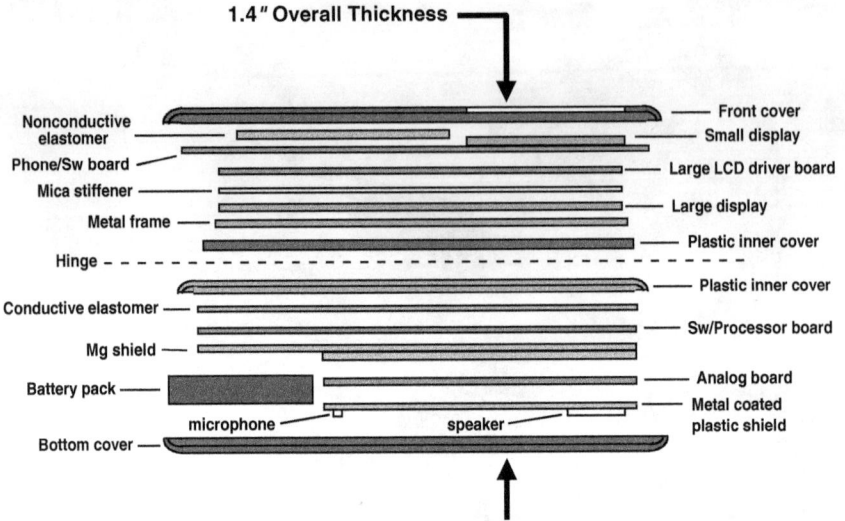

1.4" Overall Thickness

Nonconductive elastomer
Phone/Sw board
Mica stiffener
Metal frame
Hinge
Conductive elastomer
Mg shield
Battery pack
microphone
Bottom cover

Front cover
Small display
Large LCD driver board
Large display
Plastic inner cover
Plastic inner cover
Sw/Processor board
Analog board
speaker
Metal coated plastic shield

Figure 10.10 Subassembly stacking.

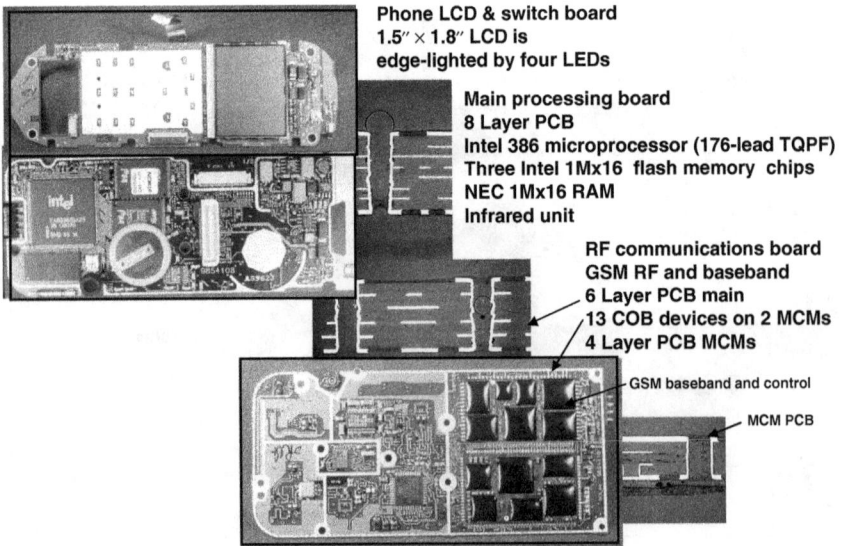

Phone LCD & switch board
1.5" × 1.8" LCD is
edge-lighted by four LEDs

Main processing board
8 Layer PCB
Intel 386 microprocessor (176-lead TQPF)
Three Intel 1Mx16 flash memory chips
NEC 1Mx16 RAM
Infrared unit

RF communications board
GSM RF and baseband
6 Layer PCB main
13 COB devices on 2 MCMs
4 Layer PCB MCMs

GSM baseband and control

MCM PCB

Figure 10.11 Nokia 9000 electronic assemblies.

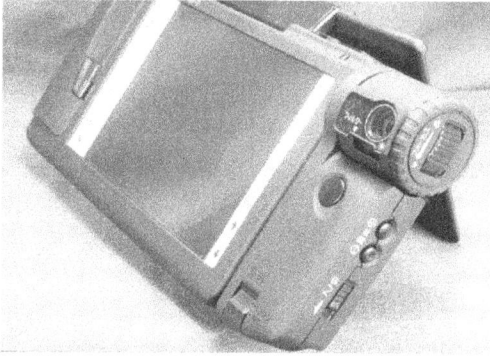

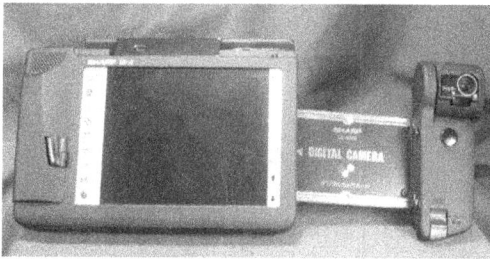

Figure 10.12 Sharp Zaurus.

6.75" × 4" × 1.1", 485 g PDA

Cost at introduction—$1380 with camera option

Standard PCMCIA slot with compatible still camera

- camera adds 1.5" and 90 g

Large touch screen/color display

- 3.1" × 4.2" LCD
- 240 × 320 RGB pixels

Handwriting recognition

Digital voice recorder

- Up to fifty 20-s messages

Communication support

- Fax, electronic mail,
- Cell phone adapter

Organizer

- Address/phone book
- Appointment book
- Finance manager
- Memo pads
- Calculator

PC link (cable and optical)

Figure 10.13   HP Jornada 928.

The original Zaurus product could achieve connectivity by connecting to an external cellular phone. Thus, it was able to function as an electronic still camera that could email images over a PHS digital cell phone link.

Combining electronic imaging and wireless connectivity with PDA functionality is a continuing trend. As a result, the PDA product classification is changing constantly and is difficult to distinguish from the *Smart Phone* product classification. In order to understand contemporary PDA design more thoroughly we will examine the HP Jornada 928 product (Fig. 10.13) in some detail.

The Jornada is a PDA that functions as a GSM/GPRS cellular phone. The product attributes are as follows:

Product type	Wireless Pocket PC
Product name	Jornada 928
Manufacturer and origin	HP, made in Singapore & PHL
Purchase date and price	August 2002, $999 MSRP, ($592) with Vodafone contract

Display	Main screen: 3.5-in, 240 × 320 pixel, 16-bit full color TFT with touch sensitive screen & CCFL front light
Caller ID display	Phone screen: 132 × 32 pixel, monochrome LCD with blue EL backlight
Processor	TI OMAP710 (133 MHz)
Memory	32 MB Flash ROM, 64 MB SDRAM
Expansion slot	One CompactFlash Type 1 extended slot
Communication	GSM/GPRS 900/1800 MHz, data rate at 9.6 & 14.4 kbps, class B GPRS multislot
External ports	Synchronized PC with USB or serial cable, infrared IrDA-SIR, stereo headset jack
Software platform	Pocket PC 2002 Phone Edition
Talk time, standby	3-h talk time, 12-h PDA, 120-h standby
Power source	Dual battery: lithium polymer 3.7v, 760 mAh (external), lithium polymer 3.7v, 760 mAh (embedded), AC adapter with charging stand
Antenna	External flip-up
Sound features	Recording, voice dialing, answering machine
Alerts	Vibrator, audible alarm, multicolor notification LED
Weight	221.6 g (with battery)
Dimensions	137 × 78 × 17 mm

The Jornada, like many of the current generation of wireless PDA products, is more bulky than the typical cellular phone. Users that would prefer this product over a cellular phone would have to place a premium on the PDA functionality in order to justify the encumbrance of the larger form-factor. The product connects to a PC for synchronization by way of a USB connector.

Figure 10.14 shows the front and back view of the product in the closed configuration. Since the product must be flipped open to access the PDA display, a secondary display is viewable on the outer enclosure. This display is for caller identification. The SEND and END buttons are also accessible in the closed configuration. The user is able to identify callers, initiate calls, and end calls without unfolding the product. This is a desirable feature, considering that it is important to keep the complexity of making a phone call to a minimum. For a product concept of this type to be successful, the product must be fully capable of replacing the user's existing cell phone.

The system uses a foldable antenna to keep the stowable form-factor to a minimum. This fold-out method of extending an antenna is potentially more rugged than designs that use a thin telescoping antenna.

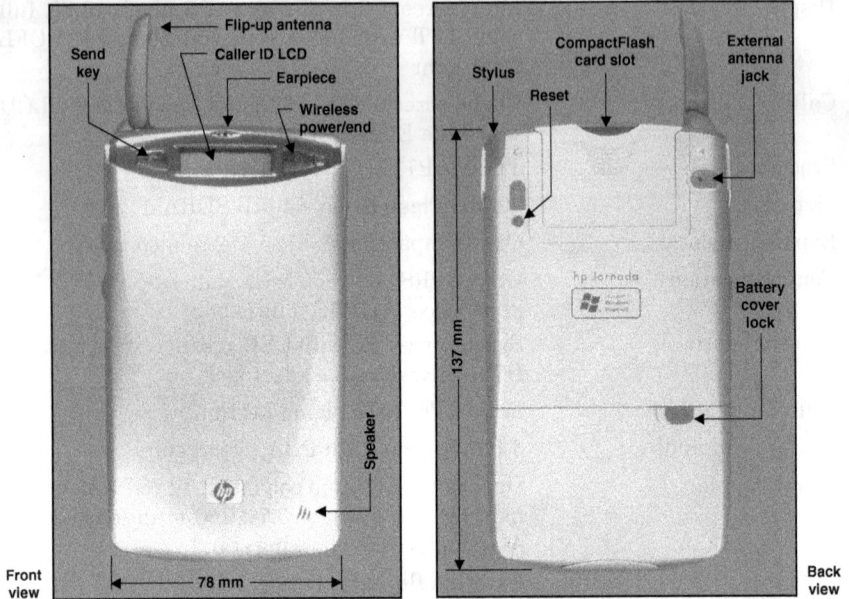

Figure 10.14 Front and back views.

The compact flash slot can be used for mass storage or potentially for other types of peripheral interface.

The protective plastic cover flips up to reveal the main LCD and system controls, as shown in Fig. 10.15. The speaker phone is automatically activated when the cover is open. The PDA has a separate power button. Separating the PDA power and RF power is a good idea since the RF section consumes considerably more power than the PDA section. A user could conceivably access contact information from the PDA many times, even after the remaining charge is not sufficient to start up the RF section. This is a useful capability, since cell phone users are often frustrated by the fact that they cannot access their stored phone numbers when their cell phone has run out of power. The display is a 3.5-in, 240×320 pixel, 16-bit color, reflective TFT with touch-sensitive screen. A reflective color display is critical to achieving a reasonable battery life.

The side views of the product are shown in Fig. 10.16. A record button on the side of the product enables the user to quickly create a voice memo. This is a particularly useful feature for PDA products which, in general, are still competing with inexpensive paper notebooks for market share. One of the drawbacks of a PDA is the inconvenience of having to turn it on and navigate to the memo function and to an appropriate file

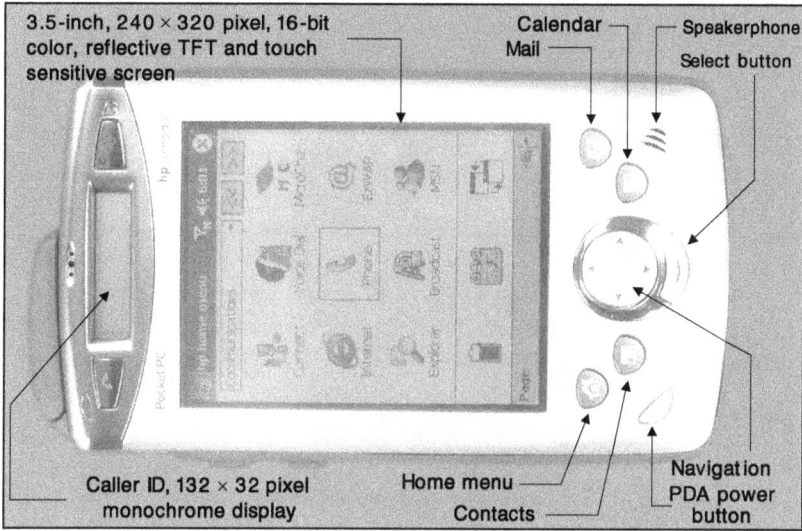

3.5-inch, 240 × 320 pixel, 16-bit color, reflective TFT and touch sensitive screen

Calendar

Mail

Speakerphone

Select button

Caller ID, 132 × 32 pixel monochrome display

Home menu

Contacts

Navigation

PDA power button

Figure 10.15 Cover open.

for memo input. Push button voice memo functionality offers a reasonable alternative for quick note taking.

Removal of the battery pack, as shown in Fig. 10.17, reveals the GSM SIM card compartment. GSM SIM cards carry the information that allows the network to identify the phone and determine if it is entitled

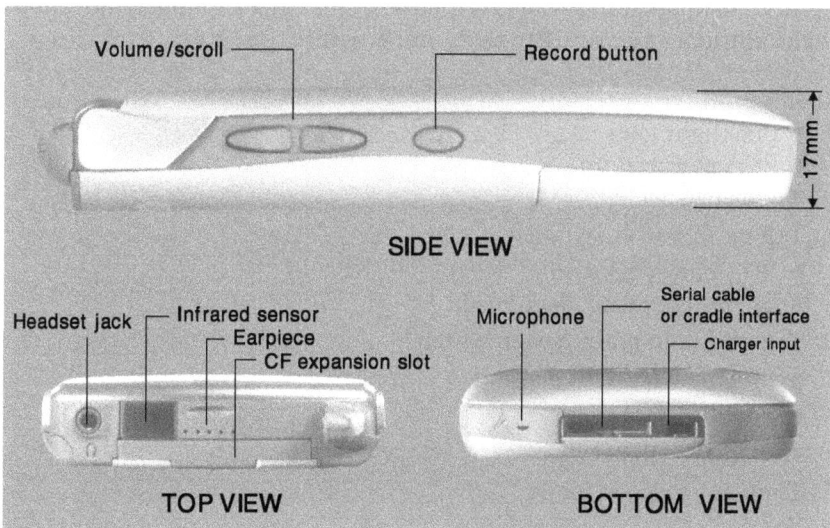

Volume/scroll

Record button

17 mm

SIDE VIEW

Headset jack

Infrared sensor

Earpiece

CF expansion slot

Microphone

Serial cable or cradle interface

Charger input

TOP VIEW

BOTTOM VIEW

Figure 10.16 Side and end views.

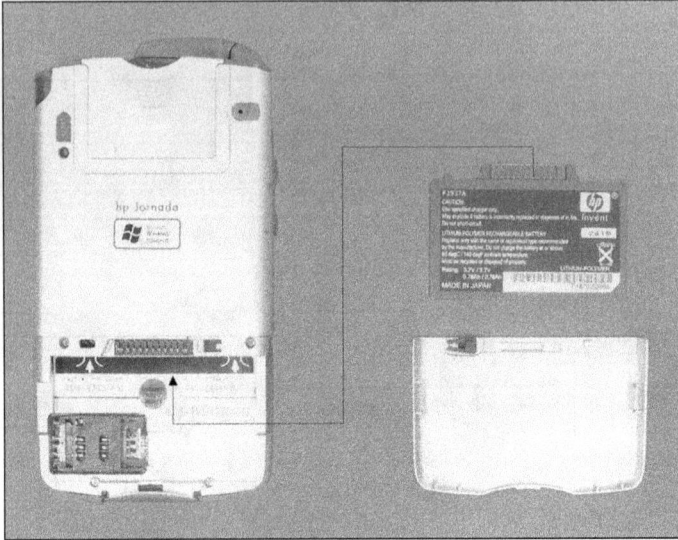

Figure 10.17   Battery removed.

to service. GSM SIM cards also store the user's phone list. SIM cards allow users to easily switch phone hardware without having to notify the network operator. Figure 10.18 shows an estimation of the simplified block diagram for this product based on the observed component set.

A power management analysis of this product indicates the power levels that might be expected in a contemporary wireless PDA product. Figure 10.19 shows the observed power consumption under the following conditions, according to the numbers in the diagram:

1. Open lid.
   - Backlight goes on.
   - PDA powered up.
   - Wireless off.
2. Tap main power switch.
   - PDA power goes off.
   - Device goes into sleep mode.
3. Tap and hold main power switch.
   - Display backlight goes off.
4. Close lid.
   - Display still active.
5. Time out to sleep mode.

From these observations it can be inferred that the main display backlight power $\sim 0.95 - 0.35 = 0.6$ W.

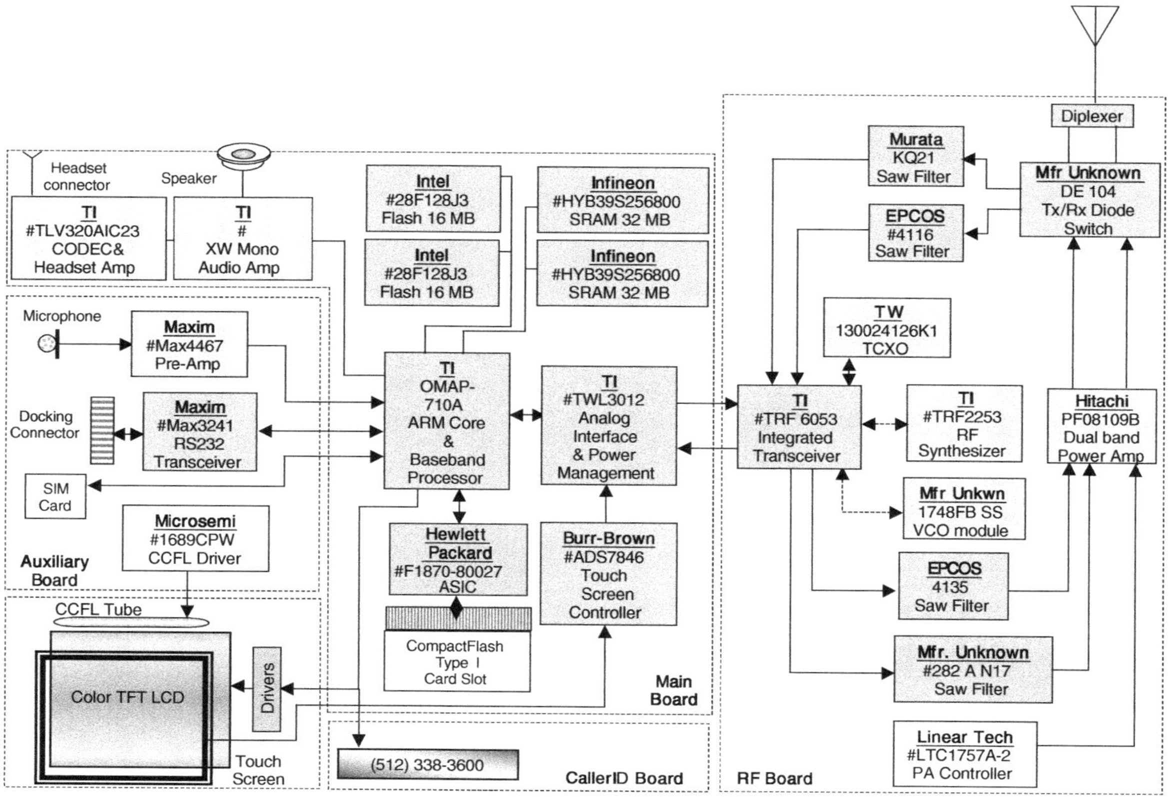

Figure 10.18 Simplified block diagram.

279

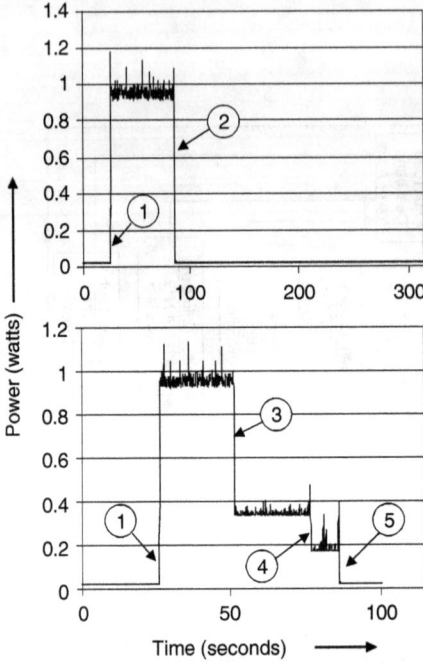

Figure 10.19 Power sequencing and backlight.

In Fig. 10.20 we see the power consumption related to viewing images stored in the PDA according to the following system conditions:

Enter test with display backlight on.

PDA powered up.

Wireless off.

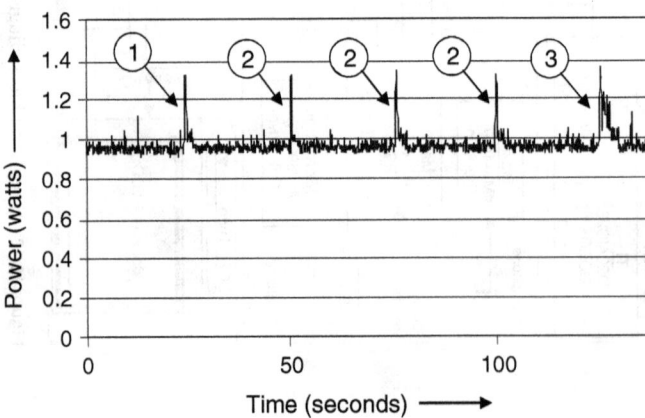

Figure 10.20 Viewing images.

Execute the following sequence:

1. Select images from main menu.

2. Choose next image.

3. Choose image with audio attachment.

From this and the previous sequence it can be concluded that the static operating power is around 1 W and that image display creates a 200-mW spike but does not incur sustained power consumption.

The power consumed by the ringer is shown in Fig. 10.21, and the power test sequence is as follows:

Enter test with display backlight off.

PDA powered up.

Wireless off.

Execute the following sequence:

1. Select ringer choice from main menu.

2. Play choice 1.

3. Play choice 2.

4. Play choice 3.

Surprisingly, system power consumption can increase significantly when the ringer is playing. Similarly, the system was tested in audio

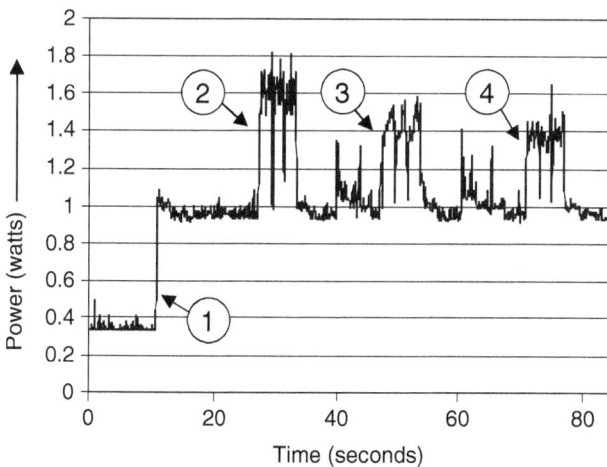

Figure 10.21 Playing ringer.

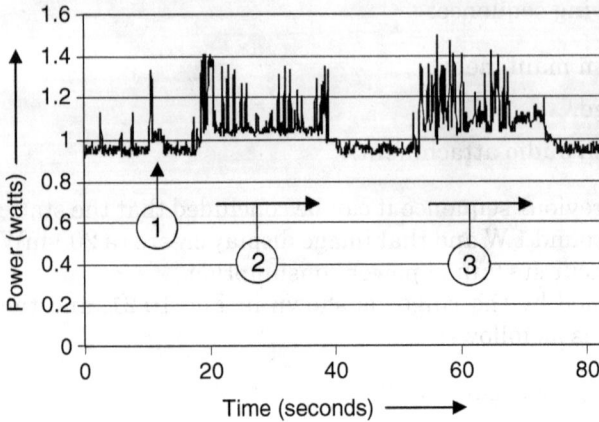

Figure 10.22 Audio record and playback.

record and playback mode with the power consumption response shown in Fig. 10.22 according to the following inputs:

Enter test with display backlight on.

PDA powered up.

Wireless off.

Execute the following sequence:

1. Select Notes from main menu.
2. Record message.
3. Playback message.

Based on the results, it appears that the audio record and playback functions, when activated, probably increase average system power consumption by about 20 percent. In each of these different power scenarios it is important to observe whether the backlight is on or off in conjunction with the particular function being observed.

The largest source of power consumption in this type of product will invariably be the cellular phone function. Figure 10.23 shows the power consumption results when trying to make a phone call with this product with the following system conditions:

Enter test with display backlight off.

PDA powered up.

Wireless off.

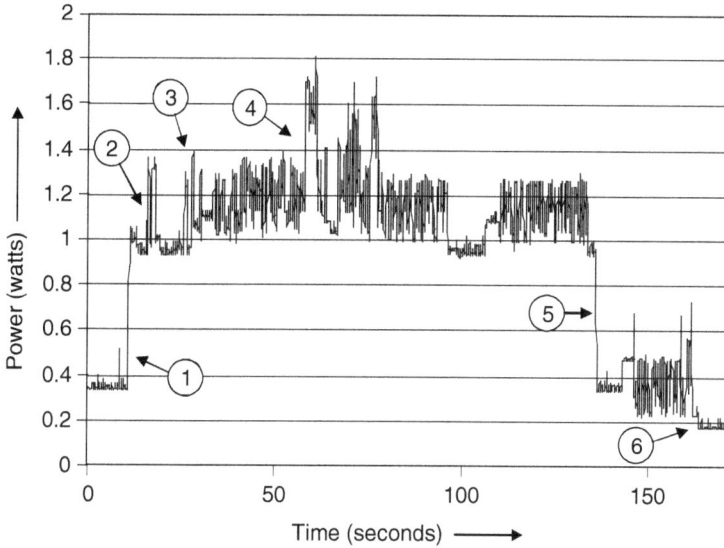

Figure 10.23 Attempted phone call.

Execute the following sequence:

1. Tap main menu.
   ▪ Backlight goes on.
2. Select phone screen.
3. Turn wireless power on.
4. Dial and attempt to place call (no network).
5. Backlight goes off.
6. Wireless power down.

Finally, the power consumption for various PDA applications is shown in Fig. 10.24. The corresponding inputs are listed below:

Enter test with display backlight on.

PDA powered up.

Wireless off.

Execute the following sequence:

1. Select settings.
2. Select display.
3. Set brightness to maximum.

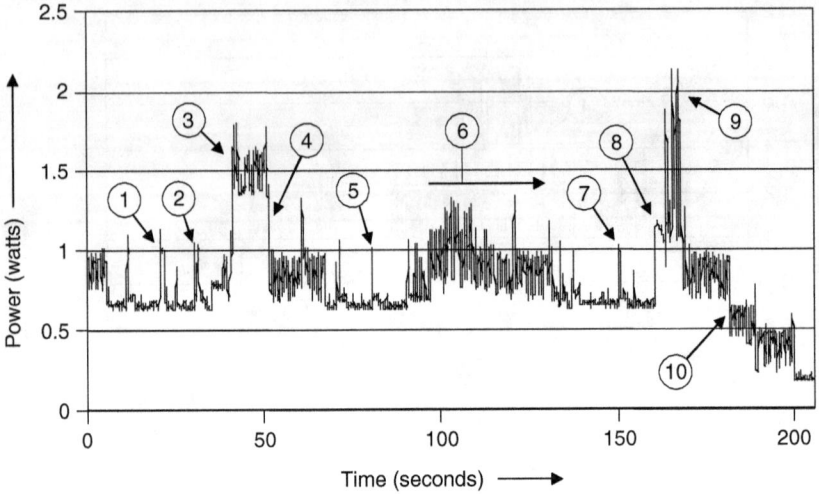

Figure 10.24   PDA applications.

4. Reduce to minimum.

5. Select "Word."

6. Use virtual keyboard.

7. Select voice dialing.

8. Attempt voice dialing.

9. Announcement "phone is off."

10. Power down.

This chart shows that the product consumes between 0.75 and 1.5 W when various PDA applications are running. This product employs two lithium-ion batteries, one external (Fig. 10.17) and one internal (Fig. 10.25). Each of these batteries has an energy storage capacity of 2.8 Wh (3.7 V × 760 mAh), for a system total of 5.6 Wh. Based on the power consumption characteristics above, we can estimate the battery life potential for the product.

Since the minimum usage mode would be for a user to operate the PDA without the backlight, the minimum usable power consumption appears to be about 0.35 W. This would imply a maximum battery life of 5.6 Wh/0.35 W = 16 h.

The maximum usage model would be represented by a user making a speakerphone call while simultaneously using PDA applications with the backlight on. The average power consumption under these conditions might be as high as 2.0 W. Under these conditions we can estimate the minimum battery life to be around 5.6 Wh/2.0 W = 2.8 h.

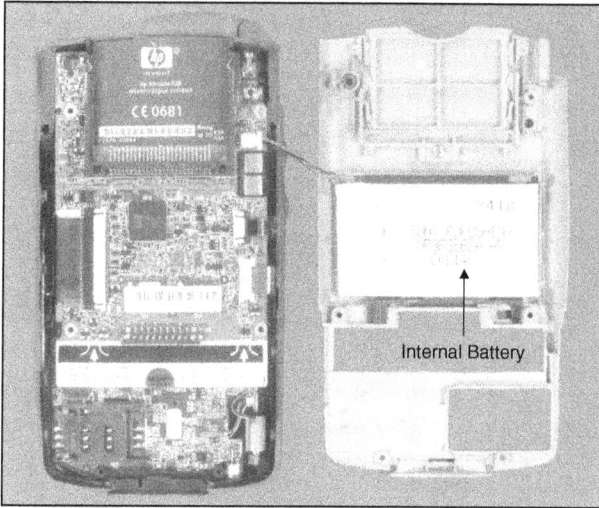

Figure 10.25 Internal battery exposed.

Clearly the battery life claims for this type of product should be based on the usage model that is expected from a typical user. The addition of electronic imaging functionality would make the variations in minimum and maximum battery life even greater.

Figure 10.26 shows the elements of the product enclosure and the battery cells deployed in this product. The most notable feature is the use

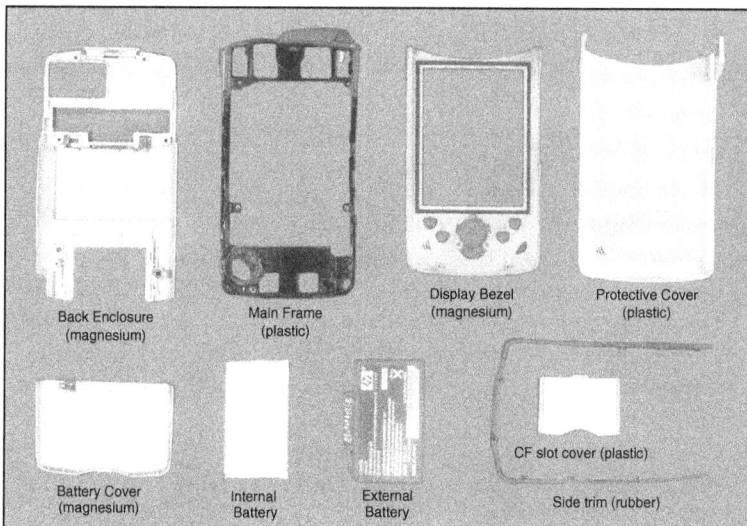

Figure 10.26 Product enclosure.

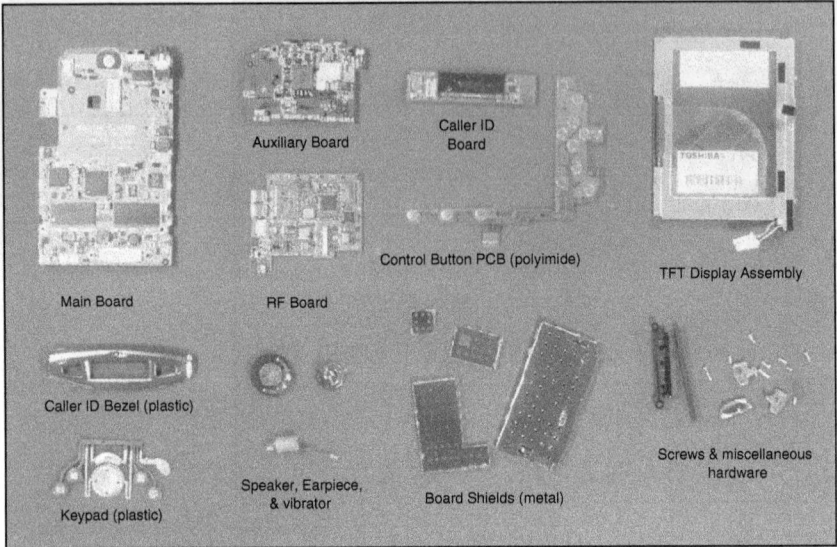

Figure 10.27 Product subassemblies.

of magnesium to form the main product enclosure. Magnesium has a very high strength-to-weight ratio and provides excellent thermal properties to quickly remove heat from the system.

Figure 10.27 shows the remaining subassemblies that comprise the product. The system electronics assemblies include a main PCB, an auxiliary daughter board, an RF board, a caller ID board, a user-interface flex circuit, and a display module. There is a single piece plastic keypad overlay that provides the physical characteristics for the control buttons deployed beneath the main display. This overlay is positioned over the flex circuit on which are patterned the contact electrodes that are switched closed when the user pushes down on the overlay button.

A complex mechanical design was required in order to package all these components into an enclosure with an overall thickness of only 17 mm. Figure 10.28 shows the mechanical stack-up of all these components. It also highlights the critical role of surface mount connectors and other miniaturized electrical connectors in achieving a compact design. The numbers in this figure show the location of the electrical connectors for each of the following components:

1. Touch screen

2. Main display

3. Control flex PCB

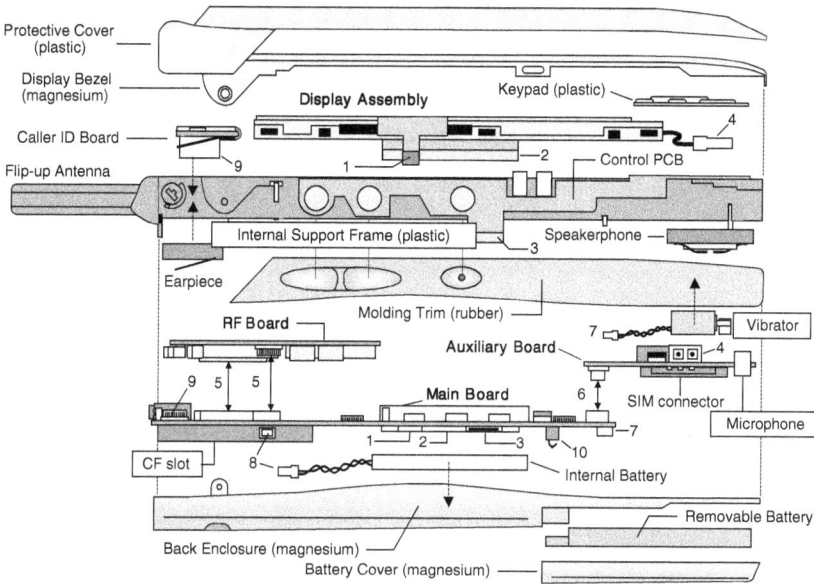

Figure 10.28 Component stack-up.

4. Backlight

5. RF to main board

6. Main to auxiliary board

7. Vibrator

8. Internal battery

9. Caller ID to main board

10. External battery

Figure 10.29 shows the main electronic assembly as it resides in the product. T1, T2, and T3 in this picture denote the location of SMT ZIF flex connectors and the corresponding flex cables that connect to them. These flex cables wrap around the main electronic assembly and connect to modules on the other side. T1 connects to the flex button interface, T2 connects to the touch screen sensor panel, and T3 connects to the main display.

The RF module connects to the main system board through two SMT connectors shown in Fig. 10.30. A close-up of the RF board is shown in Fig. 10.31. A picture of this board with the shielding can removed is shown in Fig. 10.32. Notice the tiny cylindrical surface mount coaxial connector mounted in the upper right-hand corner of the board.

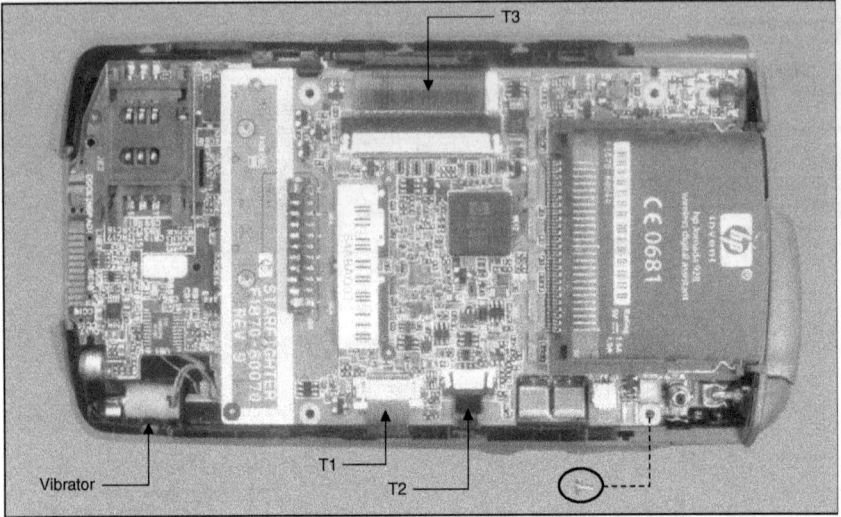

Figure 10.29 Main electronic assembly.

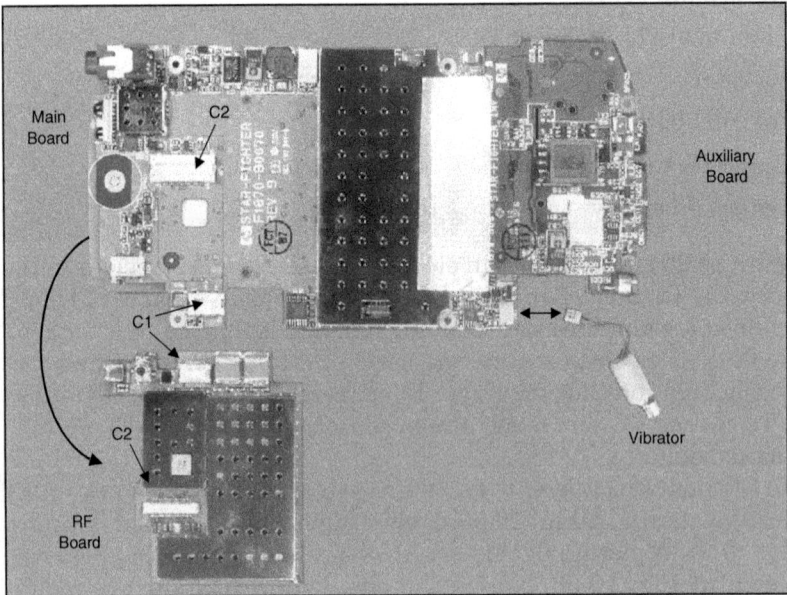

Figure 10.30 RF board removed.

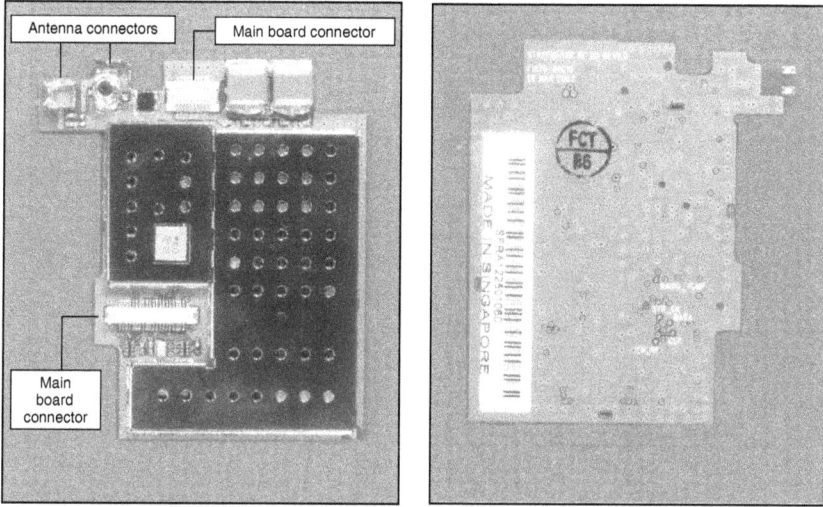

Figure 10.31 RF board close-up.

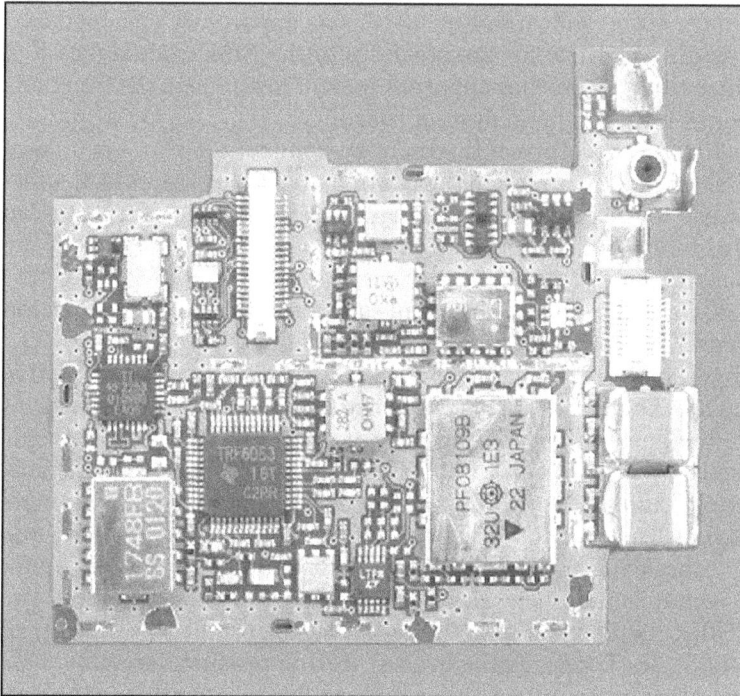

Figure 10.32 RF board (with shield removed).

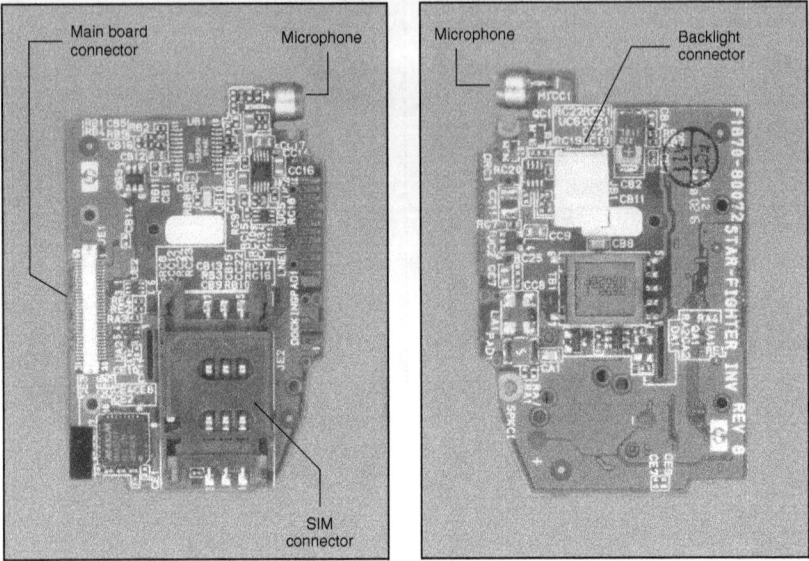

Figure 10.33   Auxiliary board.

The auxiliary board shown in Fig. 10.33 was apparently designed as a discrete module to allow for the stack-up of the SIM module (see Fig. 10.28), since there is no other apparent reason to separate the functions on this board from the main system board.

Figure 10.34 is a picture of the main system board. There is a large metal-shielding can mounted to the board and a piece of metal foil adhesively bonded to the can. Adhesive foil is sometimes used as a Bond-Aid to fix an unforeseen EMI problem in the design. The other side of the main board (Fig. 10.35) is the mounting surface for most of the system connectors, including the touch screen connector, the display connector, the compact Flash connector, and the external battery connector.

The physical user interface for this product is the touch sensitive main display, the user ID display, and the buttons. Removing the front enclosure (magnesium) of the product, as shown in Fig. 10.36, reveals how these components are seated in the product. Notice that the buttons that are underneath the main display are created with a single piece overlay. This is good design practice, as it greatly reduces the assembly complexity of dealing with individual buttons and also provides a mechanical framework for holding each button in place. The polyimide flex circuit containing the actual button contacts and electrical traces is positioned beneath the overlay during product assembly. This flex circuit also wraps around the side of the product to provide the buttons that are arranged along the product edge (Fig. 10.37). The overlays

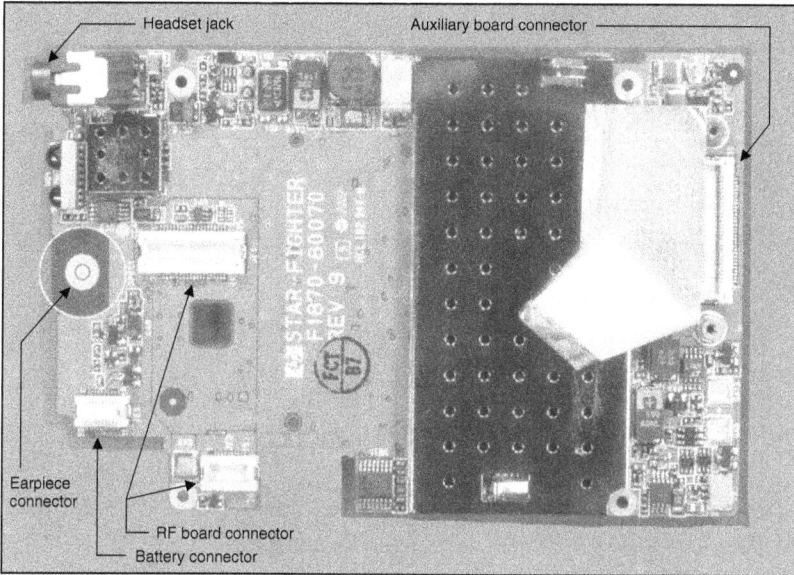

Figure 10.34 Main board (with shield).

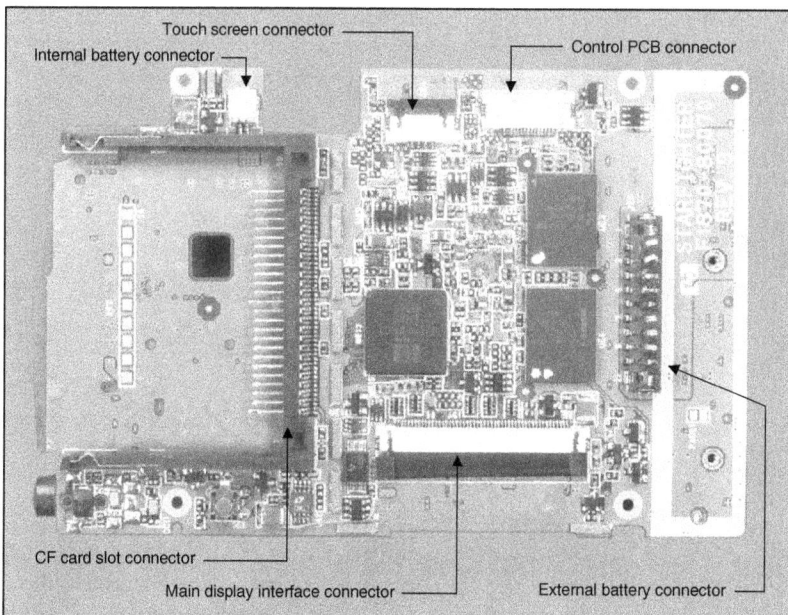

Figure 10.35 Main board (side 2, interface).

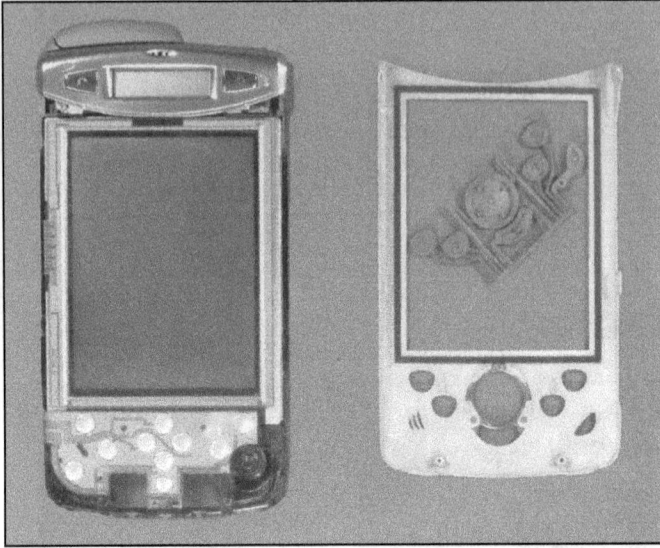

Figure 10.36 Main display bezel removed.

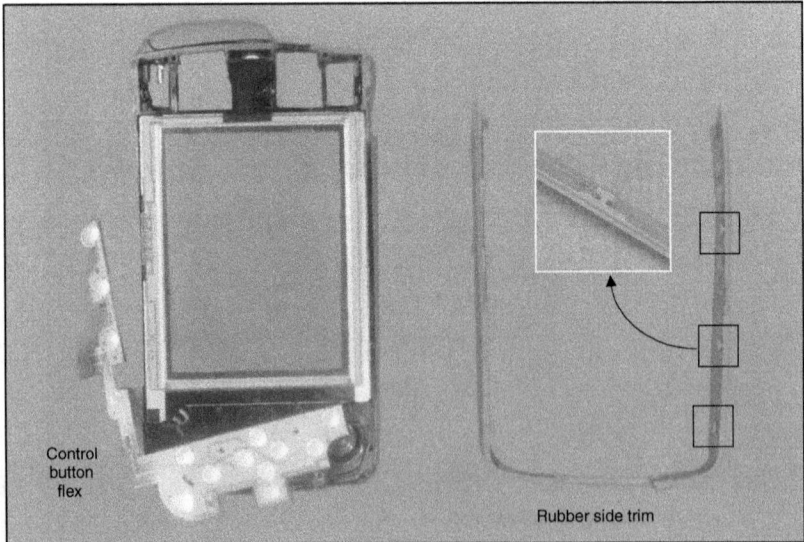

Control
button
flex

Rubber side trim

Figure 10.37 Control button PCB removed.

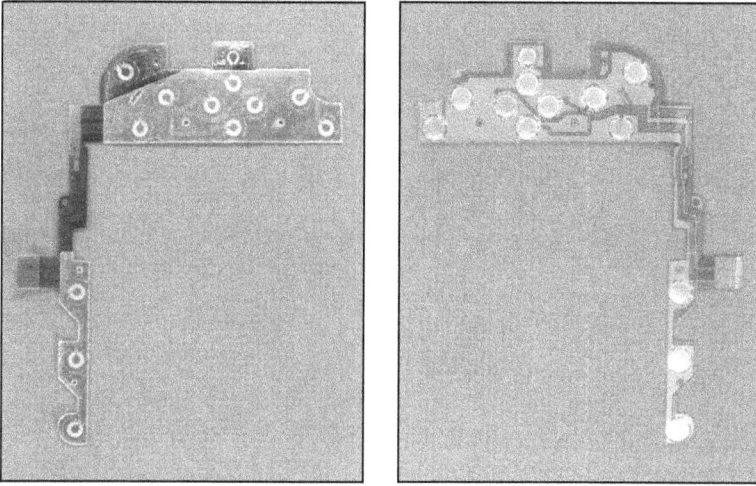

Front and back sides of the single layer copper-polyimide control button PCB

Figure 10.38 Control button flex.

for these buttons are integral to the rubber trim that is seated around the edge of the product. This trim is held in place by molded tabs that fit into slots in the product frame. Plastic stiffeners are adhesively attached to the backside of the flex circuit to provide stable mechanical support for the button contacts (Fig. 10.38).

The caller ID board (Fig. 10.39) is a self-contained module that includes the caller ID display electronics, the LCD, and an electroluminescent (EL)

Blue EL Backlight Connection

Figure 10.39 Caller ID PC board.

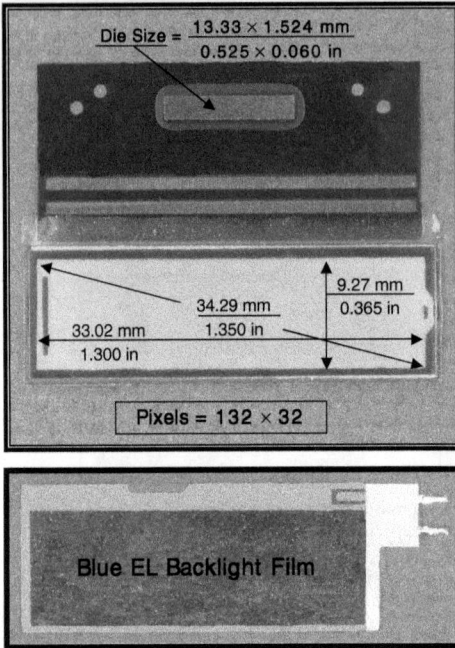

Figure 10.40   Caller ID LCD and backlight.

backlight. Figure 10.40 shows the caller ID LCD and backlight panels. The LCD driver is attached using a TAB frame that folds around the PCB and EL backlight.

The main display module is shown in Fig. 10.41. This is a full color 3.5-in diagonal active matrix display. The driver chips are attached directly to the LCD glass, using an adhesive flip-chip assembly process.

A characterization of the types of substrate technology designed into this product is shown in Fig. 10.42. Both the main board and the RF board rely on micro-Via printed circuit technology to achieve the required routing density. Via sizes can have a major impact on the practical routing density because they interfere with efficient routing schemes and reduce the effective line pitch in densely routed areas.

The weights of the various system components are shown in Fig. 10.43. The use of magnesium for the enclosure enabled the designers to build a very rugged product while keeping the total weight of the enclosure to only 30 percent of the total system weight.

The IC metrics for this product (Table 10.1) indicate that there are 29 ICs that have a combined die area of 517 mm^2.

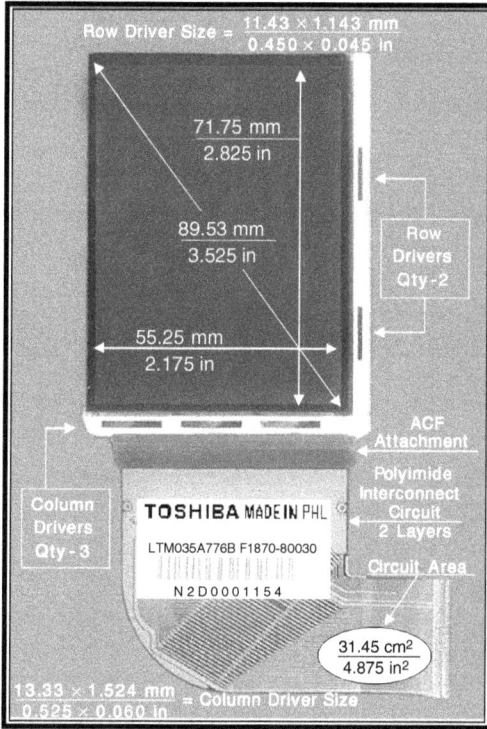

Figure 10.41 Main display module.

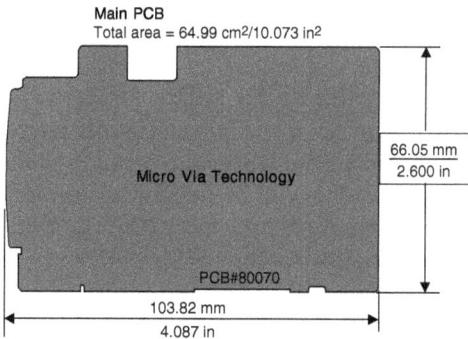

Main PCB
Total area = 64.99 cm²/10.073 in²

66.05 mm
2.600 in

Micro Via Technology

PCB#80070

103.82 mm
4.087 in

Main PCB
(*FR4 substrate*)
(Micro-Via technology)
Layers = 8
Finest pitch = 0.203 mm/0.008 in
Narrowest trace = 0.076 mm/0.003 in
Narrowest space = 0.127 mm/0.005 in
Smallest Via I.D. = 0.254 mm/0.010 in
Smallest Via O.D. = 0.457 mm/0.018 in
Micro-Via I.D. = 0.127 mm/0.005 in
Micro-Via O.D. = 0.305 mm/0.012 in
Substrate thickness = 0.762 mm/0.030 in
Assembly weight = 31.4 g

RF PCB
(*FR4 substrate*)
(Micro-Via technology)
Layers = 8
Finest pitch = 0.254 mm/0.010 in
Narrowest trace = 0.102 mm/0.004 in
Narrowest space = 0.152 mm/0.006 in
Smallest Via I.D. = 0.305 mm/0.012 in
Smallest Via O.D. = 0.610 mm/0.024 in
Micro Via I.D. = 0.203 mm/0.008 in
Micro Via O.D. = 0.406 mm/0.016 in
Substrate thickness = 0.965 mm/0.038 in
Assembly weight = 10.6 g

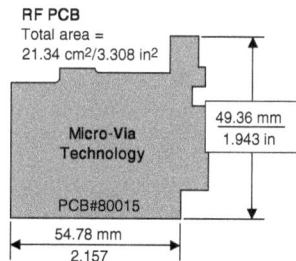

RF PCB
Total area =
21.34 cm²/3.308 in²

49.36 mm
1.943 in

Micro-Via
Technology

PCB#80015

54.78 mm
2.157

Figure 10.42 Substrate characteristics.

Item	Weight (g)	
▨ **Enclosures**	67.6	
Back enclosure (magnesium)		13.7
Protective cover (plastic)		22.1
CF Slot cover (plastic)		2.1
Battery cover (magnesium)		.2
Rubber trim		3.3
Display bezels (magnesium)		7.9
Main frame & antenna (plastic)		13.3
▨ **Electronic Boards**	56.4	
Main board		34.8
RF board		13.4
Auxiliary board		8.2
▨ **Display Assembly**	48.6	
Color TFT		33.9
Touch screen		11.1
Caller ID & board		3.6
▨ **Batteries**	36.5	
Internal		16.5
External		20
▨ **Miscellaneous**	12.5	
10 screws, tape, buttons, speaker, earpiece, stylus, vibrator, connector frame, keypad, etc		
Total	**221.6**	

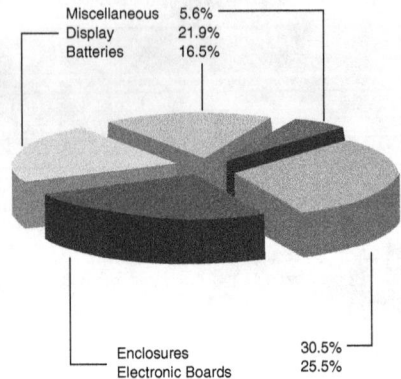

Miscellaneous 5.6%
Display 21.9%
Batteries 16.5%
Enclosures 30.5%
Electronic Boards 25.5%

Figure 10.43 Component weights.

Looking at the electronic assembly metrics (Table 10.2), we can see that the main circuit board is where most of the high-density packaging is located. The substrate is an eight-layer circuit board with an aggressive tiling density of 7.28 percent.

The connection metrics offered in Table 10.3 show that the ratio of discrete components to IC IO is 599 to 1194, or about 0.5. This ratio for the RF board (mostly analog) is a little over 1.35 but for the main board (mostly digital) is less than 0.4. These ratios are a useful reference when trying to estimate the discrete counts that would be needed to support analog and digital circuitry for a new design or for comparing the relative level of integration for a particular architecture.

TABLE 10.1   IC Metrics (IC Summary)

Board description	No. of IC's total	No. of IC IO's	Die area (mm^2)	IC footprint area (mm^2)	IC IO's/ (IC footprint area)
ID display PCB	1	8	2	15	0.5
Main PCB	21	1,030	473	1,536	1
Auxiliary PCB	4	70	23	115	1
Control PCB	—	—	—	—	—
RF PCB	3	86	19	135	1
IC total	**29**	**1,194**	**517**	**1,801**	**0**

TABLE 10.2 Electronics Assemblies Metrics (Summary of Electronic System)

Board description	Board area (cm²)	No. of PCB layers	No. of parts	No. of connections	PCB tiling density (%)	Connection density	Routing density	Part density	Average pin count
ID display PCB	8.8	2	31	98	0.2	11	18	4	3.2
Main PCB	65.0	8	426	2,222	7.28	34	40	7	5.2
Auxiliary PCB	21.1	6	78	318	1.11	15	24	4	4.1
Control PCB	18.3	1	1	26		1	18	0	26.0
RF PCB	21.3	8	134	507	0.90	24	28	6	3.8
System total	134.5	—	670	3,171	3.8	24	—	5	4.7

TABLE 10.3 Connection Metrics (Component and Connection Counts)

Board description	Opportunity count	IC's	IC IO's	Modules	Module IO's	Discretes	Discrete IO's	Connectors	Connector IO's
ID display PCB	129	1	8	3	10	26	60	1	20
Main PCB	2,648	21	1,030	7	28	386	884	12	280
Auxiliary PCB	396	4	70	1	2	70	178	3	68
Control PCB	27	—	—	1	26	—	—	—	—
RF PCB	641	3	86	9	78	117	277	5	66
Total	3,841	29	1,194	21	144	599	1,399	21	434

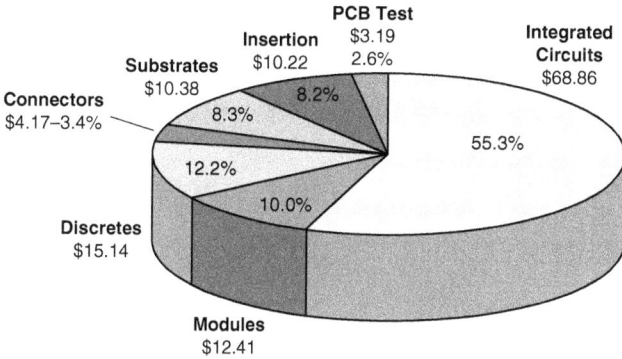

Figure 10.44   Electronic system cost estimate.

An estimate of the electronic systems cost is shown in Fig. 10.44. Not surprisingly, integrated circuits comprise over half of the electronics cost. Even though high-density substrates are critical to achieving the product form-factor, they only contribute approximately 8 percent of the product cost. Ten percent of the cost is determined by specialized RF modules needed to support the cellular phone functionality. The power amplifier module shown in Fig. 10.45 comprises several wire-bonded ICs and a number of discrete components connected to a ceramic substrate.

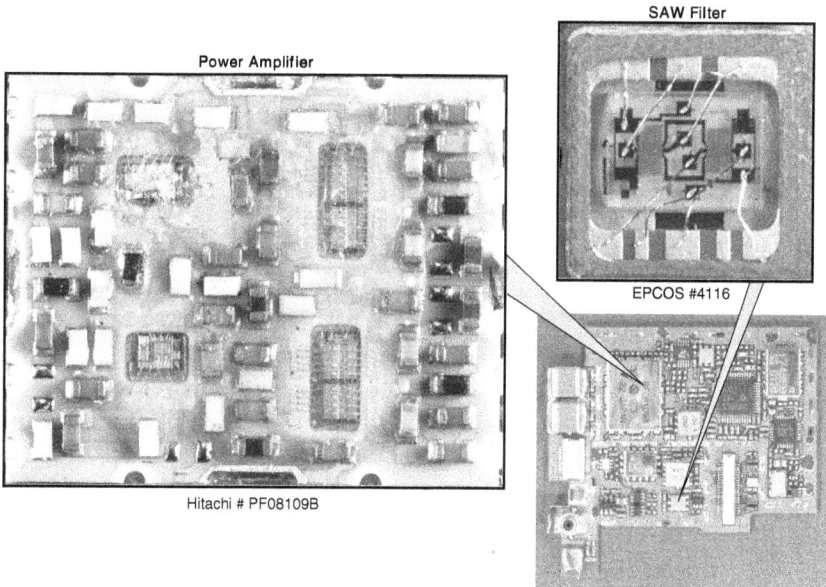

Figure 10.45   RF modules.

TABLE 10.4 Design for Manufacturing (DFM)

Item	Description	Estimated cost
Back enclosure	Magnesium—13.7 g—8 × 76 × 130 mm—paint—plastic slide frame	$1.79
Protective cover	Plastic—22.1 g—14 × 80 × 130 mm—paint both sides	$0.60
Caller ID bezel & buttons	Painted plastic—3 pcs—gasket—1.9 g	$0.45
Battery cover	Magnesium 5.2 g—6 × 50 × 76 mm—paint—latch	$0.75
Antenna	External flip-up—3.2 g—48 mm	$0.55
Molding	Rubber—3.3 g—116 × 76 mm	$0.35
Main display bezel	Magnesium—7.9 g—10 × 74 × 120 mm—paint—plastic trim—gasket	$1.10
Main frame	Plastic—13.3 g 10 × 74 × 135 mm—screw insets—cover bushing/locks	$0.72
Miscellaneous	11 screws—tape—labels—stylus—keypad—small plastic parts	$0.85
Total		**$7.16**

The cost of various mechanical parts is estimated in Table 10.4. While the estimated cost of these individual parts is small relative to the overall product cost, the design and tooling costs to produce these mechanical parts represent a substantial part of the manufacturer's financial risk.

DFA Index   **5.0%**
Total Assembly Time (s)   **589**
Total Assembly Cost ($)   **$1.64 (@$10/h, Singapore)**

Percent of total (589 s) assembly time

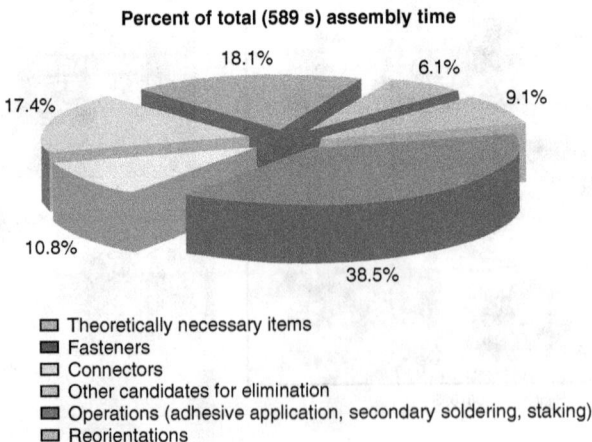

18.1%   6.1%
17.4%   9.1%
10.8%   38.5%

☐ Theoretically necessary items
■ Fasteners
☐ Connectors
☐ Other candidates for elimination
■ Operations (adhesive application, secondary soldering, staking)
☐ Reorientations

*From Boothroyd Dewhurst, Inc. DFMA release 8.0

Figure 10.46 Design for assembly (DFA).

HP Jornada 928 PDA	Cost
Electronics assemblies	$124.38
Displays #1 & #2	$30.68
Housing/hardware (DFM)	$7.16
Batteries	$6.64
Final assembly (DFA)	$1.64
Total	$170.50

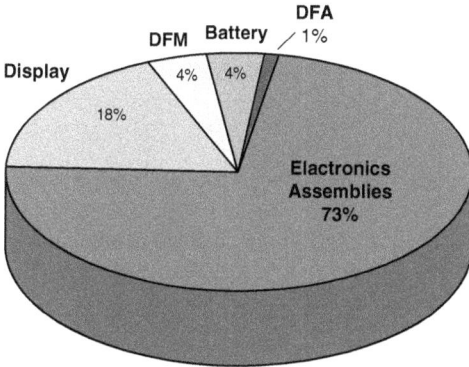

Figure 10.47 Cost summary.

The pie chart in Fig. 10.46 shows the estimated time required for final assembly of the product. The estimated cost breakdown for the entire product is shown in Fig. 10.47. Software costs for the operating system and applications used in this product are not included in this cost estimate.

Digital Imaging Products

Portable digital imaging products include camcorders, digital cameras, and imaging-enabled PDAs. The development of highly portable 8 mm camcorder products in the early 1990s marked the beginning of consumer acceptance and demand for these types of products. Japanese consumer electronics companies such as Sony and Matsushita created new high-density packaging technologies that made these products possible.

Digital cameras are a relatively new and hugely successful class of portable electronics product. The Casio QV-10 (Fig. 11.1) was a very innovative product, being one of the first digital camera products targeted directly at the consumer. The Casio QV-10 was introduced into the marketplace in the second quarter of 1995, at a retail price of around $900. It is a landmark product in that it opened up the market for consumer digital cameras.

This system contains a 250-K pixel color charge-coupled device (CCD) image sensor and a 1.8-in color TFT (Fig. 11.2) LCD display that functions as a viewfinder. The system stores up to 96 images into 16 Mbits of flash memory. These images may be downloaded to a computer for storage and manipulation, displayed on the on-board LCD or an external TV monitor, or sent to a color printer for hardcopy.

The product housing is contoured to provide an ergonomic design, and a simple set of control buttons is arranged on the top of the housing, as shown in Fig. 11.3. The I/O ports are concealed under a thin plastic cover (Fig. 11.4). Four AA battery cells (Fig. 11.5) provide the system power.

While Casio has been a major player in consumer electronics for many years with products such as calculators, clock radios, electronic musical instruments, and pocket TVs, their innovative entry into the consumer imaging market was surprising and sudden. When Casio's core

Figure 11.1 QV-10 front view.

competencies are examined, however, their entry into this business should have come as no surprise. Spotting the applicability of these competencies in advance, however, would have required that the subtle changes in the product usage paradigm be recognized first.

The QV-10 usage paradigm was derived in part from prior developments in the camcorder industry. The introduction of a 1.8-in LCD viewfinder made different (and better) user ergonomics possible. Additionally, the LCD gives the QV-10 the appeal of an "instant" camera,

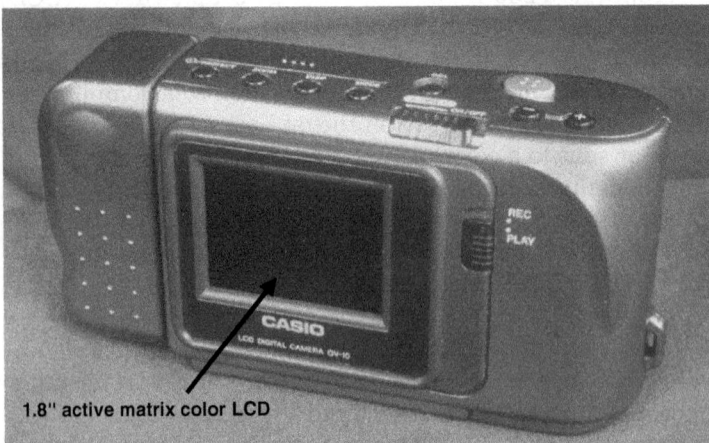

Figure 11.2 QV-10 rear view.

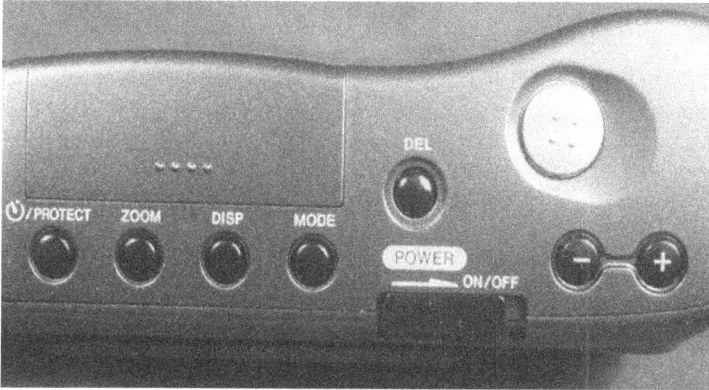

Figure 11.3 Top view.

with the cost of bad or frivolous shots being zero. Casio made some interesting trade-offs by electing to utilize nonremovable image storage. This decision recognized that most users will not require more than 96 *known good images* in a single outing and therefore will not require the added cost and bulk of removable storage.

Casio had produced small color LCD modules in high volume for several years at its Ome, Japan facility. While there were other companies that had superior LCD technology, Casio developed a highly specialized driver-chip-assembly technology that enabled them to be extremely competitive in producing small, low cost, reliable, color displays. The Casio Ome facility produced over two million such display modules each

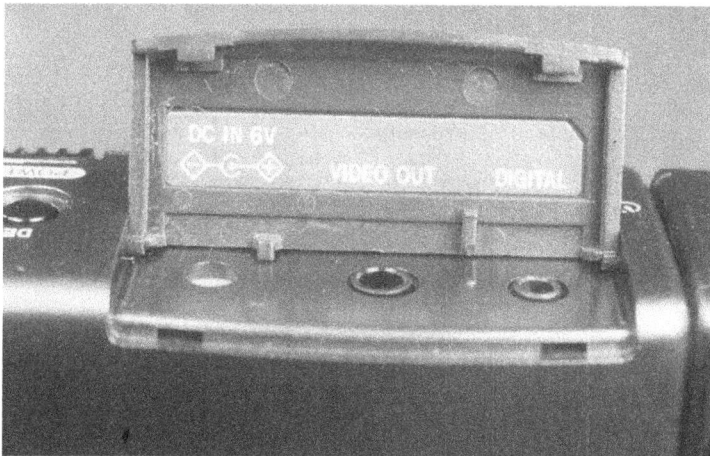

Figure 11.4 I/O ports.

Figure 11.5 Batteries.

month at peak production. This module is pictured in Fig. 11.6. Casio also utilized a CCD image sensor (Fig. 11.7) that is packaged in a low-cost, optically transparent leaded package to reduce the product manufacturing cost.

Since the introduction of the Casio QV-10, digital camera products have evolved very rapidly under stiff competition. As a detailed case study, we will examine the Sony Cyber-Shot DSC-U10 (Fig. 11.8) as an exemplary product design. This product is a highly miniaturized point-and-shoot style camera. The product specifications are as follows:

Product type	1.3 Megapixel Digital Still Camera
Product name	Cyber-Shot DSC-U10
Manufacturer & origin	Sony, Made in Japan (label)
Official release date	June, 2002 (Japan); fall, 2002 (USA)

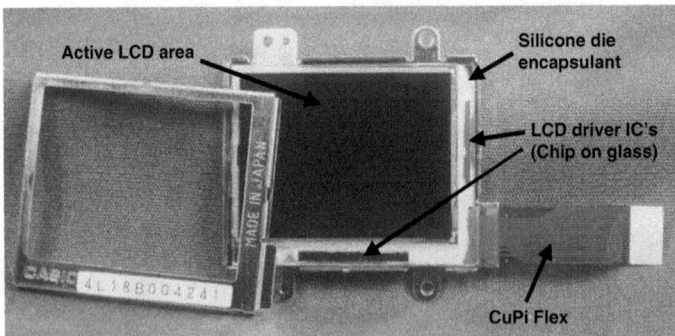

Figure 11.6 LCD module—front bezel removed.

Figure 11.7   CCD sensor—oblique view of assembled package.

Figure 11.8   Cyber-shot DSC-U10.

Purchase date & price	Purchased 10/05/02 in Japan for ~$US200
Sensor	1/2.7" 1.3 Megapixel—Super HAD CCD
Image sizes	1280 × 960 or 640 × 480 pixels
Lens	Fixed length 35 mm equivalent, f = 5.0 mm, F2.8
Focus distance	10 cm. Infinity (autofocus)
LCD monitor	2.5 cm (1.0 in), 64460 pixels (293 × 220), color TFT
Recording media	Memory stick (shipped with 8 MB stick)
Data formats	Still images: DFC compliant
	DPOF compatible. Movies: MPEG1 compliant, no audio
Viewfinder	LCD monitor (no optical viewfinder)
Flash, zoom	Built-in flash—no zoom feature
Exposure control	Automatic, three-scene selection modes
Shutter speed	Automatic, 1/30 to 1/2000 s
Power source	Two AAA nickel metal hydride cells—AC charger
Battery life	Approx. 60 min or 1600 pictures (1280 × 960 LCD backlight on)
Weight	115 g (including batteries & memory stick)
Dimensions	30 × 41 × 85 mm
Special features	Compact size—USB interface with Mac or PC—sliding lens-cover door activates power—five picture burst (640 × 480 at two frames per second)—up to 15 s movies (116 × 112 frame)

The sliding lens cover (Fig. 11.9) designed into this product also turns the product on and puts it into picture-taking mode. This design feature eliminates at least one keystroke and is a thoughtful addition to the user interface design. Simple ergonomic enhancing features such as this should be the goal of every product designer. The other buttons used to control the system are pictured in Fig. 11.10.

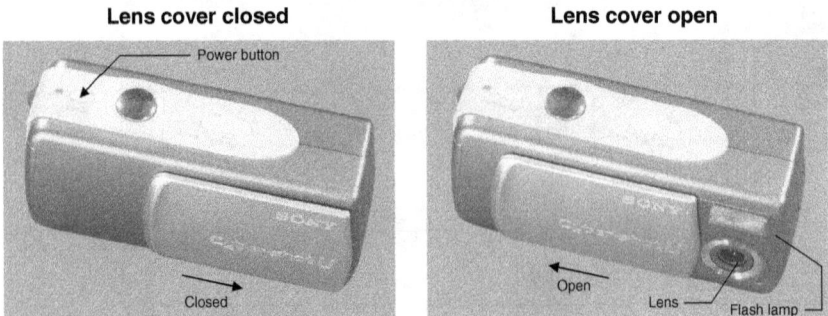

Lens cover closed   **Lens cover open**

Figure 11.9 Sliding lens cover.

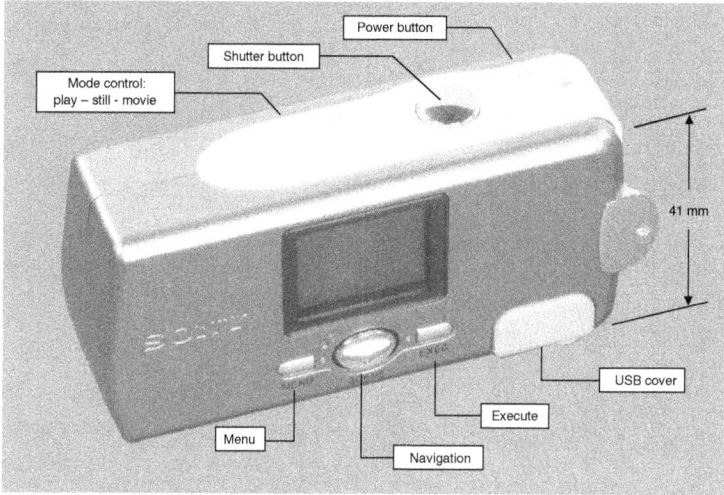

Figure 11.10 System controls.

The battery compartment is covered by a hinged door, as shown in Fig. 11.11. Opening this door also reveals the slot for the memory stick. The batteries are two NiMH (AAA cell) rated at 1.2 V and 700 mAh (each).

A simplified block diagram was estimated for this system and is shown in Fig. 11.12. The mechanical arrangement of these components is shown in Fig. 11.13. The system partitioning for this product is particularly impressive and involves some advanced electronic packaging technologies. The sketch indicating the system functional partitioning, the shape of

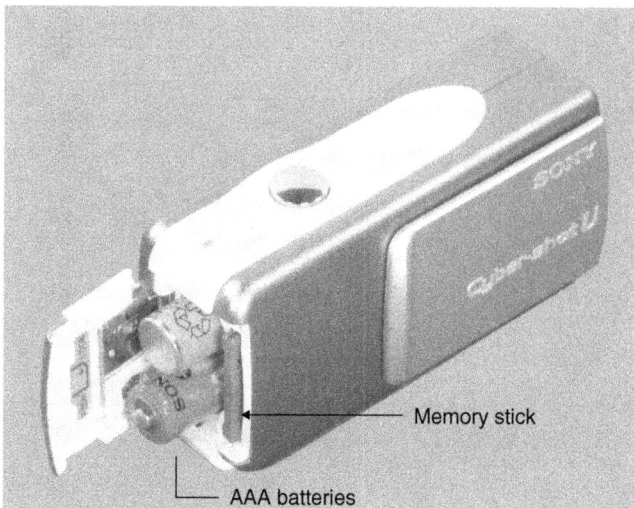

Figure 11.11 Battery removal.

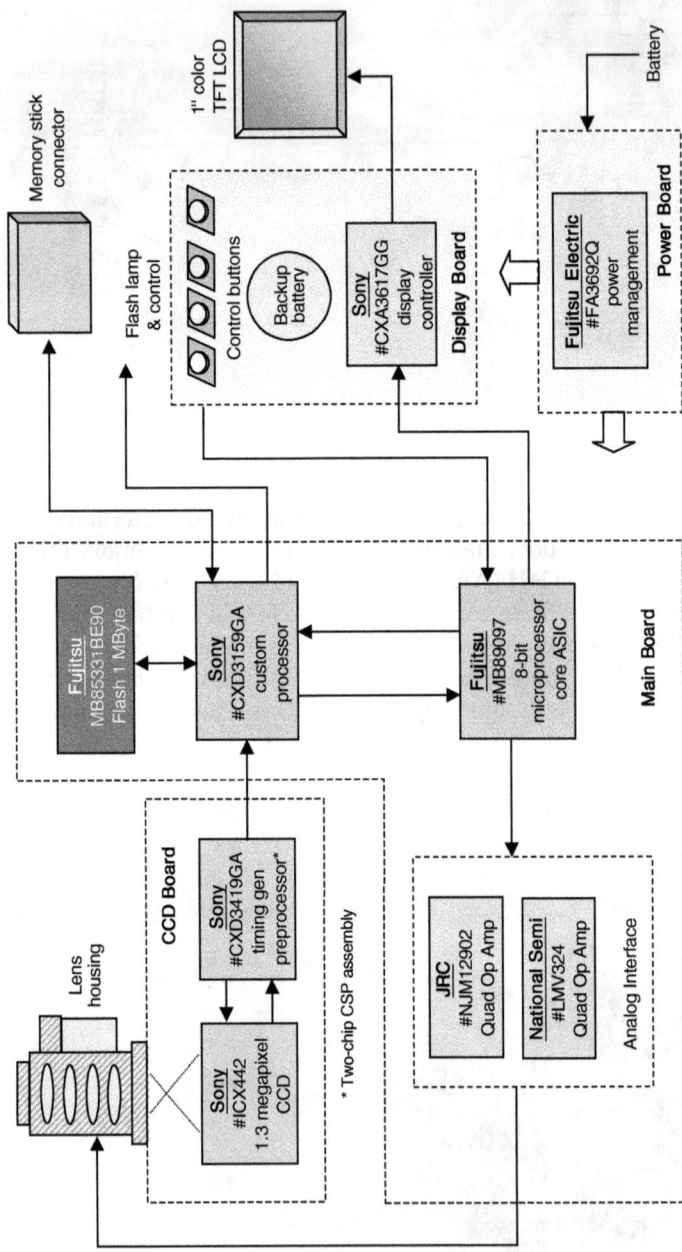

Figure 11.12 Simplified block diagram—estimated.

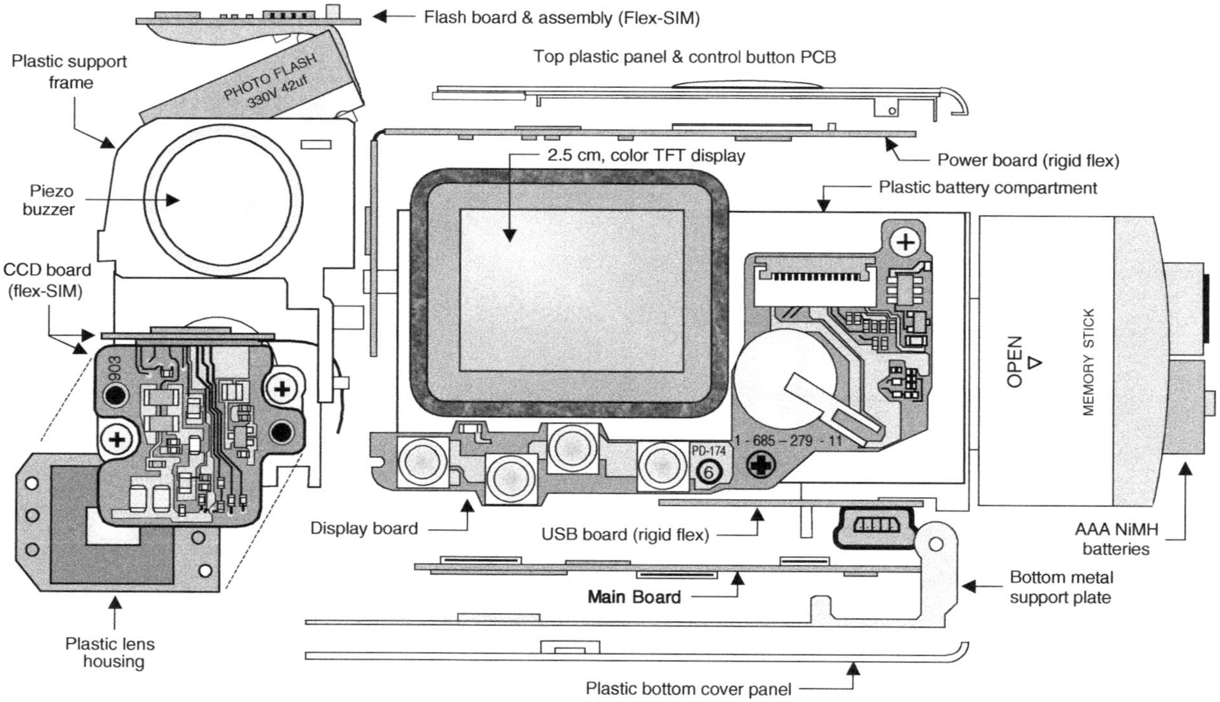

Flash board & assembly (Flex-SIM)

Top plastic panel & control button PCB

Power board (rigid flex)

Plastic battery compartment

2.5 cm, color TFT display

MEMORY STICK

OPEN
△

AAA NiMH
batteries

Bottom metal
support plate

1 · 685 — 279 · 11

PD-174
6

USB board (rigid flex)

Main Board

Plastic bottom cover panel

Display board

Plastic support
frame

PHOTO FLASH
330V 42uf

Piezo
buzzer

CCD board
(flex-SIM)

903

Plastic lens
housing

Figure 11.13 Component arrangement.

311

various printed circuit assemblies and their mutual connectivity is shown in Fig. 11.14.

Flex circuitry is essential to creating this product. The aggressive form-factor and the high number of I/O required to connect between various electronic modules requires the liberal use of polyimide flex circuitry, as shown in Fig. 11.15. Figure 11.16 shows how the display module is typically interconnected to the control electronics in this type of electronic imaging system.

Early 8-mm camcorder products drove the development of high-density SMT discrete component sizes such as the 0805, the 0603, and the 0402 package. The display controller board in this product uses 0201 discrete components, which are a mere 0.51 mm × 0.25 mm in size (Fig. 11.17).

The systems image sensor is a CCD device that is packaged with a hybrid flex-rigid board design approach. The CCD board and optics assembly are shown in Fig. 11.18. More detail on this CCD board is shown in Fig. 11.19 CCD board. The flex and rigid boards are bonded together. To achieve very high packaging density, a rather unusual multichip package is used. This multichip package barely occupies more real estate than would two separate CPSs.

The mechanical and optical design in electronic imaging products requires access to very specialized components. Figure 11.20 shows the autofocus and shutter motors after they have been removed from the main optical assembly. The components that comprise the optical assembly are pictured in Fig. 11.21.

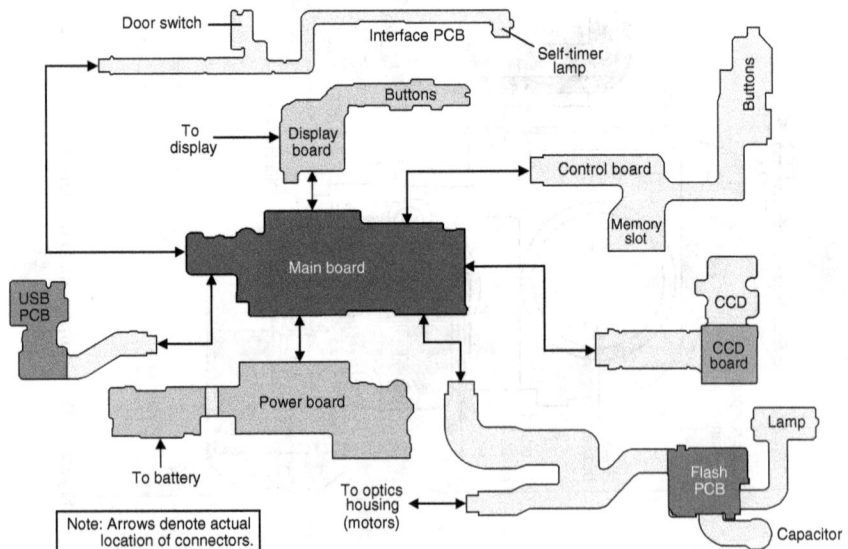

Figure 11.14   Board interconnect routing.

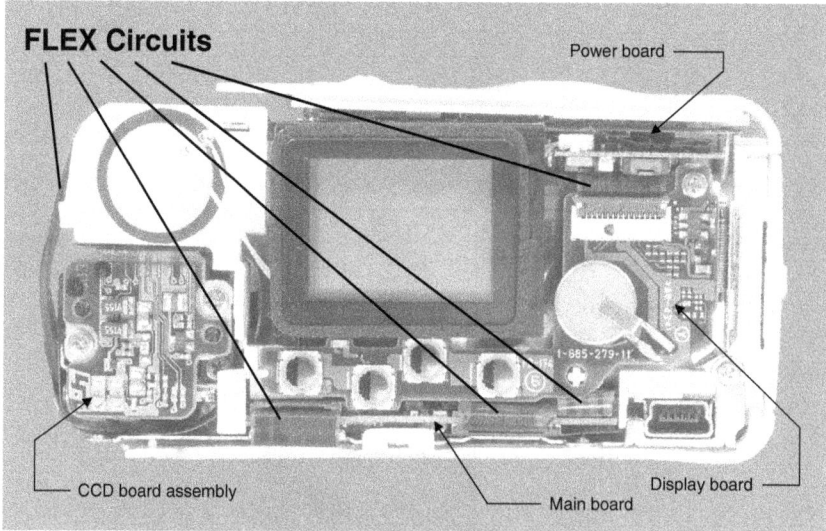

FLEX Circuits

Power board

CCD board assembly

Display board

Main board

Back side of the assembled electronics with the back enclosure removed.

Figure 11.15   Flex circuits.

Figure 11.22 shows the power consumption of this product when operating in playback mode. The numbers on this chart correspond to the system conditions below:

1. Turn power on by sliding door with "Play" mode selected.
2. Select next still picture for display.

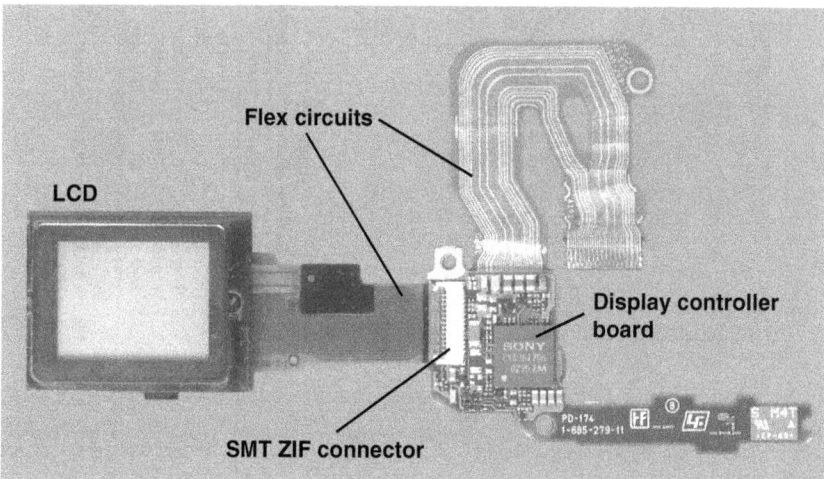

Flex circuits

LCD

Display controller board

SMT ZIF connector

Figure 11.16   Display and board assembly.

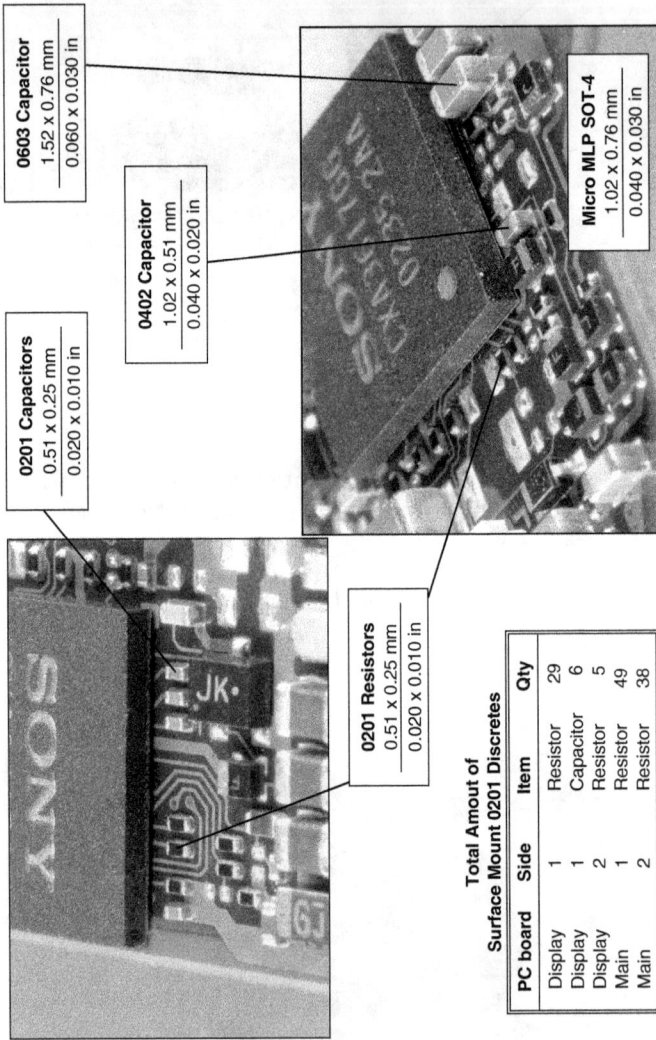

0603 Capacitor
1.52 x 0.76 mm
0.060 x 0.030 in

0402 Capacitor
1.02 x 0.51 mm
0.040 x 0.020 in

0201 Capacitors
0.51 x 0.25 mm
0.020 x 0.010 in

Micro MLP SOT-4
1.02 x 0.76 mm
0.040 x 0.030 in

0201 Resistors
0.51 x 0.25 mm
0.020 x 0.010 in

Total Amout of
Surface Mount 0201 Discretes

PC board	Side	Item	Qty
Display	1	Resistor	29
Display	1	Capacitor	6
Display	2	Resistor	5
Main	1	Resistor	49
Main	2	Resistor	38

Figure 11.17 Display PC board—0201 discretes.

314

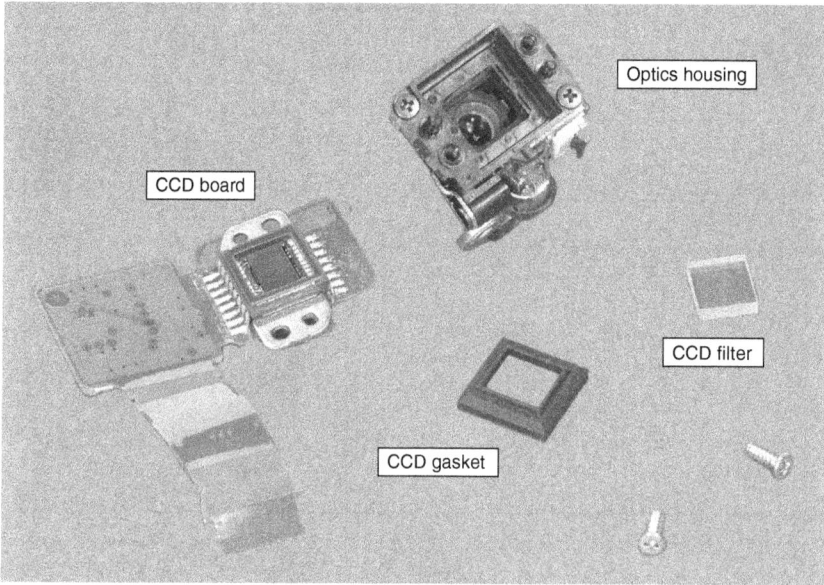

Figure 11.18 CCD and optics.

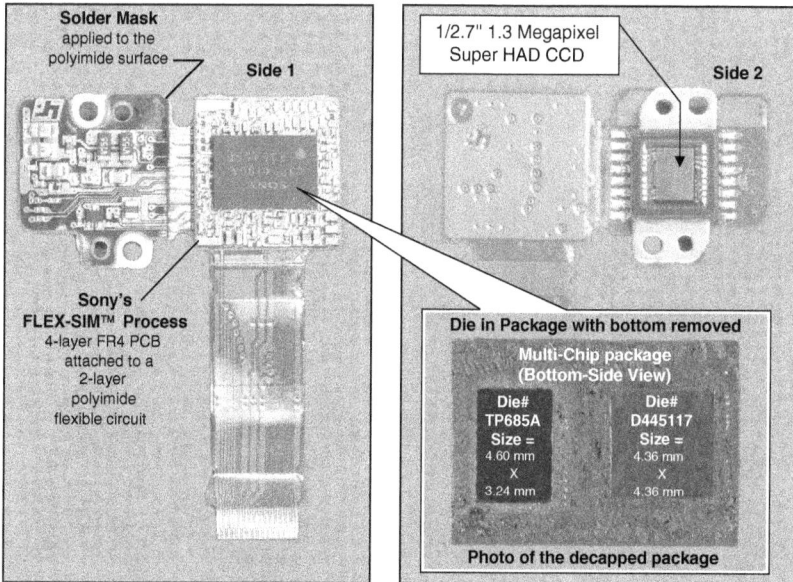

Figure 11.19 CCD board.

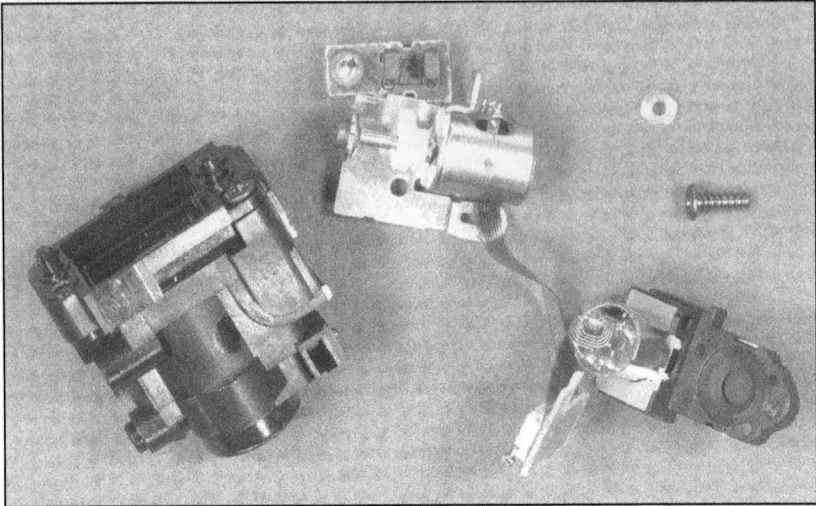

The autofocus and shutter motors were removed from the optics housing in this photo.

Figure 11.20 Focus and shutter motors removed.

3. Play movie.

4. Power off.

Figure 11.23 shows the power consumption of this product when operating in image capture mode. The numbers on this chart correspond to the system conditions below:

1. Start test in "Play" mode.

2. Switch mode to "Still."

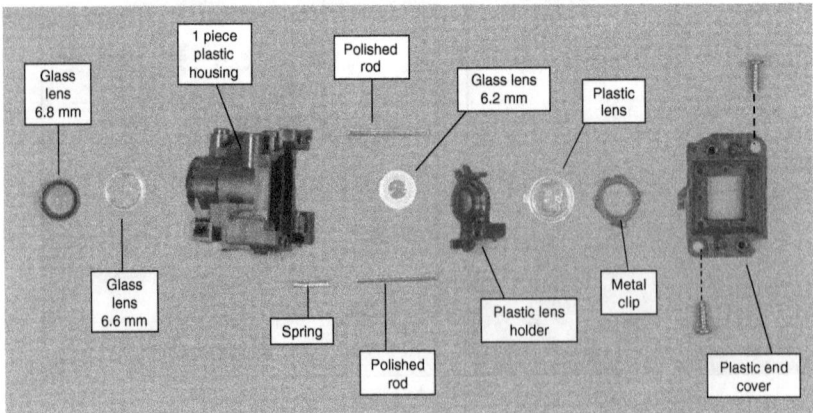

Figure 11.21 Lens housing disassembled.

3. Take picture (no flash).

4. Take picture (with flash).

5. Enter test in "Still" mode.

6. Switch mode to "Movie."

7. Take movie.

8. Power off.

All still pictures were taken with the resolution set to 1280×960 pixels. The movie uses 116×112 resolution.

Figure 11.24 shows the weight distribution within this system. The designer made the notable choice of using metal for a couple of external housing elements, instead of plastic. While comparable plastic parts would have produced an acceptable product, the selective use of metal in this case creates a look and feel of high quality. While the metal parts increase the product weight by a noticeable amount, the improved impression of quality seems to be a worthwhile trade-off.

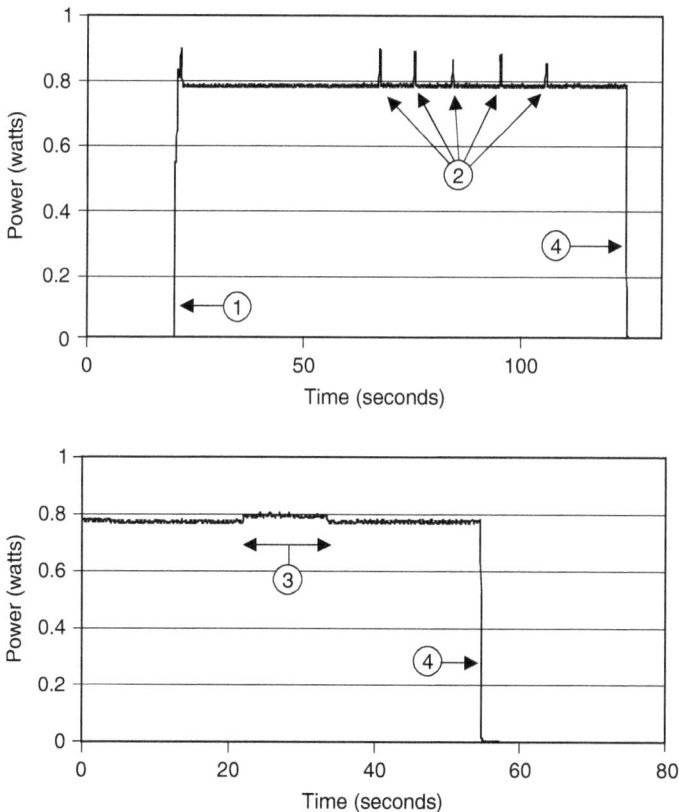

Figure 11.22 Power measurements (set 1).

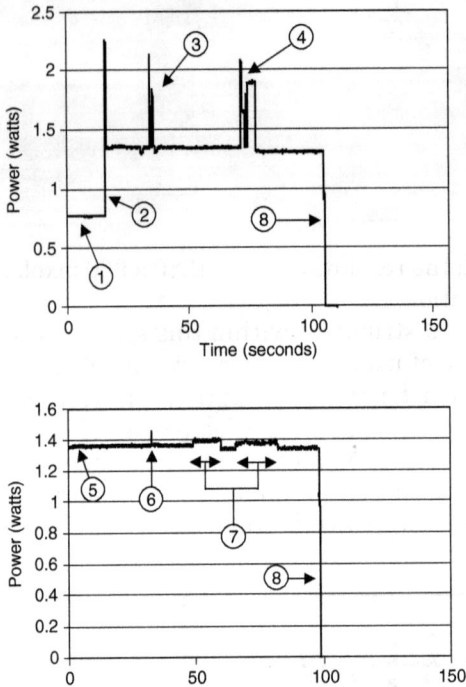

Take Still Pictures
Take Movie

1. Start test in "Play" mode
2. Switch mode to "Still"
3. Take picture (no flash)
4. Take picture (with flash)
5. Enter test in "Still" mode
6. Switch mode to "Movie"
7. Take movie
8. Power off

All still pictures were taken with the resolution set to 1280 × 960 pixels. The movie uses 116 × 112 resolution.

Figure 11.23 Power measurements (set 2).

Item	Weight (g)
Enclosures	**45.2**
Back metal enclosure	11.1
Front metal enclosure & door	12.3
Bottom plastic cover	1.7
Bottom metal support plate	4.2
Top plastic panel	2.2
Inner plastic support frame	2.2
Plastic battery compartment	11.5
Electronic Boards	**25.4**
Main board	4.5
Display board & interface	2.4
Flash board & flash circuitry	6.3
Power board	5.1
CCD board	3.1
USB board	1.1
Control button board	2.9
Battery	**23.7**
Memory Stick Support	**7.4**
Memory stick slot	4.1
Memory stick (8 MB)	3.3
Optics	**7.1**
Display Assembly	**5.2**
Miscellaneous	**1**
24 screws, tape, etc.	
Total	**115**

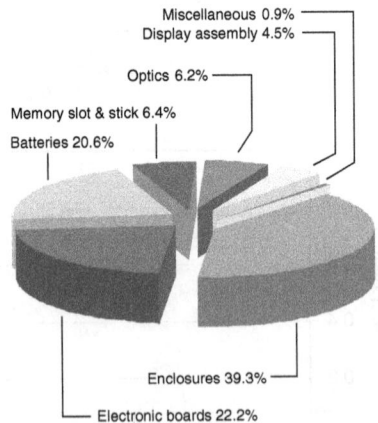

Miscellaneous 0.9%
Display assembly 4.5%
Optics 6.2%
Memory slot & stick 6.4%
Batteries 20.6%
Enclosures 39.3%
Electronic boards 22.2%

Figure 11.24 Component weights.

TABLE 11.1   IC Metrics (IC Summary)

Board description	No. of IC's total	No. of IC IO's	Die area (mm^2)	IC footprint area (mm^2)	IC IO's/(IC footprint area)
Display PCB	1	81	9	64	1.3
Main PCB	8	496	132	411	1
CCD PCB	3	112	65	219	1
Control PCB	—	—	—	—	—
Flash PCB	1	8	8	32	—
Interface PCB	—	—	—	—	—
Power PCB	4	104	25	86	1
USB PCB	—	—	—	—	—
IC total	17	801	238	811	0

This system contains 238 mm^2 of silicon, as indicated in Table 11.1. The largest silicon device in the system is the CCD which is 65 mm^2 (or about 60 percent of the area of a typical CPU found in a notebook computer). Table 11.2 shows that the main PCB has a very high silicon tiling density of over 14 percent. Due to the large proportion of flex-circuit area in the system, most of which is not populated with components, the average tiling density of the system appears deceivingly small, at around 3.1 percent. The implied routing density requirement for the main PCB, based on the part density and average pin count, is 73 cm/cm^2. This is a very high routing density as evidenced by the fact that the main PAB is an eight-layer board with line pitches as small as 0.152 mm.

Table 11.3 shows the IO distribution breakdown between ICs, modules, discretes, and connectors. The ratios between these connectors are characteristic of the particular design, but can also be used to estimate the ratios for similar products.

An estimated cost breakdown for this product is shown in Fig. 11.25. While the large percentage of cost dedicated to ICs is no surprise, it is

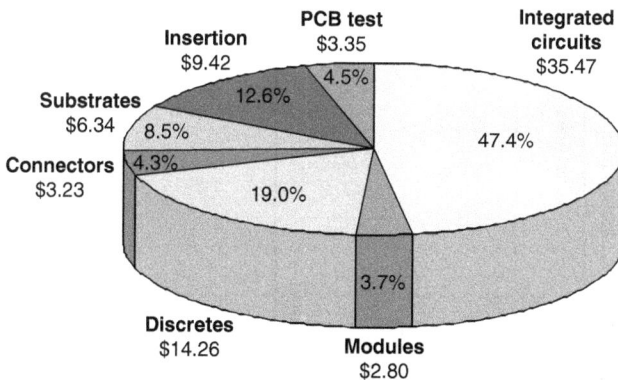

Figure 11.25   Estimated electronics cost.

TABLE 11.2 PCA Metrics (Summary of System PCA Metrics)

Board description	Board area (cm²)	No. of PCB layers	No. of parts	No. of connections	PCB tiling density (%)	Connection density	Routing density	Part density	Average pin count
Display PCB	5.7	6	93	371	1.50	65	48	16	4.0
Main PCB	9.3	8	273	1,232	14.11	132	73	29	4.5
CCD PCB	9.4	6	84	275	6.96	29	29	9	3.3
Control PCB	13.1	2	15	46		4	10	1	3.1
Flash PCB	20.0	5	56	138	0.41	7	12	3	2.5
Interface PCB	5.1	1	4	10		2	7	1	2.5
Power PCB	10.3	8	159	558	2.38	54	41	15	3.5
USB PCB	4.8	3	11	34		7	14	2	3.1
System total	**77.8**	—	**695**	**2,664**	**3.1**	**34**	—	**9**	**3.8**

TABLE 11.3 Component and Connection Metrics (Component and Connection Counts)

Board description	Opportunity count	IC's	IC IO's	Modules	Module IO's	Discretes	Discrete IO's	Connectors	Connector IO's
Display PCB	464	1	81	5	18	84	193	3	79
Main PCB	1,505	8	496	4	12	254	559	7	165
CCD PCB	359	3	112	—	—	81	163	—	—
Control PCB	61	—	—	3	18	11	22	1	6
Flash PCB	190	1	8	1	3	54	127	—	—
Interface PCB	14	—	—	1	4	3		—	—
Power PCB	717	4	104	—	—	152	354	3	100
USB PCB	45	—	—	—	—	9	23	2	11
Total	**3,359**	**17**	**801**	**14**	**55**	**648**	**1,447**	**16**	**361**

TABLE 11.4   **Design for Manufacturing (DFM)**

Item	Description	Estimated cost
Front enclosure	Anodized aluminum—sliding door—12.1 g spring & plastic hardware—1.5 × 4 × 8 cm	$0.90
Back enclosure	Anodized aluminum—plastic buttons—strap lock—1.5 × 4 × 8 cm—11.1 g	$0.55
Battery compartment	ABS—Main compartment—door & hinge— battery terminals—metal clip—11.5 g	$0.94
Support frame	ABS—2.5 × 2.5 × 3.5 cm—2.2 g	$0.30
Top cover	Plastic cover & buttons—metal support panel— 1.5 × 1 × 6.5 cm—5.3 g	$0.35
Bottom cover	Plastic cover—metal support panel— 1.5 × 2 × 8 cm—5.8 g	$0.35
Memory stick	8 MB	$4.80
Lens housing	2-piece plastic housing—CCD gasket—CCD filter—2 small motors—3 glass lenses— 1 plastic lens—2 polished rods—spring, metal clip & screws	$7.13
Miscellaneous	21 screws—tape—labels—memory slot connector cover, etc.	$0.62
Total		**$15.94**

DFA Index (%)	**3.8%**
Total Assembly Time (s)	**762***
Total Assembly Cost ($)	**$6.35 (@$30/h, Japan)**

Percent of total (762 s) assembly time

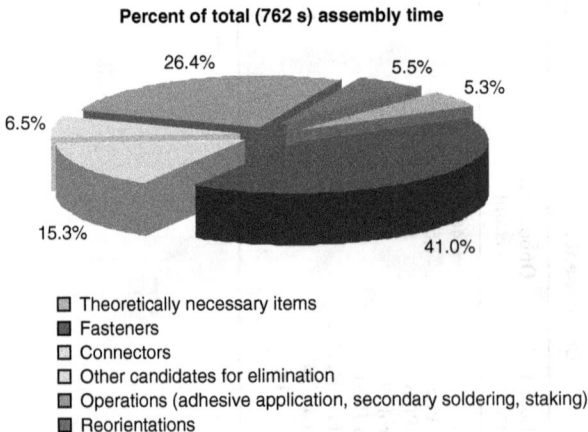

26.4%   5.5%   5.3%   6.5%   15.3%   41.0%

☐ Theoretically necessary items
■ Fasteners
☐ Connectors
☐ Other candidates for elimination
■ Operations (adhesive application, secondary soldering, staking)
■ Reorientations

Figure 11.26   Design for assembly (DFA).

Sony DSC-U10 Digital Camera	Cost
Electronics assemblies	$74.87
Display	$10.32
Housing/Hardware (DFM)	$15.94
Batteries – AAA type/Qty 2	$1.48
Final assembly (DFA)	$6.35
Total	$108.96

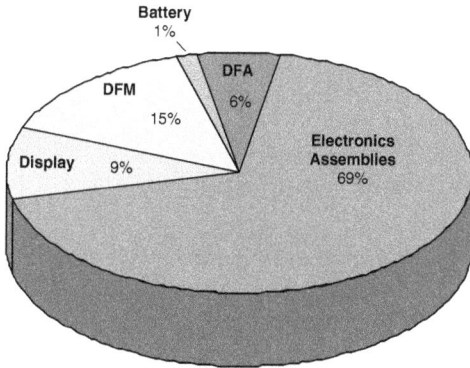

Figure 11.27   Cost summary.

unusual to see almost 20 percent of system electronics cost contributed by discrete components.

Table 11.4 shows the estimated cost breakdown for mechanical parts and other elements not included in the estimate of systems electronics cost. The optics assembly stands out as a significant cost adder in this category. Figure 11.26 shows the estimated time required for final assembly of the system elements to be 762 s. This estimate includes 110 s for the assembly of the optics module.

An estimate of the total cost to manufacture this product is shown in Fig. 11.27.

Economics

Growth in the electronics industry is driven by investment in infra-structure for new technology. State-of-the-art fabrication facilities that manufacture integrated circuits and flat panel displays can cost over a billion dollars to construct. For these facilities to be economically justi-fiable there must be immediate demand for the components that they produce. If the demand exists, then these factories generate profits that are used to finance the next generation of technology advancement.

Demand for new technology does not materialize, however, unless the new technology can add valuable functionality to the products in which it is employed. The rest of the infrastructure in which the tech-nology is used must also be able to support the increased performance of the new technology. A high-definition TV monitor, for example, is useless without communications standards to enable transmission of programming, without processor technology capable of decoding high-definition signals in real time, and without high-definition programming to view on the monitor. Furthermore, none of these other support ele-ments will be developed if there is not the expectation that the high-definition monitor will become affordable for a large number of customers at some point in the foreseeable future.

Often, "chicken-and-egg" situations develop, which slow the advance-ment of new product concepts and technologies. Manufacturers are reluctant to invest in new technology without a commitment from designers to use the technology. Designers are reluctant to use unproven technology because it may be too risky or expensive. Network infrastructure builders are reluctant to upgrade their networks with new services without the certainty that those services will be used. New services often require new products and content to be successful. Of course, the builders of the new products and content will not invest unless there is some certainty that the required network services will be available.

Standards are one tool that the industry relies on to help break these deadlocks. Standards reduce the risk of investing in new technologies by defining the system interfaces ahead of time. When many companies commit to a new standard, it signals that there is broad consensus that the new technology is worthy of investment. Standards also enable a modular approach to designing products. For some products, the system architecture can be fairly well defined simply by showing the standardized inputs and outputs on a simple block diagram. The portable electronics designer has many standards to exploit, which cover interfaces for networks, displays, memory, processor busses, and many other aspects of the system architecture.

A thorough understanding of manufacturing economics can also be helpful in justifying an investment decision. Manufacturing costs are characterized by learning curves that enable managers to project future cost reductions. Companies that understand these economics very well are able to determine how quickly the cost of a new product will be reduced. This is critical information for creating a future cash-flow analysis needed to justify a product investment. Product designers who understand learning curve dynamics are able to exploit them in their product designs and have a distinct competitive advantage. It is also important for the designer to recognize how the product design can leverage these basic economic principles. The use of product platforms is instrumental in creating a line of portable electronics that will be competitive and profitable over the long term.

12.1   High-Volume Manufacturing and Learning Curves

Knowledge of high-volume manufacturing processes and economics is important to good portable electronic design. It is only through the exploitation of high-volume manufacturing that portable electronic devices exist today. Henry Ford demonstrated with the early automobile production lines that the cost to produce a product steadily decreased with each successive unit that is produced.

This is certainly true of modern semiconductor manufacturing facilities, where billions of dollars of investment must be recovered by producing millions of units of output. Every manufacturing process in the portable electronics food chain (whether it is for components such as ICs and flat panel displays or the final product assembly) is subject to these economics. Understanding and leveraging these economics can create a strategic advantage for companies that are producing portable electronic products.

While a detailed discussion of all the individual manufacturing processes that are involved in making portable electronics is beyond the scope of this book, the principle of manufacturing *learning curves* is extremely relevant. Learning curve principles are used by leading portable and consumer electronic producers to develop costing objectives and pricing strategy for products that will be sold in high volumes. Ultimately, learning curve analysis can help the portable electronic designer determine whether a product concept is economically viable. This type of analysis is particularly important for portable product concepts that are part of a complex business model where the device itself may be a business enabler rather than the primary focus of the business. Examples of this situation are video game consoles, where the sale of game software is the primary profit generator. Another prime example is taken from the early days of cellular phone adoption. During this time, it was not unusual for cellular phones to be provided for free if the customer agreed to a cellular service contract for a certain duration. It was not always the service provider who sold cellular phones at a loss to enable their service business. Often the cell phone producer would provide underpriced phones or free phones to the service provider as part of the overall contract to supply expensive base stations and other infrastructure gear.

Learning curve theory states that the cost to produce a unit of product will decrease by a characteristic percentage with each doubling of the cumulative number of products produced. The characteristic percentage is determined by the type of manufacturing processes involved. The powerful aspect of learning curve theory is that it enables decision makers to predict future manufacturing cost reductions without requiring detailed knowledge of how those future cost reductions will be achieved.

So the learning curve is the cost of a unit of production as a function of cumulative volume. The following expression was developed to describe the learning curve:

$$K(n) = K(1)\, n^b \tag{12.1}$$

where

$$b = \log(R_e)/\log(2) \tag{12.2}$$

$K(1)$ = cost of the first unit produced
$K(n)$ = cost of the nth unit produced
n = total units produced
Re = experience rate

Another useful form of this relationship can be written as follows:

$$K(n_2) = K(n_1)\, (n_2/n_1)^b \tag{12.3}$$

where $K(n_1)$ = cost at n_1 cumulative units manufactured
     $K(n_2)$ = cost at n_2 cumulative units manufactured

We can further transform Eq. (12.2) as follows:

$$K(n_2)/K(n_1) = (n_2/n_1)^b$$

$$\log\{K(n_2)/K(n_1)\} = b \log(n_2/n_1)$$

$$b = \log\{K(n_2)/K(n_1)\}/\log(n_2/n_1) \tag{12.4}$$

Manipulating Eq. (12.3) produces

$$b \log(2) = \log(R_e)$$

$$R_e = 10^{\{b \log(2)\}} \tag{12.5}$$

Equations (12.4) and (12.5) are useful because they enable us to cal-culate an empirical experience rate based on two data points of cost and cumulative volume. Being able to extract an experience rate from exist-ing data is useful because it is very difficult to derive the experience rate theoretically. Furthermore, given the large tooling costs and other ini-tial investments involved in electronic product manufacture, it is diffi-cult to assign a meaningful cost to the first unit built. For electronic products, it is therefore recommended that Eq. (12.3) be used, rather than Eq. (12.1). This enables the analyst to base projections on more sta-bilized data since n_1 does not have to be the first unit produced but rather a somewhat later data point where production costs are less ambiguous.

Let us consider the following simple example of how we can use these equations. Suppose that a manufacturer has just completed one year of production of a new type of portable MP3 player. The initial cost esti-mates from engineering indicated that the device would cost $50 to pro-duce. After two months of production, 20,000 units were produced and the data clearly indicated that the manufacturing cost for each unit was running at $58 at the end of the two-month period. (Per unit costs during the first week would have been estimated at several thousand dollars per unit, which is a valid number but subject to huge uncer-tainty.) At the end of eight months of production, 120,000 units had been produced and the cost for each unit at the end of that period was determined to be $42.

Using Eq. (12.4) we can determine b:

$$b = \log\{42/58\}/\log(120{,}000/20{,}000) = -0.18014$$

Equation (12.5) lets us calculate the experience rate (R_e):

$$R_e = 10^{\{-0.18014 \times \log(2)\}} = 0.88 \text{ or } 88\%$$

This means that each time the cumulative volume produced is doubled, the cost to manufacture a unit decreases by 12 percent.

An experience ratio of 88 percent is not uncommon for electronic products that are composed of commodity components. Much steeper curves can be achieved for products that contain custom silicon or new components. In the early 1990s some cellular phone products had experience rates of less than 70 percent. This was because many of these products were based on custom silicon and other newly developed component technologies.

So, it is important to realize that the learning curve that a particular product will undergo is really a composite of the various learning curves for its component parts. In the same way that we calculated an aggregate learning curve for the product above, individual learning curves can be produced for key components and subcomponents to give a more detailed picture of the cost dynamics for the product.

The final assembly process for the product in the example may have been performed on a new production line. This production line would have a steep learning curve that would affect the corresponding cost contribution for final assembly. A more significant component of the cost might be determined by a few key pieces of commodity silicon. The learning curve for these components is determined not by the cumulative volumes for the MP3 player but for all products that have been supplied with that component up until that point in time.

Another method for looking at the learning curve for commodity components is to consider the industry learning curve rather than the learning curve for a particular vendor. When determining the future cost of a 14.1-in color XGA TFT display, it might be useful to look at the industry learning curve rather than the curve for a particular supplier. It is also important to keep in mind that learning curves are used to predict cost, *not* price. Price fluctuations due to changes in supply and demand can significantly affect the cost to purchase a specific component from the supplier. Very sharp price fluctuations are observed in the DRAM market and the LCD market, for example. Learning curves can be used to estimate future prices, however, if a competitive market is assumed with some additional assumptions about the producer's profit margins. Figure 12.1 shows the learning curve for a specific type of flat panel display product used in notebook computer systems.

Assuming that the proper care is taken in building the learning curve models, various strategic scenarios can be explored with the help of learning curve theory. Consider the case of Product X, which is targeted to have a market price of $199. A hypothetical company, Acme Gadgets,

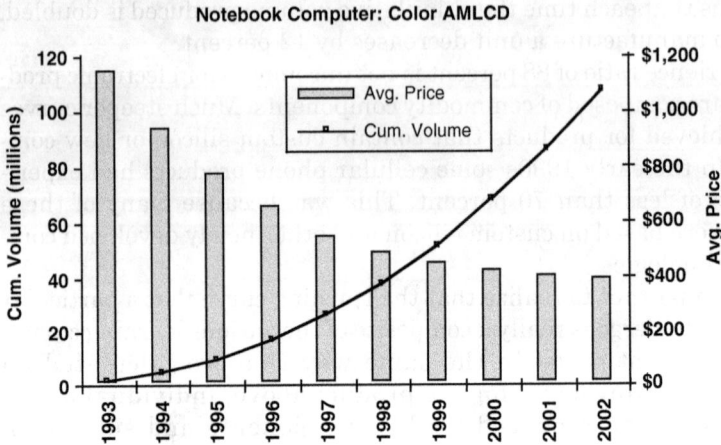

Figure 12.1   Learning curve for flat panel display.

intends to introduce the product but realizes that a successful launch
will attract the attention of some able competitors.

Based on historical price data for similar products, Acme models the
anticipated market price for the product over the first 36 months of
introduction. A learning curve model is used to shape this price profile using the initial introduction price as the starting data point and
applying an experience rate that fits the historical data for similar
products. Figure 12.2 shows the results of this model. Notice that the
experience curve equation is being used to shape the price curve, even
though, strictly speaking, the experience curve model is really
intended to model cost, not price. Again, the rational for this is the
assumption that in a competitive market, the price will roughly track
the manufacturing cost plus a delta representing the competitive
margin. In this price model, initial production volume for the entire
market is assumed to be 30,000 units per month at launch, growing
by 5 percent per month as capacity expands. Assuming an experience
rate of 83 percent, the result is a projected market price of just over $50
some 36 months after launch, with nearly 3 million cumulative units
shipped.

Figure 12.3 shows the model parameters for Acme and its competitors—
Company x and Company y. Acme estimates that its initial manufacturing cost will be $89. Historical data shows that Acme should expect a
learning curve of 80 percent for this type of product. The design rating
is a subjective metric that Acme uses to estimate the relative customer
appeal of a particular product design (100 percent being the best design
possible). In this type of analysis, two competing products would
garner share of the available market in proportion to the ratio of their

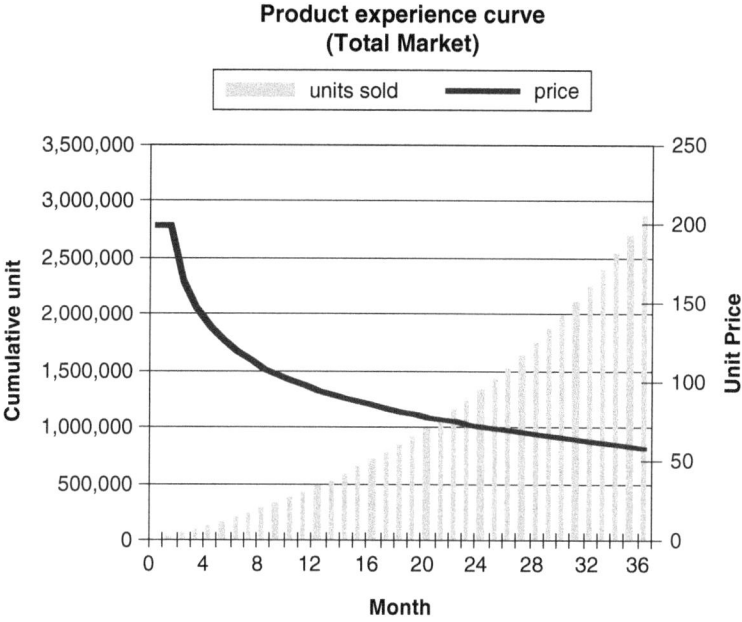

**Product experience curve
(Total Market)**

Product X	
Initial market price	$199
Initial monthly volume	30,000
Volume growth rate (monthly)	5.0%
Experience rate (price)	83%
b	−0.27

Figure 12.2   Product price curve.

Acme	
Initial unit cost	$89
Experience rate	80%
Design rating	70%
Entry point	0
b	−0.32

x	
Initial unit cost	$80
Experience rate	80%
Design rating	60%
Entry point	6
b	−0.32

y	
Initial unit cost	$70
Experience rate	60%
Design rating	30%
Entry point	9
b	−0.74

Figure 12.3   Product competitor parameters.

Market Share

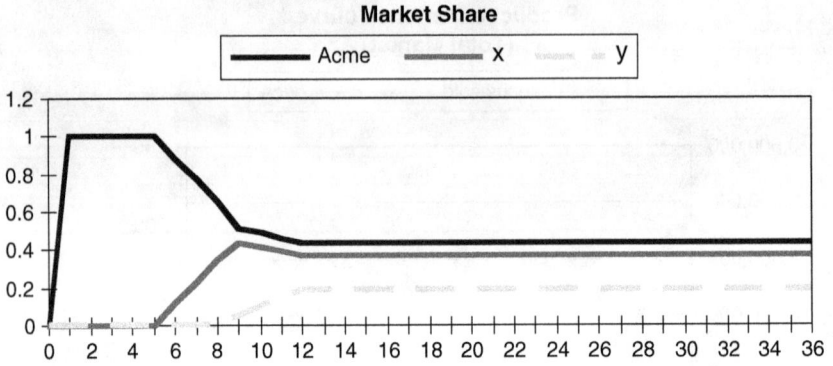

Figure 12.4 Market share distribution.

respective design ratings. The entry point indicates the months after launch of the initial product in which each competitor launches their own product.

Figure 12.4 shows the proportion of sales that each competitor captures during each month of the 36-month period. Figure 12.5 shows how each competitor's production cost changes as it produces units to meet the demand for its product. Figure 12.6 shows the cumulative gross profit for each competitor based on the difference between the projected market price and the production cost, multiplied by the appropriate unit sales for each competitor.

In the modeled scenario, Acme is the first to market with a very good design and enjoys considerable profits and 100 percent market share until Company x enters the market in month 6. Company x has a slightly less appealing design and similar manufacturing capabilities

Producers Cost

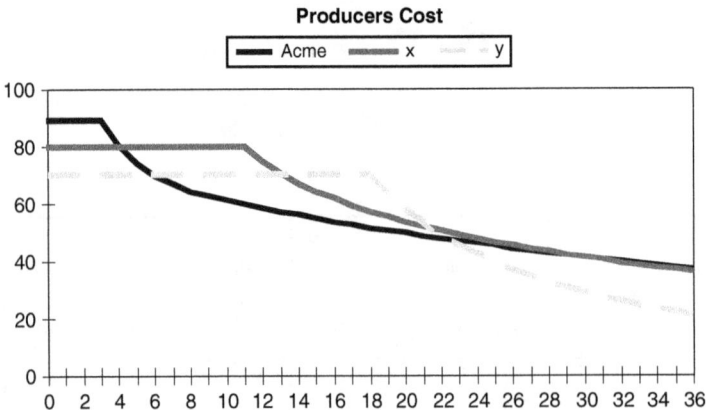

Figure 12.5 Competitor's production costs.

Cumulative Profits

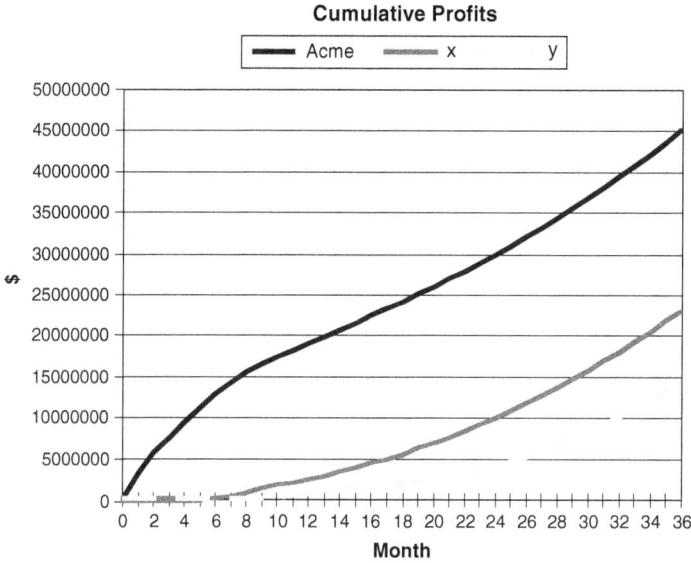

Figure 12.6 Competitor's cumulative profits.

to Acme, having the same experience rate of 80 percent. It takes Company x until month 26 to approach Acme's costs because Acme had a head start on the experience curve. Company y launches a product in month 9, with a rather clunky design. Company y has very low-cost manufacturing capabilities and is able to match Acme's production cost by month 23. Since demand is strong for all three producers, Company y elects not to drive prices down too aggressively and to reap increasing profits as its production costs continue to decline. If Company y does decide to push prices below the projected price curve, Acme may decide to exit the market.

Obviously, the scenario above is subject to many questionable assumptions. The point is that product executives should seek to use learning curve models to understand the potential market dynamics and profit opportunities.

Major consumer electronics companies such as Sony and Matsushita rely on forward-pricing models to justify new investments. When Sony enters a new electronic product market, for example, it can do so with the certainty that it will field a quality design, that it will capture some reasonable portion of market share, and that its factories will perform well enough to achieve a particular experience rate. This confidence in the learning curve gives consumer electronics executives the justification to make the large investments required to launch a major new consumer electronic product.

12.2   Leveraging Product Platforms

Revolutionary product concepts, such as the first Pilot PDA or the first Sony Walkman, are few and far between. In both cases, the products themselves did not represent a leap in technology but rather an insightful combination of existing technology components. Discovering a revolutionary product concept, however, is actually quite different from exploiting that concept with a complete line of products. Indeed, the vast majority of portable electronic products that are designed are not revolutionary products but evolutions of a proven product concept. Most portable electronic products end up being manufactured by large consumer electronics OEMs as part of a broad product line of similar products.

A prime example of this concept is the Walkman product line produced by Sony. Sony introduced the Walkman in 1979 to the Japanese market and it quickly became a worldwide hit. While many capable Japanese competitors entered the market quickly, Sony was able to maintain an astonishing 35 percent plus market share for most of the 1980s. Sony was able to achieve this by implementing a classic product platform strategy.

Industry lore has it that the Walkman was invented in 1978 by then Sony chairman Masaru Ibuka who wanted to combine some technology developments from two different Sony research groups to create a portable high-fidelity music player. One group had developed miniaturized headphones, while another had produced a miniaturized, high-quality tape recorder. The components were combined, consumers craved the product, and the Walkman was born. While the innovation itself was notable, however, Sony's ability to exploit the innovation was even more impressive.

According to Sanderson and Uzumeri (1997), Sony went to market with the original product design (TPS-L2) but also created three major platform activities to improve the product. These product platforms are classified as *generational* design efforts because they involve the integration of major new technology developments to improve the product concept. Minor tweaks to these platforms are considered *topological* or cosmetic changes and do not involve major engineering resources.

Sony's product line strategy was to invest in a small number of technology-driven platforms but a large number of topological variations. This enabled Sony to cost-effectively cover many market niches and deprive its competitors of profitable opportunities. By driving and controlling market segmentation, Sony was also able to maintain high margins for selective portions of its own product line.

In 1981, Sony launched the WM2 product platform that weighed less than 10 oz with a 1-oz headset. Enabled by a thin, lightweight tape transport mechanism and other innovations, this platform was the basis for a wide range of Walkman products.

A couple of years later Sony introduced products based on the WM20 platform. This platform included a new, thin motor technology and reduced power consumption. Rather than drop the WM2-based products, however, Sony continued to use the platform for low-priced products targeted at cost conscious consumers. The thinner profile and longer battery life of WM20 products commanded a premium.

Concurrently, Sony was developing the WMDD platform series which used an improved drive mechanism to reduce sound distortion resulting from vibration and variation in the motor speed. This platform was targeted at the high end of the market and the first product based on this platform was introduced into the market in 1982.

Thus, the WM2 series provided a basic platform for the lower-pricing tier, the WM20 for the middle tier, and the WMDD for the upper tier. Based on these platforms, Sony introduced over 250 product variations during the 1980s, averaging about 20 per year. Variations in color, plastics, button location, or the addition of simple electronic enhancements allowed Sony to fragment the market into many niches, which it could service very cost-effectively.

Sony developed Walkman products tailored to appeal to particular demographics and lifestyles. Japanese models were equipped with remote controls in the headphone cord, allowing for the control of the system in the confines of a crowded train. Many U.S. models were equipped with a radio tuner, and ruggedized versions were introduced in the United States for people that lead an athletic lifestyle. European models were more classically styled with the emphasis on sound quality.

With the platform strategy developed by Sony, they could reportedly break even on some product variations with the sale of as few as 30,000 units. Adding simple electronic features to these platforms was also a low-cost way to differentiate the product. Sony led the industry in the introduction of these new features. In addition to creating the initial Walkman product in 1979, Sony (and its subsidiary Aiwa) also introduced the first models with the following features: AM/FM stereo radio, stereo recording, auto-reverse, FM headphone radio, Dolby, shortwave tuner, remote control, separate speakers, water resistance, graphic equalizer, solar power, dual cassette, TV audio band, children's model, and enhanced bass. (Sanderson and Uzumeri, 1997).

Each of these functional enhancements required relatively little engineering investment compared with the development of the overall product platforms. In one example, enhance bass was added to the feature set by adding a switch and a single capacitor. This enhancement cost less than 20 cents in manufacturing but added $20 to the system retail price.

According to Sanderson and Uzumeri (1997), Sony also enjoyed another considerable advantage over its competitors in the Walkman space during the 1980s. The average product life for a Sony Walkman product was almost two years, which is nearly double the product life of most of the competitors. The advantages to achieving longer product life are enormous. First, the manufacturer is able to realize a very high return on investment, since the initial investments in design and tooling were paid for early in the product life. Second, additional production efficiencies continue to reduce the product manufacturing cost according to the principle of manufacturing learning curves. Third, Sony continues to introduce additional new models into the market, meaning that at any point in time Sony has more models available than its competitors.

The reason for Sony's superior product longevity also merits examination. Sony does appear to invest more in developing a high-quality design than does its competitors. Sony products consistently demonstrate very ergonomic features and are constructed with a high-quality finish. Sony's brand recognition virtually guarantees market share provided they can field a competitive product in a proven market. Thus, investing in high-quality designs is a low-risk proposition for Sony. This may not be the case for many of Sony's competitors. Perhaps most importantly, since Sony was able to focus on many niche markets simultaneously, each niche product embodied a highly optimized design for the niche that it serviced. These highly optimized designs simply did not need to be refreshed as often as a more generic product.

In the heyday of the Walkman, Sony was organized around the product platform concept. Product engineering was centralized in Japan. This is where all research into new technologies was conducted and where new product platforms were developed. Production was also coordinated from Japan. Two Japanese production facilities produced higher-end products, while factories in Taiwan and Malaysia produced lower-cost products and assemblies. Industrial design and marketing were handled regionally, however, with organizations in the United States, Japan, and Europe. These regional organizations were able to better develop strategies for customizing the platforms to the requirements of the local market.

So focused did Sony become around the efficient production of these platforms, that Sony even developed its own automation equipment which was later used to assemble VCR and camcorder products. Sony automation later developed this equipment into a product line that it sold to other manufacturers.

Sanderson and Uzumeri summarize their analysis of Sony's success with the following conclusions:

1. Sony's variety-intensive product strategy did both, denied competitors of profitable niches and increased Sony's product longevity.

2. Sony's regionally distributed design and marketing organizations ensured that product design could quickly adapt to regional market differences and changes.

3. Sony's emphasis on industrial design enabled Sony to rapidly respond to a changing market through topological variations rather than through the reengineering of an entire platform.

4. Sony's emphasis on minimizing design costs through its platform strategy enabled designers to exercise greater creativity of design in the marketplace by reducing the financial risk of experimental designs.

13

The Past, Present, and Future of Portable Electronics

13.1 A Brief History of Portable Electronics

The history of portable electronic devices is very brief. It is only since the development of the transistor, and subsequently the transistor radio in the 1950s, that portable electronics have even been possible. Further developments lead to the planar integrated circuit in 1959, which made possible highly integrated electronics. The first highly integrated portable electronic devices were calculator products developed by Texas Instruments. These devices combined integrated circuit technology for processing and LED technology for the display. Even though these first portable electronic devices were rather expensive, at several hundred dollars each, they became very popular because they greatly reduced the time required to perform routine calculations.

While Texas Instruments was first-to-market with these new devices, innovative Japanese companies such as Sharp were quick to apply their advancing manufacturing capabilities to electronic calculators. Efficient Japanese production methods and low operating costs enabled the price of electronic calculators to fall rapidly during the 1960s and 1970s. By the mid-70s, almost any student could afford an electronic calculator to perform their math homework. Affordable, programmable scientific calculators also began to appear in this time frame. These programmable calculators were the precursors to the first personal computers.

Digital watches also became popular in the 1970s. The first digital watches were also developed by Texas Instruments and used integrated circuits plus an LED display. Since watches require a smaller form-factor and must be left on all of the time, however, battery life was more of an issue with these products. Early LED displays were not particularly power efficient, so early digital watches required the user to push a button to activate the display and check the time. In addition, these early watch displays had poor daylight readability and were rather unaesthetic.

Development of the LCD was a major boost to the digital-watch industry. LCDs require very little power when compared to LEDs, are very thin, and have good contrast for daylight readability. Sharp and Casio were quick to adopt LCD technology and develop a broad selection of compact, affordable, and stylish digital watches.

Around 1979, Sony introduced the first Walkman product to the Japanese market. Consumers found the combination of high-quality sound and small form-factor irresistible. The first 30,000 units sold out in three months. Within two years, Sony had sold over 1.5 million units.

Beginning in the late 1970s and early 1980s the first personal computers began to appear on the market. While these products were not portable, per se, they were greatly diminished in form-factor when compared to the computers that preceded them. This burgeoning industry drove the development of high-volume surface mount technology infrastructure which was to enable truly portable electronic products just a few years later.

By the late 1980s the portable electronics industry began to pick up a real head of steam. The first notebook computers, portable cellular phones, electronic organizers, Walkmans, and point-and-shoot 35-mm cameras appeared in the marketplace during this period. Japanese companies were dominating these product categories.

Sharp and Toshiba introduced the first affordable LCD-based portable computers. Their strategic focus LCD technology had paid off. The ensuing competition drove increasing functionality and decreasing form-factors simultaneously. North American competitors joined in the fray, but not without an Asian partner to supply high-volume, high-density electronics manufacturing capabilities.

With the increasing levels of functional integration being driven by notebook computing, the existing manufacturing infrastructure already in place for calculators, and the advancing LCD technology, Japanese companies began to introduce electronic organizer products such as the Sharp Wizard and the Casio Boss. These products incorporated simple but useful functionality into a very small form-factor. Typically, these products held a personal calendar, reminder alarm, contact information, memo, calculator, and clock. These products were also significant because they drove new manufacturing technology, particularly for packaging small displays and keyboards.

Compact portable cellular phones were also becoming popular in the late 1980s. Companies such as Mitsubishi and Toshiba came seemingly out of nowhere with extremely competitive form-factors for cellular phones.

Motorola, with its own high-volume automotive electronics manufacturing infrastructure, was holding its own against the Japanese competitors, but just barely. Cellular phone products differed from calculators, watches, and organizers in that they contained a large amount of analog circuitry that could not be integrated into silicon. Smaller discrete components such as the 0402 were developed to enable the design and production of small form-factor cellular phones. With hundreds of small, discrete devices that needed to be connected in a very small area, new printed circuit board processes and technologies were developed to enable smaller cellular phones.

Portable cameras have been available since the turn of the century. It was not until the 1980s, however, that these products started to incorporate significant amounts of electronics. By the late 1980s, low-cost point-and-shoot 35-mm cameras were being built by Kodak, Fuji, Canon, and others. These systems made heavy use of flexible printed circuits (FPC) in their design, as well as highly miniaturized photosensors, aspherical optics, integrated circuits, and various electromechanical components.

As the world's most successful consumer electronics producer, Sony was not idle in the portable electronics segment during the 1980s. Besides fielding a few token products in the categories just mentioned, Sony was busily producing Walkman products by the millions. While other manufacturers were trying to find a new killer product, Sony had already found one and was reaping the benefits on a large scale.

Walkman products also drove high-volume manufacturing technology, although the required circuit densities and functional integration was not as high as with other types of products. Walkmans were unique at that time, however, in that they required a highly miniaturized and cost-effective tape transport system to handle cassette audio tapes. Sony had developed specialized robotic systems for assembling layer upon layer of tiny mechanical gears and linkages into an integrated audio tape record/playback system. This product category also drove Sony to pay careful attention to battery technology. Investments in these capabilities would prove to be a great asset for Sony in the 1990s.

In the early 1990s, Sony introduced a line of compact handheld video Camcorder products. These 8-mm camcorders, named after the compact video tape cassette format on which the video is recorded, were significantly smaller and lighter than previous camcorder products. The consumer market for video exploded with the introduction of this category of product and other Japanese companies followed suit with compact 8-mm video camcorder products. Eight-mm camcorders exploited integrated circuit technology, high-volume surface mount manufacturing, miniaturized discrete components, high-density circuit boards, flexible circuits, improved battery technology, and miniaturized electromechanical assemblies. Eight-mm camcorders represented the culmination of technologies developed for other portable product categories during the previous decade.

The pace of change did not slow, however, with the introduction of the 8-mm camcorder. Indeed, the adoption of the portable camcorder as a successful consumer electronic device only increased the interest in new types of portable electronic devices. The mid-1990s saw the development of an ever wider selection of portable electronic devices.

Casio introduced the QV-10 digital still camera. This product was a wake-up call for film manufacturers and film camera makers. The QV-10 was not the first electronic still camera, but it was the first product that captured the interest of large numbers of consumers. The color display enabled the user to share images instantly in a social setting while storing the images in a format that could be readily saved or emailed using a PC, without the cost of film development.

Nokia moved to the forefront of the cellular phone industry by embracing digital cellular technology and producing some very good product designs and user interfaces. Nokia phones of this era set the standard for future cellular phone interfaces in terms of ease of use. Nokia phones also adopted small form-factors with ergonomic design, in contrast to the boxlike appearance of many competitors.

Several manufacturers also began to introduce a class of products known as PDAs. Apple and Sony both introduced high-profile PDA products, but these offerings were not widely accepted.

The Personal Computer Memory Card Industry Association (PCMCIA) standard also flourished during the mid-1990s. PCMCIA cards were developed to enable the user to upgrade notebook PCs with new functions by plugging in a standard format PC card. These cards are typically used for adding new communications capabilities to notebook PCs such as a wireless network connection or an Ethernet port. Some manufacturers have also created electronic imaging modules, personal organizers (like the Rollodex REX), or other novel functions that expand the basic functionality of the PC. PCMCIA products were, and continue to be, produced in high volumes to support the portable PC industry. Since these cards are less than 4 mm thick, thin electronic packaging technologies were developed to make it possible to fit a double-sided PCA inside the PCMCIA housing.

Also during the mid-1990s, high-volume production of CD drives for notebook computers drove lower costs and reduced form-factor for CDs. This resulted in the Walkman market changing from cassette-based products to CD-based products.

The late 1990s saw even more product evolution and diversification. The growth of the Internet changed the dynamics of portable product development considerably.

As notebook computer components became more modular and integrated, much of the production moved from Japan to Taiwan. OEM manufacturers consolidated their supply chains to take advantage of a common pool of component suppliers and system-level ODM manufacturers. Network connectivity became an expected feature for every notebook. Interest in wireless began to develop. Expensive thin and light notebook PCs were introduced for the enterprise market, enabled by lithium-ion batteries and low-profile disk drives. Traditional camera manufacturers such as Canon and Minolta fought back against electronics savvy competitors, such as Casio and Epson, with their own high-quality consumer digital still camera products. Low-cost CMOS sensor technology and higher pixel counts improved digital photography quality immensely.

U.S. Robotics was the first company to develop a successful PDA product concept, with the introduction of the Pilot. This product was based on a functionality that had proven value in the enterprise market, namely, the Microsoft Outlook application.

Digital cellular phones became the standard offering as analog phones became outdated. Digital cellular standards enabled network operators to carry more traffic on their networks. Digital cellular phones also enabled users to get much longer battery life. Wireless access protocol (WAP) was introduced to enable cellular phones to browse text-based content from the Internet. Short messaging service (SMS) was introduced in Europe to enable users to exchange short text messages via cell phone. Pagers, which had enjoyed niche markets for much of the previous decade, were pushed into obsolescence. The RIM Blackberry emerged as a popular product for enterprise customers. The RIM is a small form-factor device that has Microsoft Outlook functionality and leverages the legacy pager network to provide wireless email connectivity.

The most advanced cellular-related portable product introduced during this time was the Nokia Communicator. The Communicator combined a cellular phone with abbreviated personal computer functionality, including email and web browsing. Talk of third-generation (3G) networks had everyone excited about high-bandwidth, wide area networks that would enable ubiquitous streaming video services to handheld devices. How such products would achieve a useful battery life was inexplicably left out of the equation.

As the twenty-first century dawned, industry momentum slowed but it has by no means come to a halt. Investment in high-bandwidth, wide area networks has dried up as telecoms struggle to recoup past investments. Projected revenues based on "push advertising," banner-ads, "attracting millions of eyeballs," and other creative notions have proved elusive.

Over the past three years, portable electronic products have continued to evolve in spite of a general reduction in development resources.

Notebook computers have become the fastest growing segment of the PC market, as notebook prices have dropped dramatically and the performance gap between desktop and notebook systems has narrowed.

802.11 WLAN technologies have enjoyed widespread adoption. 802.11 networks have become commonplace in corporations and are beginning to spread into homes and public places. This makes the demand for notebook computing systems even more compelling, given that there is greater opportunity to get connected when needed.

Cellular phone evolution seems to have slowed somewhat as people have become comfortable with the familiar feature set and user interfaces. Nextel's walkie-talkie functionality is a notable exception to the standard functionality and it has enjoyed success in vertical markets where mission critical communications are required. While many phones have Internet browsing capability, customer interest in this capability is low, since it compares poorly to PC-based web browsing. Voicemail (stored on the service provider's network) on the other hand, is very well received. Products that combine cellular phone and electronic still camera functionality are being introduced, but their level of success is still in question.

Additional products that follow the lead of the Nokia communicator have been introduced but this class of products has yet to prove that it has widespread appeal, or even a strong niche value proposition. Windows CE machines, designed to function as lightweight, ultracompact notebook PCs have been positioned against Palm products and a number of competitors. Market growth for these products is currently muted, however, against growing interest in fully functional, wireless, thin, and light notebook computer products.

Most interesting to the portable electronics designer should be the continued introduction and acceptance of single function products with a clearly articulated value proposition. Portable MP3 and DVD players have had tremendous market growth recently. High-density storage media is becoming a mainstream product as individuals begin to accumulate libraries of digital images, music, video, and documents. Keychain-sized solid-state USB FLASH sticks are as likely to be in the possession of a junior high school student as in the possession of an IT professional.

13.2 Cardinal Functions

Portable electronics are wonderful products. They encapsulate all the power of the computer age into a device that performs a powerful function. Portability in a device means that the device is personal; it is an extension of the individual. The smaller and more discreet the device, the greater the perception that it is the individual, not the device itself, that has become more powerful.

To understand the future of portable electronic devices, it is helpful to have a model of what the ultimate portable electronic device might look like. Science fiction has often foretold reality with respect to technology development. The original Star Trek television series introduced the concept of the *transporter* that can instantly relocate matter to a different point in space. The Tri-corder and the communicator devices, also introduced in this series, were archetypes for the portable computer and the cellular phone. Already, many of the devices portrayed in recent science fiction seem out of date because of the incredible rate at which technology is advancing.

To understand where the portable electronics products are headed, perhaps we should look to sources that transcend scientific and technological themes. In J.R.R. Tolkien's *Lord of the Rings* trilogy, Tolkien has created a fantasy world, *Middle Earth*, containing magical devices that convey great power to the individuals that possess them. The *palantir* is a small globe that enables individuals to see and hear each other across great distances. The wizards of Middle Earth can create force fields, lightning bolts, and other manipulations of natural forces with a portable staff; and of course, there are the Rings of Power, which give their bearers ultimate authority and dominion over others. Extending our vision beyond science fiction, into the world of fantasy, we might discover the form of the ultimate portable electronic devices. On this basis, I would propose the following functionalities as the *cardinal functions* for which it is the ultimate goal of portable electronic designers to produce.

First there is *telepathy*. Telepathy is the communication between minds by some means other than sensory perception. Cellular telephony begins to approach this mark in that it enables two people to instantly communicate from two different locations virtually anywhere on the globe. Cell phones fall short of true telepathy because they rely on the sense of hearing to convey the information and because they convey what a person chooses to say, not what they are thinking. A cellular phone that enabled true telepathy would represent a communications usage paradigm on which it would be rather difficult to improve.

Next there is *teleportation*. We will define teleportation as the purposeful and instant relocation of a body of matter to any desired point in space. The popularity of corporate jets is evidence of the value that the rich and powerful place on attempting to achieve teleportation, although these are hardly portable electronic devices. Video-conferencing products are also chasing this cardinal function by using available technology to create a working environment that achieves the advantages of co-location among geographically distributed teams.

Transmutation is the next cardinal function on the list. Transmutation is defined as the changing of matter from one form to another. Modern manufacturing industries are intensely concerned with achieving this goal. The transformation of sand into silicon into a portable electronic device is one of the countless transformations that happen every day. Speeding up this process and reducing the cost has been the goal of manufacturing companies for many decades. As cycle times are reduced, the elapsed time between someone's desire for a particular object and the fulfillment of that object is shrinking drastically. It is not difficult to envision a future where an individual visualizes (designs) a complex product and that product is produced almost immediately. All desires for physical objects could be immediately fulfilled (economics ignored). What types of portable electronic devices will be involved in that process?

Recreation should not be overlooked as one of our cardinal functions. People need to entertain themselves. Portable electronic devices that focus solely on this function, such as portable video games and MP3 and DVD players, are already hugely successful. Advancing technology will only make the recreational experiences richer and more compelling.

Finally, there is *omniscience*—the state of having complete or infinite knowledge of all things. Certainly, this is beyond human capacity, yet there is a basic human desire to have access to more information than anyone else. Information is power. The ability to have all of the information in the Library of Congress contained in a credit-card-sized device will certainly happen.

So, the proposed cardinal functions are telepathy, teleportation, transmutation, recreation, and omniscience. Arguably, any portable electronic device has been designed in an attempt to achieve one or more of these functions to some degree. Any functionality that can be described can be aligned to one of these vectors.

In considering the critical factors of a portable electronic device, we have just discussed the ultimate possibilities in product functionality. The notion of performance for these cardinal functions is straightforward. The performance is simply the extent to which the cardinal function is performed in its purest state. That is to say, the functionality is carried out instantly and the fidelity of the fulfillment is perfect to the extent of human perception. Where this is not the case, there will be room for improvement in the performance.

The ultimate user interface would be a mental interface whereby the user controls the device by thought alone. Short of that, devices that respond to verbal commands, handwriting, and simple gestures would give the user a compelling sense of power and control.

The ultimate form-factor is a device so small that it can be embedded in a piece of jewelry or in the user's body. The sense of power is greater when the functions seem attributable to the user rather than the device.

The issues of battery life and reliability need to be eliminated. There is no room for battery replacement or a nonfunctioning device in this future vision of powerful, magical devices. The user should not be concerned with energy management or environmental issues that might impact reliability. Future portable electronic products should recharge transparently from sources in the user's environment. Ultimate reliability means the device should function in any environment in which the user can survive.

Cost and time-to-market issues do not seem as relevant in a discussion of future product directions. As devices become more powerful, it is difficult to say how these products will be valued by society and what will be their pace of technical advancement. Will there be a powerful overclass of "Wizards" who can afford to obtain these powerful devices? Or will the technology be broadbased and accessible by all, helping to create a society with greater equality of opportunity?

13.3   Powerful Thin Clients

The key to making portable electronics more powerful is to view them as remote controls for vastly more powerful devices and resources. This is known as the thin client model. This is the path of least resistance for advancing portable electronic products, both economically and technologically. With this thin client philosophy in mind, there are a number of product concepts that can be used to demonstrate the likely direction of future portable electronic devices.

With the ever-increasing gate densities of advancing silicon technology it is tempting to force more and more functionality into the portable electronic product. This temptation should be resisted, however, in favor of keeping the functionality of the portable electronic device focused on the pure product concept. Exploiting the network infrastructure is the most highly leveraged way to expand the power of a device. Partitioning internal and external device functionality is becoming the essence of portable electronic design.

Processing, mass storage, and information retrieval can be easily offloaded from a portable electronic device to a more powerful stationary device somewhere on the network. The limitations are network bandwidth, latency, availability, and coverage.

Let us examine this concept by considering the hypothetical product concept of the "ear phone." The ear phone, shown in Fig. 13.1, is a device that is worn in the ear. It consists of a speaker, a microphone, a radio transceiver and a microcontroller with a small amount of memory. The Jabra earpiece is pictured as well as partial evidence of feasibility, although the Jabra does not include a radio transceiver, memory, or microcontroller. Jabra does include microphone technology that allows

- Hearing aid form factor
- Integrated microphone & speaker
- Low power/short range RF (like Bluetooth)
- Embedded IP address
- Voice processing: external to unit, controlled by software agents

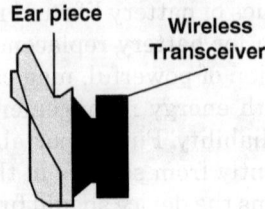

Ear piece

Wireless Transceiver

Jabra™ wired earphone

Figure 13.1 Ear phone.

the voice of the user to be picked up through the ear rather than placing the microphone in front of the mouth.

We will assume that the ear phone has a wireless connection to the network capable of supporting high-quality, two-way voice communication. In this respect, the ear phone is nothing more than a cellular phone. In fact, it is less. The ear phone makes no provisions for dialing, storing phone numbers, providing a choice of musical ringing tones, playing underpowered video games, or any of the other features that modern cellular phones offer.

The ear phone maintains a constant connection to the network; therefore, presenting a dial tone to the user is unnecessary. The connection that is maintained by the ear phone is a voice channel directly to a voice-processing server somewhere on the network. The "receive" audio channel is quiet when not in use so that the user is not distracted. The voice-processing server waits for a keyword or key phrase from the user as a signal to begin receiving instructions.

When the user says the keyword, such as "ear phone," the voice server interprets the subsequent dialog as commands. At this point the user can issue various commands based on the services that are available. The most obvious service would be to place a phone call to another individual. The user might say "Dial John Smith," and the server would take the necessary actions to complete a voice call to the John Smith referenced in the user's network-based phone list. The user can now conduct a hands-free voice call with John Smith.

John Smith, on the other end of the call with an ear phone of his own, hears a soft beep indicating that he has an incoming call. Since John is walking through an international airport, heavily encumbered with baggage, he elects to state the command "Ear phone identify" rather than to ignore the call or accept it from an unknown source. Realizing that the phone call is important, John speaks the "Accept" command and is able to conduct a hands-free discussion as he makes a dash to the next terminal.

In this example, the ear phone has accomplished nothing more than a cellular phone call, although the ease of use is significantly improved compared to a traditional cellular handset. This voice channel to the network is capable of performing other powerful functions, however. Suppose John Smith learns during his phone call that he needs to fly to a different city than he had intended. John issues the command "Ear phone dial travel agent." The voice server connects John to a voice-based travel server and he conducts a dialog with the system to change travel arrangements. This raises the notion of people spending their time speaking to computer-controlled systems rather than to other people. While this is unpleasant to contemplate, based on experiences with current technology, the fact is that natural language processing will improve dramatically in the coming years. Even if this technology requires powerful servers to become easy to use, it will still have a profound impact on portable electronic devices.

Let us say that the user of an ear phone wants to send an email but is sitting in traffic on the way home from work. The user can say "Ear phone write email." The voice server routes the voice channel to a service that converts voice to text. After dictating a message, the server reads the message back with improved grammar and composition and asks for approval. The user approves and directs the server to whom to send the message. In a similar way, a user might access specialized services that let him conduct banking transactions, order groceries, call a cab, make restaurant reservations, tell his personal video recorder to record a program, or activate countless other voice-controlled events.

In another situation, a user may walk up to a wall-mounted flat panel display in a hotel conference room where she is about to give a presentation. The user issues the command "Ear phone, access desktop on local display." (Keep in mind that any number of logical, natural language commands of this type would achieve the desired result.) The voice server is able to positively identify the user because her ear phone has an embedded authentication code (in memory) and IP address. The network is aware of the user's proximity so that the appropriate local display is deduced. The server creates a secure link between the display and "hosted desktop" where the user's business files and applications reside. The user is now able to share any documents accessible from her desktop with the occupants of the conference room.

To interact with the desktop, the user might use a wireless keyboard residing in the hotel conference room. Another possibility is that she might bring her own user-interface devices with her. The ear phone itself could be used to control the desktop via the voice server on the network.

The user could issue commands, such as "Desktop, open sales presentation" or "Desktop, next slide," to control what is being displayed. Let us now examine some other hypothetical devices that leverage this same thin-client concept.

In order to control the conference room display the user might rely on a product such as the "magic wand," shown in Fig. 13.2. The magic wand consists of an array of miniaturized motion and orientation sensors, a radio transceiver, a microcontroller, and a small amount of memory. Like the ear phone, it establishes a network connection whenever it is turned on. It automatically connects to a server that authenticates the device and its owner, and associates the device with a user profile. The internal sensors determine which way the device is pointed. With a built-in laser pointer, the user might *calibrate* the device for the local display by tracing the four edges of the display. The magic wand can then be used as a type of mouse to control the desktop cursor.

A similar idea might be the use of one or more "power rings," worn on the fingers, which could be used to monitor the user's gestures. Different finger motions would represent the user input. Snapping a finger might be the same as a traditional mouse click. Pointing could be used to move the cursor.

Nippon Telephone and Telegraph (NTT) pioneered the concept of using rings as a keyboard. The user, wearing rings on each finger, can "type" on any surface, without an actual keyboard. The rings transmit

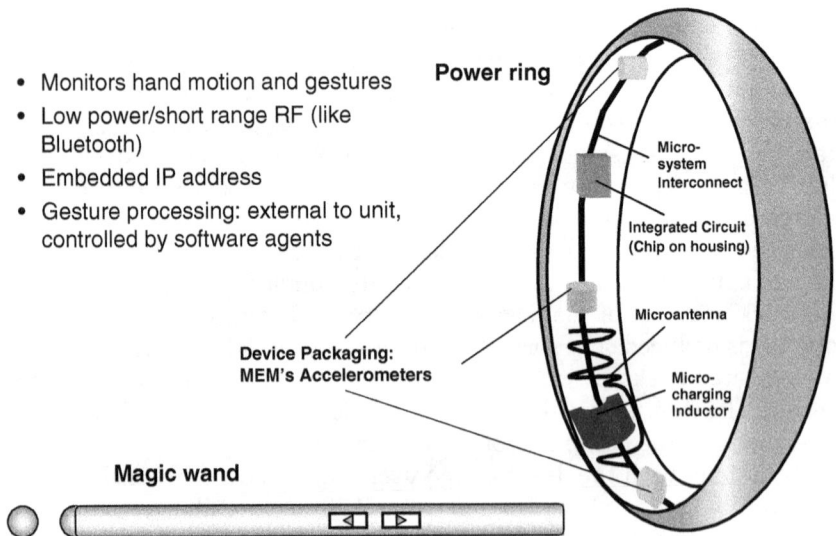

- Monitors hand motion and gestures
- Low power/short range RF (like Bluetooth)
- Embedded IP address
- Gesture processing: external to unit, controlled by software agents

Power ring

Micro-system Interconnect

Integrated Circuit (Chip on housing)

Microantenna

Device Packaging: MEM's Accelerometers

Micro-charging Inductor

Magic wand

Figure 13.2 Power ring and magic wand.

data about the finger motion to the network server. The network server uses the shock data from the motion sensors to determine which alphanumeric keys were intended by the user. *Voila!* Text input without a keyboard.

Now imagine the following scenario. One of the people in the hotel conference room wants to take notes on the presentation. Using the ear phone he instructs his hosted desktop to create a file named to reflect the topic, location, and date. He then commands that the text data streaming in from his power rings be written into that file. He can now discreetly take notes of the meeting by typing on the conference room table top, without the need for a notebook computer.

Another attendee at the meeting may take a slightly different approach. Using a device such as the display pad, shown in Fig. 13.3, the meeting attendee can take notes in a more traditional fashion. The display pad is essentially a thin-client tablet PC. Like the large flat panel display being used for the presentation, the display pad is being driven over the network by a hosted desktop. User identification and authentication are embedded in the device and can be strengthened by detecting the presence of a power ring or other authenticating device. The supporting applications are run on the network, which minimizes the amount of processing that must be done onboard the display pad. The user interface for the display pad is a pad of paper and a pen that is equipped with MEMS-based inertial sensors. These sensors convert the movement of the pen into two-dimensional vectors that are passed to the display pad via a local wireless RF connection. The display pad

- Color reflective polymer display
- Embedded IP address
- Local Wireless link to network (like Bluetooth)
- Minimal on board electronics, software, data, processing, reside on network controlled by agents
- Cursor control and pen input using inertial "smart pen"
- Smart Pen writes on paper but digitizes motion and transmits wirelessly

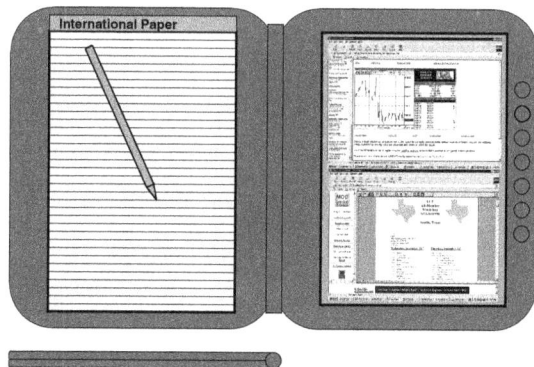

Figure 13.3   Display pad.

stores these vectors as a two-dimensional image. When the user is finished writing on a page, he taps a button on the display pad to store the page. Pages are then passed to the hosted desktop for long-term storage and additional processing such as the conversion of the notes to text.

The benefit of this interface over the tablet PC concept is that it relies on a time-tested user interface—the pen and paper. Tablet PCs that require the user to write on the display have frictional and display characteristics that make the writing and sketching on them less comfortable than writing on paper. In addition, anything that the user writes down with the pen is stored over the wireless link into the display pad. Names, phone numbers, and sketches written on scrap paper are all being recorded by the display pad for later retrieval.

Another advanced product concept that we will introduce is called the Camulet and is shown in Fig. 13.4. The Camulet has the ability to capture and store still-and-motion video. It also has a short-range radio link that enables it to store captured image data on the network. It can also display stored or streaming data containing video and still images. The Camulet is worn as a necklace and is designed to look like a piece of fashionable jewelry. While being worn in use it can display still pictures (like a locket) or video as a piece of "dynamic jewelry." The user can also point the device and push a button which causes the Camulet to capture a still image. Holding the button down causes the Camulet to capture a video clip. These captured data are buffered locally while being downloaded automatically to virtual storage on the network. Since Camulet has limited surface area for controls, only one button is available and it is used to capture video. All other control over the device is exercised through the network using an ear phone interface. For example, "Ear phone, play video undersea on Camulet" would cause the hosted desktop to stream a video file of undersea footage to the Camulet.

- Electronic still and motion w/ playback functionality
- Can function as display with communications link
- Low power/short range RF (like Bluetooth)
- Limited on board image storage, uploads and downloads to network
- Embedded IP address
- Polymer reflective display

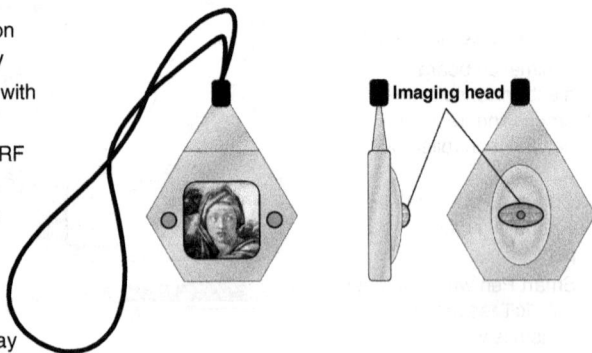

Figure 13.4 Camulet.

This thin-client model enables a different set of technology trade-offs than are currently available with the product processor centric architectural model. Achieving a thin-client world requires high bandwidth, low power, and widely deployed radio linkages to the network. Current WLAN technologies such as 802.11a/g are getting close to providing the necessary bandwidth and coverage but require too much power for highly miniaturized products such as ear phone and power ring. Power issues aside, a typical 802.11 network could support many of these thin-client concepts.

The concept of a "server pack" (Fig. 13.5) is a potential solution to bridging this problem in the near term. Server pack is a cigarette pack form-factor device containing a short-range, low-power, high-bandwidth digital radio transceiver for communicating with personal thin-client devices such as the ear phone and the Camulet. It also contains a high-bandwidth, longer-range transceiver for communications with WLAN and WWAN networks. The form-factor enables a larger battery than the ultraminiaturized thin-client devices, so the power consumption for longer-range radio connections can be handled by server pack and not by the thin clients. Server pack can be worn on the belt or carried in a purse, pocket, or briefcase.

Besides these communication devices, the server pack has basic personal computer functionality minus any kind of user-interface hardware. The server pack is used as an intermediary device for connecting ultraminiaturized products to the network. With its own onboard storage and processing power, server pack can provide local support to the network-hosted desktop to enhance performance and improve the user experience. When a WAN of WWAN is not available for connection,

- High-bandwidth wireless communications (voice and data), Macrocellular
- Full PC functionality, embedded IP address, works in tandem with agents internal and external
- 10 GB storage
- Has local RF mode (like Bluetooth) for connection to peripherals (display pad, power ring, ear phone, etc.)
- Built-in charging circuitry

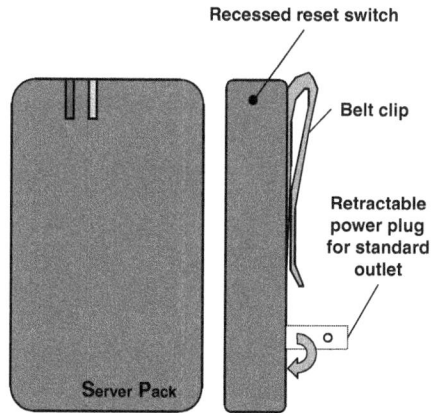

Figure 13.5   Server pack.

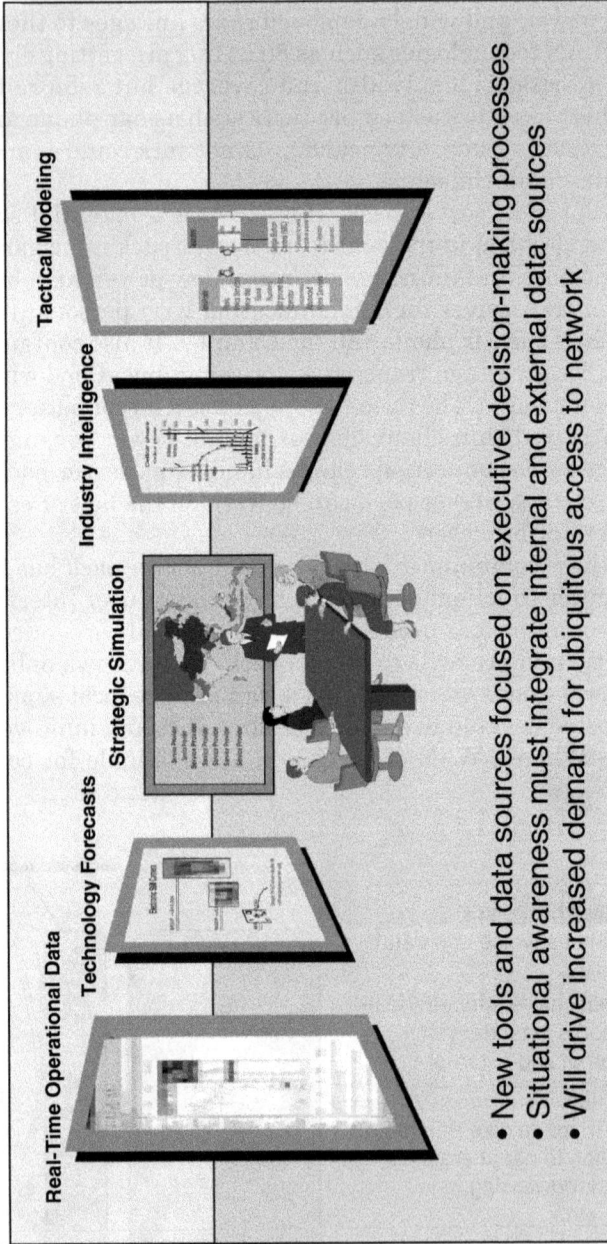

Situation awareness and decision support

Real-Time Operational Data
Technology Forecasts
Strategic Simulation
Industry Intelligence
Tactical Modeling

- New tools and data sources focused on executive decision-making processes
- Situational awareness must integrate internal and external data sources
- Will drive increased demand for ubiquitous access to network

Figure 13.6   Enterprise environment for thin-client devices.

server pack can provide some level of desktop support until the user moves back into network coverage.

We could expand the vision of our earlier scenario, involving people attending a meeting at a hotel conference room. With the widespread adoption of thin-client devices and broadly deployed flat panel displays in corporate conference rooms, it is feasible to imagine a new type of work environment for the enterprise—one in which individuals are able to be fully mobile while being able to collaborate at any location. This model lends support to the notion of the executive control room, which is focused on providing situation awareness and decision support for large enterprises. This concept is exemplified in Fig. 13.6.

Having espoused the virtues of the thin-client philosophy, there is no doubt that fat clients offer a powerful value proposition for many product categories. Wireless network limitations in service coverage, bandwidth, and latency continue to lag the performance that is possible with onboard mass storage and processing.

As wireless network capabilities continue to improve, certain classes of applications will move from the fat-client domain to the thin-client domain. Today, the Blackberry (wireless email), the cellular phone, and voice mail represent products in the thin-client domain. DVD players and electronic still cameras are squarely in the fat-client domain. As low-cost, high-bandwidth data services come into being, streaming high-quality video from a network server may enable a more compelling product than one that is limited to local storage of content (Fig. 13.7).

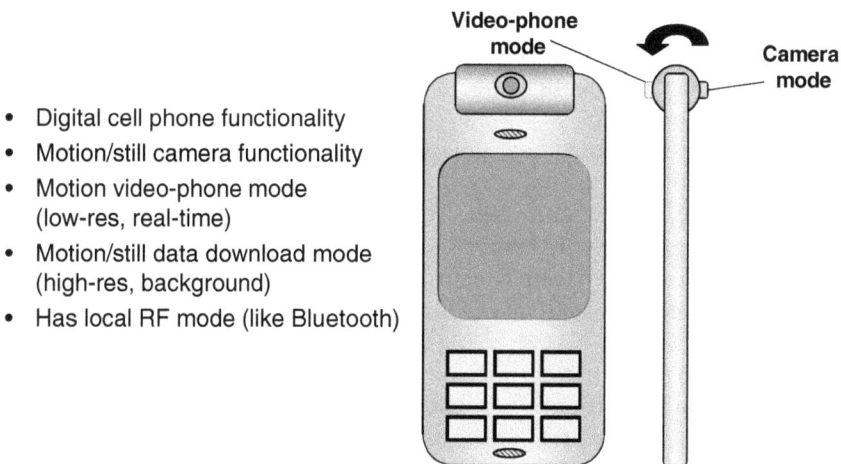

- Digital cell phone functionality
- Motion/still camera functionality
- Motion video-phone mode (low-res, real-time)
- Motion/still data download mode (high-res, background)
- Has local RF mode (like Bluetooth)

Figure 13.7 Cam Com.

Some general comments about technology developments and application requirements that impact thin-client vs. fat-client thinking are as follows:

Cheaper, more compact mass storage favors fat client architecture.

Improved network performance favors thin-client architecture.

Improved information compression technology favors thin clients.

Diminishing performance returns, with respect to memory latency and processing, favors thin-client architecture for a given application space.

Time-sensitive access to remote sources of data is favored by the thin-client approach.

Of course, thin client and fat client are time-sensitive terms and subject to individual interpretation. Today's portable thin-client devices have more processing power than fat-client devices of just a few years earlier. In general, both thin- and fat-client architectures will continue to coexist within the market, indeed within the same product, for the foreseeable future.

13.4   Concluding Comments

The goal of this book has been to encourage and enable technologists from diverse technical backgrounds to become designers of portable electronic devices. Given the broad range of technologies needed to field a portable electronic product, no single technical discipline is sufficient to create a new product. Resources must be drawn on from many technical specializations in order to create a successful product. The portable electronic product designer's mission is to combine these technologies into a product that meets a specific need in a manner that is delightful to the end user.

More often than not, new portable devices are being developed to deliver a technology-based service, such as cellular telephony or downloaded music, to the end user. In these instances, it is the service provider's branding with which the user identifies the product origin. Thus, the likelihood that the product designer will be working on behalf of a service provider rather than a hardware manufacturer is increasing.

As such, the product essence of portable product design centers on the functional specification of the product. It is the identification of the usage paradigm and value proposition to the end user that is essential to a successful product. The subsequent technical design then comprises establishing an appropriate product architecture that can then be parsed out to the appropriate subsystem and component designers.

It is this system level approach to portable electronics design that makes the creation of complex products possible. The portable electronic designer does not need to be a master of all disciplines, but he or she must become an expert at leveraging existing infrastructure. System-on-chip solutions are lowering the cost of product development substantially and making the hardware design for many classes of products relatively easy. Many portable electronic products in the market today are built by contract manufacturing houses. Many of these contract manufacturers have the ability to perform complete electronic and physical design, if the product designer can provide the functional requirements.

A few fundamental principles will govern the success of portable electronic products moving forward. Establishing a focused and valuable functionality has been emphasized in this text as the most important principle. One source of inspiration for product functionality is the notion of enabling the "usual things in unusual places." This concept speaks to the true value of portability, as we seek to enable end users to access network-based services from any location.

Some in the industry have adopted the philosophy that a portable electronic product should either *save time* or it should *kill time*. The natural tension between productivity and entertainment is perhaps one of the greatest opportunities for future product designers. While current usage paradigms treat work and play as separate and, perhaps, competing requirements, future products may transform work into play. Intelligent automation of redundant tasks can reduce the monotony of many types of work, leaving more time for the creative portion of a particular job function. Portable electronic devices that can assist the user with problem visualization might make even the creative portion of certain types of jobs seem more like a video game than work.

Another important tenet is that *transparency* of the underlying technology to the end user is paramount. The extent with which the product performs its functionality with simplicity contributes to the feeling of satisfaction on the part of the user. The early days of the silicon age started with simple products that left the user with an awestruck feeling of complexity and sophistication. People's amazement with technology is wearing thin, however. Now people want products that perform complex and powerful tasks with a feeling of simplicity.

The recent technology recession may have left many people jaded about the prospects of future opportunities in the area of portable electronics.

Aspiring product designers should not be discouraged but should have hope! Recent developments in low-cost broadband wireless technology and standards are only just now beginning to unleash the full potential of portable electronic design. I sincerely hope that the reader of this book will have found some insights that will shape the next generation of portable electronic products.

A List of Acronyms and Abbreviations

3G (Third Generation) Third generation cellular incorporates digital data channels into a cellular connection that are capable of data rates above 128 kbps

A/D (Analog to Digital) The conversion of a signal from the analog domain to the digital domain

ABS (Acrylonitrile Butadiene Styrene) A common form of plastic used for product housings in consumer electronics

AC (Alternating Current)

ACF (Anisotropic Conductive Adhesive) An adhesive film filled with conductive particles that provide for electrical conductivity in the z axis only

ACPI (Advanced Configuration and Power Interface) A standardized interface that enables power management in notebook computer systems

ALU (Arithmetic Logic Unit) A basic element of a microprocessor system

AMLCD (Active Matrix Liquid Crystal Display) An LCD in which each pixel is controlled by a transistor fabricated next to the pixel on the glass substrate, enabling faster refresh rates than passive matrix LCD devices

AMPS (Advanced Mobile Phone Service) The predominant analog cellular standard that was widely deployed in North America prior to more recent digital standards

APC (Average Pin Count) The average number of input/output pins per device for a set of electronic devices

API (Application Programming Interface) A layer of software that abstracts the details of an operating system and enables a programmer to more easily develop applications for a particular hardware platform in terms of end-user functionality

ASIC (Application Specific Integrated Circuit) An integrated circuit designed for a specific application

BD (Boothroyd Dewherst) A company that produces manufacturability analysis software

BGA (Ball Grid Array) A type of integrated circuit package in which the input/output pins are implemented as an array of solder balls distributed in a grid formation on the underside of the package

BIOS   (Basic Input-Output System) The lowest level of system software that enables critical system devices to communicate with each other

BOM   (Bill of Materials) A complete list of parts required to make a particular product

CAD   (Computer-Aided Design) The process of using computers to design a product or component

CCD   (Charge-Coupled Device) A type of integrated circuit used in electronic cameras to detect and digitize an image

CCFL   (Cold Cathode Fluorescent Light) A miniature fluorescent light bulb used as a backlight in portable electronic devices

CD   (Compact Disc) A common format for optical data storage

CDMA   (Code Division Multiple Access) A digital cellular phone standard that enables multiple phone calls to share the same frequency band by assigning each call a unique code

CDPD   (Cellular Digital Packet Data) A cellular data standard that enables data connections to be established over the cellular network on unused analog or digital voice channels

CE   (Consumer Electronics) Electronic products that are targeted at the consumer market

CFD   (Computation Fluid Dynamics) An engineering method that enables digital simulation of the thermal characteristics of fluids and gases (within a portable electronic device)

CISC   (Complex Instruction Set Computing) A system of instruction coding that enables a CPU to execute more complex operations

CMOS   (Complementary Metal-Oxide Semiconductor) The most commonly used and least expensive material system used in the production of integrated circuits

COB   (Chip-on-Board) The direct attachment of an integrated circuit to a printed circuit board with electrical connections made using wire bonds

CPU   (Central Processing Unit) The basic "engine" of any digital computing system

CRT   (Cathode Ray Tube) Prior to flat panel displays, the predominant technology for making television and computer monitors

CSP   (Chip Scale Package) A type of integrated circuit package that is only slightly larger than the integrated circuit itself

CYM (subpixels)   (Cyan, Yellow, Magenta) Primary colors for a display pixel

D/A   (Digital to Analog) The conversion of a signal from the digital domain to the analog domain

DDR   (Double Data Rate) A memory bus standard that enables a higher rate of data transfer between the CPU and memory

DFA (Design for Assembly) The process of designing the mechanical attributes of a part to reduce the time required to assemble the part into a product

DFM (Design for Manufacturing) The process of designing a product with consideration for minimizing the manufacturing cost

DFMA (Design for Manufacturing Analysis) The assessment of an existing product in terms of how well it is designed with respect to minimizing the manufacturing cost

DIMM (Dual Inline Memory Module) A type of memory package

DRAM (Direct Random Access Memory)

DSP (Digital Signal Processor) A specialized processor optimized for executing a specific class of algorithms

DVD (Digital Video Disk) A common standardized optical storage media developed primarily for video storage

EDA (Electronic Design Automation) Computer-based tools and processes that speed up the product design process

EEPROM (Electrical Erasable Programmable Read Only Memory)

EL (Electroluminescent) A material that gives off light when under the influence of an electric current

EMI (Electromagnetic Interference) Interference in product functionality or performance resulting from the presence of incidental sources of radiation

EPROM (Erasable Programmable Read Only Memory) A type of memory device used in various portable electronic applications

FC (Flip-Chip) A method of attaching an integrated circuit directly to a circuit board by soldering the integrated circuit inputs and outputs directly to connections on the circuit board

FEA (Finite Element Analysis) An engineering method for digitally analyzing the structural characteristics of a physical part

FPC (Flexible Printed Circuit) A printed circuit substrate that is fabricated on a thin flexible material that can be bent to fit into a confined space

FPD (Flat Panel Display) A flat display device

GPRS (General Packet Radio Service) A cellular standard that enables a high-speed internet connection using a packet-data-based system

GPS (Global Positioning System) A digital device that uses radio communications with satellites to determine a precise geographic location

GSM (Global System for Mobile Communications) A globally adopted second generation digital cellular standard

HAL (Hardware Abstraction Layer) A software layer that simplifies the programmatic control of various system hardware elements

HDTV   (High-Definition Television) A standard for very high-resolution television broadcast and display

I/O   (Input/Output) The nodes that represent an individual electrical connection (or a set of electrical connections) through which information encoded signals are transferred

IC   (Integrated Circuit) A device in which multiple electronic logic elements are fabricated as a monolithic structure

IEEE   (Institute of Electrical and Electronics Engineers) The industry organization responsible for many of the standards that govern the portable electronics industry

IF   (Intermediate Frequency) A processing stage in a wireless communications system in which the signal is converted to a particular frequency as a means of preparing that signal for further processing

IO   (Input/Output) The nodes that represent an individual electrical connection (or a set of electrical connections) through which information encoded signals are transferred

IP   (Internet Protocol) A standard for routing information to the correct location on the Internet

IR   (Infrared) A frequency of radiation used for short range, line of sight, wireless communications.

IRDA   (Infrared Data Association) An industry organization which created the standard that enables portable devices to communicate using an infrared tranceiver

ITO   (Indium Tin Oxide) A transparent conductor commonly used to connect electrical signals to pixels in a flat panel display

JEDEC   (Joint Electron Device Engineering Council) An industry organization that produces standards related to integrated circuit packaging

LAN   (Local Area Network) A communications network that connects devices within a single building or campus

LCD   (Liquid Crystal Display) The predominant form of flat panel display technology being used today

LED   (Light-Emitting Diode) A device used for fabricating bright, single color, character displays in portable electronic product

LFB   (Local Frame Buffer) Solid-state memory used to store video data in support of a graphics processor

LIF   (Low Insertion Force) A type of connector used in conjunction with flex circuits

MCC   (Microelectronics and Computer Technology Corporation) A technology consortium in operation between 1983 and 2001 (based in Austin, Texas), which conducted research and consulting in the field of portable electronics (among other things)

MEMS (Microelectrical Mechanical Systems) Miniaturized mechanical devices fabricated using photolithographic processes similar to those used to fabricate integrated circuits

MIPS (Millions of Instructions per Second) A measure of processor performance

MOE (Metal on Elastomer) A type of z axis, board to board, electrical connector

MPEG (Moving Pictures Users Group) An industry organization that created standards for the compression of audio-visual data

MP3 (MPEG Audio Layer III) A widely adopted audio data compression standard

Ni-Cd (Nickel-Cadmium) A common chemistry system for batteries used in portable electronic products

Ni-MH (Nickel-Metal Hydride) A common chemistry system for batteries used in portable electronic products

NRE (Nonrecurring Engineering) One-time engineering and tooling costs required at the start of a project

NTSC (National Television System Committee) The organization for which the legacy North American analog television broadcast standard is named

ODM (Original Design Manufacturer) A company that actually designs and manufactures a product, somewhat synonymous with "contract manufacturer"

OEM (Original Equipment Manufacturer) The company who puts the brand on the product, regardless of the source of manufacturing

OS (Operating System) The layer of software that allows the user to access system devices and control various program and data resources

OSI (Open System Interconnection) An architectural standard that enables computing device to communicate over a network

PCA (Printed Circuit Assembly) A printed circuit board with a complete set of electronic components attached

PCB (Printed Circuit Board) The bare wiring board (substrate) used to connect electronic components

PCI (Peripheral Component Interconnect) A standardized architecture that provides a high-speed data path between peripherals and the CPU

PCMCIA (Personal Computer Memory Card Industry Association) The industry organization for which the PC card (cards that are plugged into a notebook computer by the end user) standard is named

PDA (Personal Digital Assistant) A class of handheld products that enable the user to manage personal information and scheduling

PHS (Personal Handy-phone System) A digital cellular phone standard deployed only in Japan

PVR   (Personal Video Recorder) A device that enables the user to conveniently record and manage multiple hours of television programming

QFP   (Quad Flat Pack) A common type of integrated circuit package

QOS   (Quality of Service) A metric used to convey the extent to which a stream of audio-visual data can be delivered to the end user without interruption or degradation in quality and continuity

RAM   (Random Access Memory) System memory that is used in direct support of CPU operations

RF   (Radio Frequency) The range in the electromagnetic spectrum loosely defined from 30 MHz to 3 GHz, where most wireless communications take place

RGB   (Red Green Blue) Primary colors for display subpixels

RISC   (Reduced Instruction Set Computing) A simplified CPU instruction set that enables higher CPU performance

RTOS   (Real Time Operating System) A type of embedded operating system associated with portable electronic devices

RW   (Read Write) Signifies the capacity of a memory media to be used for multiple recordings and playback

SLC   (Surface Laminar Circuitry (IBM Trade Name)) A type of high density substrate technology

SMT   (Surface Mount Technology) The predominant electronic assembly technology in which components are soldered to the surface of the printed circuit board

SOC   (System-on-Chip) A type of integrated circuit that integrates many elements of the system such as CPU, memory, and peripheral control into a single device

SOP   (Small Outline Package) A type if integrated circuit package

SOT   (Small Outline Transistor) A type of packaging format for a discrete transistor device

STN   (Super-Twisted Nematic) A variety of LCD technology that enables faster refresh rates and sharper contrast

TAB   (Tape-Automated Bonding) An integrated circuit packaging and assembly technique used primarily in flat panel display modules

TDP   (Thermal Design Power) A CPU characteristic metric that communicates the type of thermal management solution required for product implementation

TFT   (Thin Film Transistor) A pixel control transistor that is fabricated directly onto an LCD panel

TIM   (Thermal Interface Film) A material placed between a CPU and a heat sink to improve the rate of heat transfer

TN   (Twisted Nematic) A variety of LCD technology

TQFP   (Thin Quad Flat Pack) A type of integrated circuit package that is typically less than 2 mm thick and has peripheral leads around four sides of the package

TSOP   (Thin Small Outline Package) A type of integrated circuit package that is typically less than 2 mm thick and has peripheral leads on two opposing sides of the package

TTM   (Time-to-Market) The period of time required to develop a product and deliver it to the marketplace

UMA   (Unified Memory Architecture) A personal computer architecture in which the graphics processor and the CPU share the same support memory (RAM)

USB   (Universal Serial Bus) A widely adopted standard for connecting peripheral devices to notebook computers and other portable electronic devices

VGA   (Video Graphics Array) A display standard referring to a video adapter capable of a resolution of up to 640 by 480 pixels

WAN   (Wide Area Network) A network that connects to systems over a dispersed geographic area such as neighborhood or city

WAP   (Wireless Access Protocol) A standard for formatting "Web pages" that are accessible on small display devices such as cell phones

WLAN   (Wireless Local Area Network) A LAN that uses wirelss technology to connect to the client devices

WWAN   (Wireless Wide Area Network) A WAN that uses wireless technology to connect to the client devices

ZIF   (Zero Insertion Force) A type of connector used in conjunction with flex circuits

References

Henry Petroski, *The Evolution of Useful Things*, Vintage Books (Random House), New York, 1994.

Alan Cooper, *The Inmates are Running the Asylum*, Sams Publishing (Macmillan Computer Publishing), Indianapolis, 1999.

Mark Grossman, *The SG Cowen Semiconductor Primer*, 4th ed, 2003.

S. Sanderson and M. Uzumeri, *Managing Product Families*, Erwin Publisher (Times Mirror), Chicago, 1997.

Portelligent, Inc., Report 117-03-030114-1b_Audiovox_CDM-8300, 2003.

Portelligent, Inc., Report 118-16-021010-1c_Nokia7210, 2002.

Portelligent, Inc., Report 120-000807-1b_IBM-T20_Notebook, 2000.

Portelligent, Inc., Report 120-010125-1j_Dell_C800_Laptop, 2001.

Portelligent, Inc., Report 131-011111-1f_Pentax_Tottemo_EG, 2001.

Portelligent, Inc., Report 132-021011-1e-Sony_DSC_U10_CyberShot, 2002.

Portelligent, Inc., Report 141-020424-1d_Sony_Clie_PEG-NR70V, 2002.

Portelligent, Inc., Report 142-020822-1f_HP_Jornada_928_PDA, 2002.

Portelligent, Inc., Report 155-020528-1d_Motorola_Bluetooth_Headset, 2002.

Portelligent, Inc., Report 240-020730-1d_PDA_Metrics_2Q02, 2002.

Portelligent, Inc., "Flexible Circuit Connectors," Report LCP-090-95(P), 1995.

Portelligent, Inc., "Cardio-486, Model SCE8643513," Report LCP-091-95(P), 1995.

Portelligent, Inc., "Substrate Technology Routing Strategies for CSP," Report LCP-095-95(P), 1995.

Portelligent, Inc., "Wristwatch Study," Report LCP-096-95(P), 1995.

Portelligent, Inc., "Digital Imaging Products," Report LCP-089-96(P), 1996.

Portelligent, Inc., "Summary of Eight Personal Handy Phone Systems (PHS)," Report LCP-090-96(P), 1996.

Portelligent, Inc., "Flat Panel Display Roadmap: Overview," Report LCP-091-96(P), 1996.

Portelligent, Inc., "Personal Digital Assistants (PDAs)," Report LCP-092-96(P), 1996.

Portelligent, Inc., "Analyis of the Paldio 101P Panasonic PHS Phone," Report LCP-008-97(P), 1997.

Portelligent, Inc., "Analysis of the DDI Sanyo P101 PHS Phone," Report LCP-009-97(P), 1997.

Portelligent, Inc., "Analysis of the Aiwa DDi #PT-H50 PHS Phone," Report LCP-010-97(P). 1997.

Portelligent, Inc., "Analysis of the Paldio 101S Sharp PHS Phone," Report LCP-011-97(P), 1997.

Portelligent, Inc., "Analysis of the Kyocera DDI #PS-501 PHS Phone," Report LCP-012-97(P), 1997.

Portelligent, Inc., "Analysis of the Paldio 102H NEC PHS Phone," Report LCP-013-97(P), 1997.

Portelligent, Inc., "Analysis of the Toshiba (Carrots) DDI #DL-S22PL," Report LCP-014-97(P), 1997.

Portelligent, Inc., "Analysis of the Paldio 102S Sharp PHS Phone," Report LCP-015-97(P), 1997.

Portelligent, Inc., "Analysis of the Sharp Zaurus MI-10 PDA," Report LCP-017-97(P), 1997.

Portelligent, Inc., "Analysis of the US Robotics Pilot PDA," Report LCP-018-97(P), 1997.

Portelligent, Inc., "Analysis of the Sony CM-RX100 Cellular Phone," Report LCP-019-97(P), 1997.

Portelligent, Inc., "Analysis of the Nokia 9000 Communicator,"
    Report LCP-020-97(P), 1997.
Portelligent, Inc., "Sony Handycam DCR-PC7 Digital Camcorder Analysis,"
    Report LCP-052-97(P), 1997.
Portelligent, Inc., "Chip Scale Packaging Technology in Japanese Products,"
    Report LCP-061-97(P), 1997.
Portelligent, Inc., "Digital Imaging Update," Report LCP-062-97(P), March 1997.
Portelligent, Inc., "Ultra Low Cost Electronics Study," Report LCP-063-97(P),1997.
Portelligent, Inc., "Analysis of Sony's Consumer Electronics Strategy,"
    Report LCP-064-97(P), 1997.
Portelligent, Inc., "Flat Panel Display Roadmap," Report LCP-065-97(P), March 1997.
Portelligent, Inc., "Battery Roadmap," Report LCP-066-97(P), March 1997.
Portelligent, Inc., "Memory Roadmap," Report LCP-067-97(P), March 1997.
Portelligent, Inc., "DSP Roadmap," Report LCP-068-97(P), March 1997.
Portelligent, Inc., "Thermal and Power Studies," Report LCP-069-97(P), 1997.
Portelligent, Inc., "Board Level Packaging Advisor," Report LCP-070-97(P), 1997.
Portelligent, Inc., "Chip Scale Packaging Technology Status Report,"
    Report LCP-071-97(P), 1997.
Portelligent, Inc., "Miniaturization in Portable Electronic Products,"
    Report LCP-086-97, 1997.
Portelligent, Inc., "Nokia 232 Cellular Phone Analysis,"
    Report LCP-087-97(P), 1997.
Portelligent, Inc., "Sony DSC-F1 Cybershot Digital Still Camera Analysis,"
    Report LCP-088-97(P), 1997.
Portelligent, Inc., "Hardware Analysis of Five Notebook Computers,"
    Report LCP-116-97(P), 1997.
Portelligent, Inc., "Sony's Product Strategy for Portable Electronics,"
    Report LCP-118-97(P), 1997.
Portelligent, Inc., "Overview of Secondary Battery Market," Report LCP-081-98(P), 1998.
Portelligent, Inc., "Flat Panel Display Market Update," Report LCP-082-98(P), 1998.

Index

ABOUT THE AUTHOR

BERT HASKELL is President & General Manager of Wireless Age, Inc., Austin, Texas, a company that specializes in the deployment of wireless network technologies that enable user mobility. He was formerly Vice President for Product Development at Stellar Display Corporation, where he was responsible for the development of flat panel display products. Prior to this he was Director of Marketing for Interactive TV and Wireless Solutions at Concero, Inc. He has also served as Vice President for Portable and Consumer Electronics at MCC Corporation. He is currently sits on the board of Portelligent, Inc., the leading supplier of technical intelligence for the portable electronics industry. He has bachelor's and master's degrees in mechanical engineering from the University of Rochester and was a contributor to the textbook *Anytime, Anywhere Computing*. He lives in Austin.

www.ingramcontent.com/pod-product-compliance
Lightning Source LLC
Chambersburg PA
CBHW060756220326
41598CB00022B/2452